310
Current Topics in Microbiology and Immunology

W. Doerfler and P. Böhm (Eds.)

DNA Methylation: Development, Genetic Disease and Cancer

With 25 Figures and 10 Tables

Springer

Prof. Dr. Walter Doerfler
Universität zu Köln
Institut für Genetik
Zülpicher Str. 47
50674 Köln
and
Institut für Klinische und Molekulare Virologie
Universität Erlangen
Schlossgarten 4
91054 Erlangen
Germany
e-mail: walter.doerfler@uni-koeln.de,
walter.doerfler@viro.med.uni-erlangen.de

Petra Böhm
Universität zu Köln
Institut für Genetik
Zülpicher Str. 47
50674 Köln
Germany
e-mail: p.boehm@uni-koeln.de

Cover Illustration:
Random distribution of parental chromosomes during somatic cell division. Fluorescence in situ hybridization with differentially labeled centromeric satellite DNAs was used to distinguish between maternal Mus musculus and paternal M. spretus chromosomes in a mouse hybrid metaphase.

Library of Congress Catalog Number 72-152360

ISSN 0070-217X
ISBN-10 3-540-31180-7 Springer Berlin Heidelberg New York
ISBN-13 978-3-540-31180-5 Springer Berlin Heidelberg New York

Springer is a part of Springer Science+Business Media
springeronline.com

Printed in Germany

Editor: Simon Rallison, Heidelberg
Desk editor: Anne Clauss, Heidelberg
Production editor: Nadja Kroke, Leipzig
Cover design: design & production GmbH, Heidelberg
Typesetting: LE-TEX Jelonek, Schmidt & Vöckler GbR, Leipzig
Printed on acid-free paper SPIN 11536994 27/3150/YL - 5 4 3 2 1 0

Preface

Volume 301 of *Current Topics in Microbiology and Immunology, DNA Methylation: Basic Mechanisms*, which was published earlier this year, provided an introduction to the field and summarized ongoing research on elementary mechanisms related to DNA methylation. Due to the delayed availability of some of the manuscripts scheduled to appear in the book, the chapters dealing with the significance of this genetic signal for development, genetic disease, and cancer were assigned to this follow-up volume.

Obviously, the study of DNA methylation has had a profound impact on research in many areas of biomedicine. By January 2006, there were well over 10,000 citations in PubMed on DNA methylation and its ramifications. It has become clear that we need to re-sequence the human and other genomes at least partly to obtain a true account of the distribution of the fifth nucleotide in functional genomes. At this time, we are far from understanding the regulatory and structural functions that the fifth nucleotide exerts on chromatin.

What, for many years, seemed to be the almost-forgotten fifth nucleotide in DNA, 5-methyldeoxycytidine (5-mC), has now been generally recognized as one of the important topological signals in molecular genetics. In an intricate way and in conjunction with histone modifications, 5-mC residues in highly specific locations in a nucleotide sequence play crucial roles in long-term gene silencing and in influencing chromatin structure. The two mechanisms are probably intimately connected. Moreover, the genomes of many organisms are characterized by unique patterns of DNA methylation that can differ from genome segment to genome segment and cell type to cell type. These patterns can be instrumental in determining cell type and function. Studies on the role of DNA methylation have now moved center stage in many fields of biology and medicine such as developmental biology, genetic imprinting, genetic disease, tumor biology, gene therapy, cloning of organisms, and others. Again, basic research in molecular biology has opened new vistas for biomedical problems.

In this volume, the 12 contributions written by experts in the fields of development, genetic disease, and cancer biology deal with the role of DNA methylation in biology and pathogenesis. The series *Current Topics in Microbiology and Immunology* continues a long-standing tradition in that it offers indispensable reading in many fields of biology and medicine for novice and expert alike.

Erlangen/Köln, January 2006 *Walter Doerfler*

List of Contents

List of Contributors

(Addresses stated at the beginning of respective chapters)

CTMI (2006) 310:1–12

The Regulation of Chromatin and DNA-Methylation Patterns in Blood Cell Development

C. Bonifer (✉) · P. Lefevre · H. Tagoh

Division of Experimental Haematology, Institute for Molecular Medicine, Epidemiology and Cancer Research, University of Leeds, St James's University Hospital, Leeds LS9 7TF, UK
c.bonifer@leeds.ac.uk

Abstract All developmental processes in metazoans require the establishment of different genetic programs to generate functionally specialised cells. Differential gene expression is also the basis for the alterations in the developmental potential of differentiating cells. However, the molecular details concerning how this is achieved are still poorly understood. The haematopoietic system has for many years served as an excellent model system to study how developmental processes are regulated at the epigenetic level. In this article we will summarise recent results from others and from our own laboratory that have yielded profound insights into the general principles of how cell-fate decisions are regulated in the cell nucleus. We summarise (1) how the interplay of sequence-specific transcription factors and chromatin components is responsible for the cell type and cell stage-specific activation of specific genes and (2) how these findings impact on current concepts of epigenetic regulation of developmental processes.

1
Cell-fate Decisions in the Haematopoietic System

All metazoans originate from totipotent fertilised eggs which undergo a series of cell divisions before establishing cell types with altered developmental potential. Germ cells are set aside very early in development and retain their totipotency, whereas all other differentiating cells undergo a gradual loss of developmental potential in favour of functional specialisation. The elucidation of the molecular mechanism of this process has been a long-standing fundamental question in biology. Studies of what drives differentiation of individual cell lineages have been instrumental in experimentally addressing this question. A cell lineage forms part of a developmental pathway, which is defined as a hierarchy of differentiating cells. These cells originate from a defined precursor cell type that has the potential to give rise to related but functionally specialised cells. One of the best-characterised developmental pathways is blood cell development. All blood cells originate from haematopoietic stem cells (HSCs), which in the adult mammalian organism reside in the bone marrow and have the capacity to self-renew as well as differentiate (Weissman et al. 2001). The different cell lineages are depicted in Fig. 1. The first major developmental decision occurs at the branch point between lymphoid and myeloid cells. This generates the first functionally restricted precursors: common lymphoid precursors, which only generate B cells, T cells and natural killer (NK) cells, and common myeloid precursors, which only give rise to granulocytes, erythrocytes, mast cells, megakaryocytes and monocytes. These two precursor types undergo further restrictions in developmental potential until terminally differentiated cells are generated.

It is now abundantly clear that the crosstalk between extracellular signals—triggering surface receptor responses—and the nucleus regulates the survival and the differentiation of specific blood cell types. It is also clear that these signals co-operate with cell type-specific sets of sequence-specific transcription factors. These factors interact with genes organised in specific chromatin architectures, and the assembly of transcription factor complexes on specific *cis*-regulatory elements sets chromatin remodelling and modification events in motion that ultimately lead to the stable establishment of specific genetic programs. This "opening-up" of silent chromatin is not an all-or-none event, but a multi-step process that takes place much earlier than previously thought. A number of experiments over the last several years have shown an ever-increasing number of epigenetic alterations accompanying the activation of genes during haematopoietic development, all of which occur in a co-ordinated and regulated fashion.

In our laboratory we are interested in the following questions: What is the order of events taking place during the developmental activation of specific genes, and which factors participate in this process? At which developmental stage do such processes start?

Using myeloid-specific genes and macrophage differentiation as a model, our lab has uncovered a number of important insights that are summarized

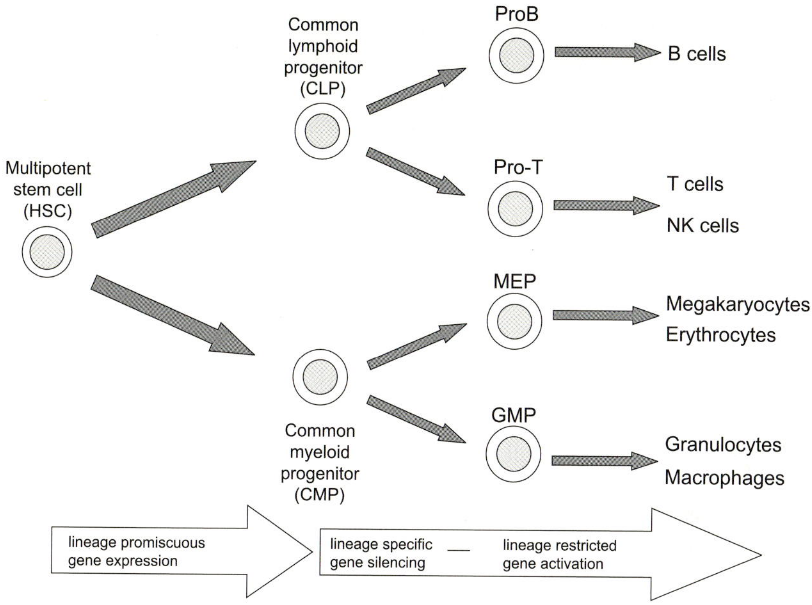

Fig. 1 Haematopoietic hierarchy and the specification of a lineage-specific gene expression program. This figure shows the development of the different blood cell types originating from multipotent haematopoietic stem cells (HSCs), which in a first step differentiate into functionally restricted common lymphoid precursors (CLPs) and common myeloid precursors (CMPs). HSCs—but to some extent also CMPs and CLPs—show lineage-promiscuous gene expression, which is then silenced in inappropriate lineages. In addition, genes not expressed in HSCs are activated during cell-lineage differentiation

below regarding the order of events taking place during the activation of gene expression from the silent state. The picture is far from complete, but the experiments aimed at answering these questions have yielded a first glimpse of the principles governing the epigenetic basis of development.

2 Transcription Factors Co-operate with Chromatin Components and Reorganise Chromatin Structure Prior to the Activation of Gene Expression

Permanently inactive genes are organised in silent chromatin, which can be characterised as being tightly packed and poorly accessible to the action of DNA binding proteins, and lacking active histone marks, such as acetylated histones (Gilbert et al. 2004). Instead, such genes present inactive histone marks such as methylated histone H3 lysine 9 (Fig. 2). In addition, the DNA within silent chromatin is often methylated at CpG dinucleotides. This renders such genes poorly accessible to the action of DNA binding proteins. An extensive description of all known heterochromatin proteins interacting with DNA and with each other is not the subject of this article and has been extensively reviewed (see, for example: Dillon and Festenstein 2002; Jaenisch and Bird 2003). However, in the context of the regulation of cell-fate decisions, it is important to mention three important facts: (1) Heterochromatin can spread via protein complexes that bind to methylated DNA and modified histones; (2) heterochromatin formation at a specific DNA region is a self-sustaining process; and (3) an open chromatin structure has to be actively maintained. Once established, patterns of heterochromatin are faithfully maintained throughout cell division, until a gene-activating stimulus is received (reviewed in Maison and Almouzni 2004). The nature of this stimulus is different for each gene but in all cases involves sequence-specific transcription factors as the endpoints of such signalling processes.

Heterochromatin restricts access of transcription factors in a number of ways, and for transcription to occur it has to be reorganised so that the transcription machinery can gain access to promoters. In addition, methyl groups on the DNA need to be removed. Methylated DNA does not only recruit the gene-silencing machinery, but many transcription factors are unable to bind to their recognition sequences if these contain methylated CpG dinucleotides. Moreover, demethylating DNA does not only remove the signal for the recruitment of repressive complexes, but also serves as a stable memory mark—for the next cell generation—that the gene is destined for activation (Thomassin et al. 2001). To gain further insight into these processes, we studied the onset of chromatin remodelling and the dynamics of DNA demethylation during cell differentiation, using the chicken lysozyme gene as a model.

Within the haematopoietic system, the chicken lysozyme gene is expressed specifically in granulocytes and macrophages and is first transcriptionally activated at the granulocyte-macrophage progenitor (GMP) stage (Jägle et al. 1997). It reaches its highest level of expression in lipopolysaccharide (LPS)-stimulated macrophages (Huber et al. 1995). We analysed the methylation status of key regulatory regions of the chicken lysozyme locus in early haematopoiesis—where the gene is silent—and the kinetics of demethylation at all stages of macrophage differentiation up to the terminally differentiated state. We found that one of the first steps in gene locus activation was the selective demethylation of specific CpGs (Lefevre et al. 2003; Tagoh et al. 2004a). This occurred already in early multipotent precursor cells long before the gene started to be expressed. However, when we measured the kinetics

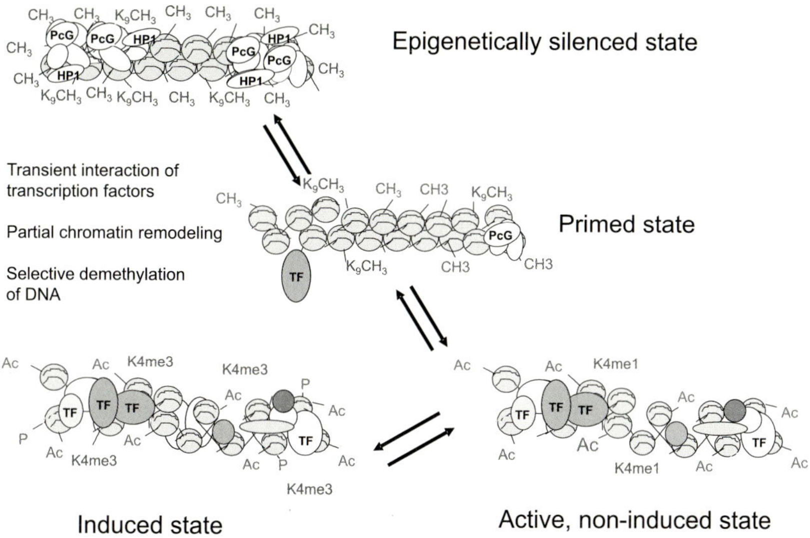

Fig. 2 Gene locus activation during haematopoietic development occurs in distinct steps. This model for the activation of a gene locus starts from the epigenetically silent, heterochromatinised state, which is characterised by compacted chromatin, methylated DNA (CH_3), histone H3 methylated at lysine 9 (K_9CH_3) and the binding of silencing factors such as heterochromatin protein 1 (*HP1*) and polycomb proteins (*PcG*). The transient action of transcription factors (*TF*, indicated as *heterogeneous shapes*) then leads to a dynamic primed state. The gene is still not transcribed, but is already marked by selective DNA demethylation and a partially reorganised chromatin structure. The next step is the active but non-induced state. This state is characterised by the removal of all heterochromatin proteins, the formation of transcription factor complexes, further chromatin remodelling, the removal of inactive histone marks, the partial acquisition of active histone marks, such as low-level acetylation of histone H3 lysine 9 (*Ac*) or monomethylation of histone H3 lysine 4 (K4me1) and low-level transcription. High-level transcription in the active, induced state goes along with full and stable transcription factor complex occupancy and DNase I hypersensitive site formation as well as histone hyperacetylation, histone H3 serine 10 phosphorylation (*P*) and histone H3 lysine 4 trimethylation (*K4me3*)

of demethylation we found that it was not uniform. Specific CpGs within individual *cis*-elements were demethylated with differential kinetics. Central CpGs within the core of important transcription factor binding sites were demethylated very early in development, but methyl groups at surrounding CpGs disappeared much slower. DNA demethylation was not the only epigenetic modification present in the chromatin of precursor cells. We saw changes in DNA topology as measured by UV photofootprinting, indicating that partial chromatin remodelling had taken place (Kontaraki et al. 2000; Tagoh et al. 2004a). We inferred from these results that it was the transcription factors themselves which were responsible for the selective removal of the methyl group. This is in concordance with a number of experiments by others. For example, it was shown that the demethylation of the immunoglobulin κ-chain locus during B cell maturation requires the presence of the transcription factor nuclear factor (NF)κB (Kirillov et al. 1996). Another example is the B cell-specific mb-1 gene, which is hypermethylated in haematopoietic stem cells but becomes demethylated during B cell differentiation. This process is directly dependent on the presence or absence of the transcription factor early B cell factor (EBF). Transfection of this factor into B cells can induce DNA demethylation at specific sites within the mb-1 gene (Maier et al. 2004).

In order to test the hypothesis that transcription factor binding was directly responsible for DNA demethylation in early multipotent precursor cells and to identify the factors responsible for the differential loss of methyl groups during macrophage differentiation, we performed in vivo footprinting experiments (Tagoh et al. 2004a). This technique allows the identification of DNA sequences that are occupied by transcription factors in vivo. We used the dimethyl sulphate (DMS) footprinting method described by Kontaraki et al. (2000), which depends on the fact that transcription factor binding affects the reactivity of DNA with DMS, which can rapidly penetrate intact cells. After in vivo formation of alkylated bases, the position of these lesions is determined at nucleotide resolution by use of ligation-mediated PCR (LM-PCR). In contrast to our expectations, we saw little or no binding of transcription factors in early multipotent haematopoietic precursor cells. During macrophage differentiation we started to see significant transcription factor interaction with DNA only at the GMP stage, coinciding with the onset of lysozyme gene expression. The DMS footprinting pattern in GMPs was indistinguishable from that of mature cells, indicating that the same transcription factor complexes interact with their recognition sequences at all differentiation stages. However, throughout a number of differentiation stages, transcription factor binding appeared to be unstable as indicated by partial footprints. Complete occupancy as indicated by maximal footprinting signals was only seen in fully differentiated macrophages. But if there was little or no binding of transcription factors to DNA in early multipotent progenitors and if transcription factor binding was unstable for a number of cell generations in differentiating cells, how was DNA-demethylation and partial chromatin remodelling achieved?

This question was answered by experiments with multipotent progenitor cell lines, which provided more material for biochemical assays. In these cell lines, we obtained the same results as in the primary cells. In vivo footprinting experiments showed no transcription factor binding at all, and we saw the same type of partial DNA-methylation pattern. In addition, we neither saw formation of DNase I hypersensitive sites nor the establishment active histone marks such as acetylated histone H3 lysine 9 or trimethylated H3 lysine 4 (Huber et al. 1995; Kontaraki et al. 2000; Lefevre et al. 2003). However, chromatin immunoprecipitation assays showed that certain transcription factors, although not bound stably enough to cause alterations in DMS reactivity, could transiently bind to lysozyme chromatin (Lefevre et al. 2003). Transient transcription factor interaction is not a phenomenon unique to the lysozyme locus. A similar result was reported with the mouse mammary tumour virus enhancer. Here it was shown that chromatin structure at this element was altered in the absence of stable binding of transcription factors (Belikov et al. 2004). The question now arose: What are the consequences of these findings on how gene locus activation is regulated in precursor cells?

3 The Role of Transcription Factors and Chromatin Components in the Regulation of Cell-fate Decisions

It has been known for many years that shifting the balance of specific transcription factors in haematopoietic precursor cells can dictate the outcome of cell differentiation (reviewed in Graf 2002; Orkin 2000), and this principle holds true for all differentiation decisions in all multicellular organisms that have been studied so far. To quote a few examples: The overexpression of the transcription factor PU.1 in multipotent progenitor cells shifts differentiation towards myeloid cells (DeKoter and Singh 2000; McIvor et al. 2003; Nerlov and Graf 1998; Yamada et al. 2001). PU.1 functions in opposition to GATA-1, which regulates erythropoiesis and inhibits GATA-1 action (Heyworth et al. 2002; Kulessa et al. 1995; Nerlov et al. 2000; Zhang et al. 2000). The overexpression of myeloid-specific transcription factors in mature B cells can reprogram these cells into macrophages (Xie et al. 2004). In turn, the conditional elimination of the B cell-specific transcription factor Pax5 in mature B cells leads to the re-expression of myeloid-specific genes in B cells, indicating that not only activators, but also repressors are required for the establishment of a specific genetic program (Mikkola et al. 2002; Tagoh et al. 2004b). In addition, Pax5-null cells in the bone marrow of knock-out mice are blocked in B cell differentiation (Hayashi et al. 2003; Nutt et al. 1997). The reason for this is that Pax5 is required for the activation of B cell-specific genes (Nutt et al. 1998), indicating that the same factors can act as activators and repressors in one cell type.

It is now firmly established that the extended developmental potential of haematopoietic stem cells goes along with a promiscuous genetic program that promotes expression of genes destined to become lineage specific (Enver and Greaves 1998). Such "lineage priming" is the likely reason for the influence of transcription factor balance on cell differentiation. However, transcription factors are not the entire story. As outlined above, it takes two to tango: Transcription factors encounter a specific chromatin architecture and cooperate with chromatin-modifying complexes to establish genetic programs. From genetic studies with worms and flies it has been long known that basic chromatin components are involved in the regulation of cell-fate decisions, it was therefore not surprising that recent studies showed the same to be true for cell differentiation processes in the mammalian haematopoietic system. Knock-out of the gene encoding the methyl-binding protein MBD2, which is part of the gene-silencing machinery recruited by methylated DNA, leads to early onset of gene expression and delays gene silencing (Hutchins et al. 2002). Similar results were obtained with mice that carried a conditional mutation of the DNA methyltransferase 1 (dnmt1) gene (Makar and Wilson 2004). But how are transcription factors and chromatin components working together in precursor cells?

The results of our experiments described above point to a scenario in which transcription factors can influence the activation of genetic programs by a "hit and run" mechanism depicted in Fig. 2. This mechanism relies on transiently interacting transcription factors that recruit chromatin modification complexes, which in turn leave a modified chromatin structure behind and block de novo DNA methylation. Such a mechanism could direct progressive demethylation after each cell division. This may occur immediately after DNA synthesis, when the epigenotype has to be newly established. This is consistent with experiments demonstrating that lineage-determining activators are induced in bipotent progenitors in a cell cycle-independent fashion but require entry in to S-phase to execute a program of differentiation (Mullen et al. 2001). In addition, it was shown that the interaction of high-affinity DNA binding proteins with DNA can lead to the progressive demethylation of CpG sequences. Moreover, it was demonstrated that transcription factor-mediated demethylation requires DNA replication, (Matsuo et al. 1998). However, recently published experiments describe rapid, DNA replication-independent removal of DNA methylation at the interleukin 2 gene following T cell activation (Bruniquel and Schwartz 2003). The mechanism by which this occurs is currently unknown, because no enzymatic activity for DNA demethylation has been found yet.

DNA demethylation and partial chromatin remodelling events bring the gene closer to the active conformation until transcription factor complexes are able to assemble, recruit RNA polymerase and initiate gene expression. But who does the first step? It is likely that not all factors are able to transiently interact with their binding sites. For example, we can exclude all factors that are unable to bind to methylated DNA or are unable to recognise their binding sites when they are masked by a positioned nucleosomes or compacted chromatin. It has been postulated that distinct "pioneer factors" have to act first which are unaffected by silent chromatin. This was based on findings describing (1) factors that bind with high affinity to condensed chromatin and activate genes early in development (Cirillo and Zaret 1999) and (2) other factors that prefer to bind to methylated rather than unmethylated DNA (Bhende et al. 2004).

4
Outlook: Epigenetic Plasticity and Reprogramming

A number of studies have recently shown that apparently committed cells still have the potential to differentiate into cells of another lineage if subjected to experimental manipulation (reviewed in Graf 2002). As described above, altering the dosage of transcription factors in mature B cells was sufficient to promote the upregulation of myeloid-specific genes. This indicates that most of the *trans*-acting factors required to express alternate lineage genes are still present in those cells. However, very few studies have addressed the chromatin structure of specific genes upon which those factors act. To address this issue, we studied epigenetic silencing of the gene for colony-stimulating factor 1 receptor (*csf1-r* or *c-fms)* during B lymphopoiesis (Tagoh et al. 2004b). *c-fms* is a target of the transcription factor PU.1 in macrophages and is already expressed in at a low level in HSCs. This gene is upregulated during macrophage differentiation and is silenced in B lymphopoiesis, despite the presence of PU.1 in both macrophages and B cells. As described above, *c-fms* can be reactivated in B lymphoid cells, and we could show that this ability correlates with a partially active chromatin structure of this gene. Although no transcription factors were bound to *c-fms cis*-regulatory elements in mature B cells, its chromatin was still DNase I accessible, nucleosomes were positioned in the active conformation and the DNA of the cores of *cis*-elements were unmethylated. We inferred from these results that it is difficult to convert an active chromatin structure into an inactive one while transcriptional activators for lineage specific genes are still present. Similar to what we described for the lysozyme locus, it is tempting to speculate that transcription factors capable of activating *c-fms*, such as PU.1, transiently interact with recognition sequences and protect these sequences from being methylated during cell division and prevent chromatin from being completely silenced. This notion is consistent with our observation that in T cells, which do not express PU.1, *c-fms* is fully methylated and chromatin has reverted to the inactive state.

As discussed above, DNA methylation is a stable epigenetic mark that has to be removed to allow for gene expression. We would extend this statement to say that efficient reactivation of gene expression by shifting transcription factor balance is difficult if a gene is organised in silent heterochromatin and is fully methylated. Once permanently shut down, genes require additional treatment with methyltransferase inhibitors such as 5-azacytidine (5-AzaC) to be effectively reactivated. In support of this idea, it was recently reported that even unrelated cells such as neural stem cells can be reprogrammed into haematopoietic cells by treatment with 5-AzaC and the histone acetylase inhibitor trichostatin A, followed by transplantation into the bone marrow of lethally irradiated mice, which provides the necessary signalling environment (Schmittwolf et al. 2005).

In summary, the experiments described above indicate that cell-fate decisions occur gradually at the epigenetic level. Up to a certain developmental stage early haematopoietic cells still retain the ability up to respond to manipulation of extrinsic and intrinsic signals and change cell fate. Our data also indicate that a specific, partially accessible chromatin structure correlates with the ability to reactivate gene expression. This could indicate that the future may hold significant promise with the design of chromatin-modifying compounds tailored to activate or repress specific genes in a disease context.

Acknowledgements Research in C. Bonifer's laboratory is supported by grants from the Wellcome Trust, the BBSRC, the Leukaemia Research Fund, the City of Hope Medical Centre and Yorkshire Cancer Research. H. Tagoh is a Kay Kendall Leukaemia Fund Fellow. The authors thank Peter Cockerill for critically reading the manuscript.

References

Belikov S, Holmqvist PH, Astrand C, Wrange O (2004) Nuclear factor 1 and octamer transcription factor 1 binding preset the chromatin structure of the mouse mammary tumor virus promoter for hormone induction. J Biol Chem 279:49857–49867

Bhende PM, Seaman WT, Delecluse HJ, Kenney SC (2004) The EBV lytic switch protein, Z, preferentially binds to and activates the methylated viral genome. Nat Genet 36:1099–1104

Bruniquel D, Schwartz RH (2003) Selective, stable demethylation of the interleukin-2 gene enhances transcription by an active process. Nat Immunol 4:235–240

Cirillo LA, Zaret KS (1999) An early developmental transcription factor complex that is more stable on nucleosome core particles than on free DNA. Mol Cell 4:961–969

DeKoter RP, Singh H (2000) Regulation of B lymphocyte and macrophage development by graded expression of PU.1. Science 288:1439–1441

Dillon N, Festenstein R (2002) Unravelling heterochromatin: competition between positive and negative factors regulates accessibility. Trends Genet 18:252–258

Enver T, Greaves M (1998) Loops, lineage, and leukemia. Cell 94:9–12

Gilbert N, Boyle S, Fiegler H, Woodfine K, Carter NP, Bickmore WA (2004) Chromatin architecture of the human genome: gene-rich domains are enriched in open chromatin fibers. Cell 118:555–566

Graf T (2002) Differentiation plasticity of hematopoietic cells. Blood 99:3089–3101

Hayashi K, Yamamoto M, Nojima T, Goitsuka R, Kitamura D (2003) Distinct signaling requirements for Dmu selection, IgH allelic exclusion, pre-B cell transition, and tumor suppression in B cell progenitors. Immunity 18:825–836

Heyworth C, Pearson S, May G, Enver T (2002) Transcription factor-mediated lineage switching reveals plasticity in primary committed progenitor cells. EMBO J 21:3770–3781

Huber MC, Graf T, Sippel AE, Bonifer C (1995) Dynamic changes in the chromatin of the chicken lysozyme gene domain during differentiation of multipotent progenitors to macrophages. DNA Cell Biol 14:397–402

Hutchins AS, Mullen AC, Lee HW, Sykes KJ, High FA, Hendrich BD, Bird AP, Reiner SL (2002) Gene silencing quantitatively controls the function of a developmental trans-activator. Mol Cell 10:81–91

Jaenisch R, Bird A (2003) Epigenetic regulation of gene expression: how the genome integrates intrinsic and environmental signals. Nat Genet 33 Suppl:245–254

Jägle U, Muller AM, Kohler H, Bonifer C (1997) Role of positive and negative cis-regulatory elements in the transcriptional activation of the lysozyme locus in developing macrophages of transgenic mice. J Biol Chem 272:5871–5879

Kirillov A, Kistler B, Mostoslavsky R, Cedar H, Wirth T, Bergman Y (1996) A role for nuclear NF-kappaB in B-cell-specific demethylation of the Igkappa locus. Nat Genet 13:435–441

Kontaraki J, Chen HH, Riggs A, Bonifer C (2000) Chromatin fine structure profiles for a developmentally regulated gene: reorganization of the lysozyme locus before trans-activator binding and gene expression. Genes Dev 14:2106–2122

Kulessa H, Frampton J, Graf T (1995) GATA-1 reprograms avian myelomonocytic cell lines into eosinophils, thromboblasts, and erythroblasts. Genes Dev 9:1250–1262

Lefevre P, Melnik S, Wilson N, Riggs AD, Bonifer C (2003) Developmentally regulated recruitment of transcription factors and chromatin modification activities to chicken lysozyme cis-regulatory elements in vivo. Mol Cell Biol 23:4386–4400

Maier H, Ostraat R, Gao H, Fields S, Shinton SA, Medina KL, Ikawa T, Murre C, Singh H, Hardy RR, Hagman J (2004) Early B cell factor cooperates with Runx1 and mediates epigenetic changes associated with mb-1 transcription. Nat Immunol 5:1069–1077

Maison C, Almouzni G (2004) HP1 and the dynamics of heterochromatin maintenance. Nat Rev Mol Cell Biol 5:296–304

Makar KW, Wilson CB (2004) DNA methylation is a nonredundant repressor of the Th2 effector program. J Immunol 173:4402–4406

Matsuo K, Silke J, Georgiev O, Marti P, Giovannini N, Rungger D (1998) An embryonic demethylation mechanism involving binding of transcription factors to replicating DNA. EMBO J 17:1446–1453

McIvor Z, Hein S, Fiegler H, Schroeder T, Stocking C, Just U, Cross M (2003) Transient expression of PU.1 commits multipotent progenitors to a myeloid fate whereas continued expression favors macrophage over granulocyte differentiation. Exp Hematol 31:39–47

Mikkola I, Heavey B, Horcher M, Busslinger M (2002) Reversion of B cell commitment upon loss of Pax5 expression. Science 297:110–113

Mullen AC, Hutchins AS, Villarino AV, Lee HW, High FA, Cereb N, Yang SY, Hua X, Reiner SL (2001) Cell cycle controlling the silencing and functioning of mammalian activators. Curr Biol 11:1695–1699

Nerlov C, Graf T (1998) PU.1 induces myeloid lineage commitment in multipotent hematopoietic progenitors. Genes Dev 12:2403–2412

Nerlov C, Querfurth E, Kulessa H, Graf T (2000) GATA-1 interacts with the myeloid PU.1 transcription factor and represses PU.1-dependent transcription. Blood 95:2543–2551

Nutt SL, Urbanek P, Rolink A, Busslinger M (1997) Essential functions of Pax5 (BSAP) in pro-B cell development: difference between fetal and adult B lymphopoiesis and reduced V-to-DJ recombination at the IgH locus. Genes Dev 11:476–491

Nutt SL, Morrison AM, Dorfler P, Rolink A, Busslinger M (1998) Identification of BSAP (Pax-5) target genes in early B-cell development by loss- and gain-of-function experiments. EMBO J 17:2319–2333

Orkin SH (2000) Diversification of haematopoietic stem cells to specific lineages. Nat Rev Genet 1:57–64

Schmittwolf C, Kirchhof N, Jauch A, Durr M, Harder F, Zenke M, Muller AM (2005) In vivo haematopoietic activity is induced in neurosphere cells by chromatin-modifying agents. EMBO J 24:554–566

Tagoh H, Melnik S, Lefevre P, Chong S, Riggs AD, Bonifer C (2004a) Dynamic reorganization of chromatin structure and selective DNA demethylation prior to stable enhancer complex formation during differentiation of primary hematopoietic cells in vitro. Blood 103:2950–2955

Tagoh H, Schebesta A, Lefevre P, Wilson N, Hume D, Busslinger M, Bonifer C (2004b) Epigenetic silencing of the c-fms locus during B-lymphopoiesis occurs in discrete steps and is reversible. EMBO J 23:4275–4285

Thomassin H, Flavin M, Espinas ML, Grange T (2001) Glucocorticoid-induced DNA demethylation and gene memory during development. EMBO J 20:1974–1983

Weissman IL, Anderson DJ, Gage F (2001) Stem and progenitor cells: origins, phenotypes, lineage commitments, and transdifferentiations. Annu Rev Cell Dev Biol 17:387–403

Xie H, Ye M, Feng R, Graf T (2004) Stepwise reprogramming of B cells into macrophages. Cell 117:663–676

Yamada T, Abe M, Higashi T, Yamamoto H, Kihara-Negishi F, Sakurai T, Shirai T, Oikawa T (2001) Lineage switch induced by overexpression of Ets family transcription factor PU.1 in murine erythroleukemia cells. Blood 97:2300–2307

Zhang P, Zhang X, Iwama A, Yu C, Smith KA, Mueller BU, Narravula S, Torbett BE, Orkin SH, Tenen DG (2000) PU.1 inhibits GATA-1 function and erythroid differentiation by blocking GATA-1 DNA binding. Blood 96:2641–2648

CTMI (2006) 310:13–22

Methylation Dynamics in the Early Mammalian Embryo: Implications of Genome Reprogramming Defects for Development

T. Haaf (✉)

Johannes Gutenberg-Universität Mainz, Langenbeckstrasse 1, Bau 601,
55131 Mainz, Germany
haaf@humgen.klinik.uni-mainz.de

Abstract In mouse and most other mammalian species, the paternal and maternal genomes undergo parent-specific epigenetic reprogramming during preimplantation development. The paternal genome is actively demethylated within a few hours after fertilization in the mouse, rat, pig, bovine, and human zygote, whereas the maternal genome is passively demethylated by a replication-dependent mechanism after the two-cell embryo stage. These genome-wide demethylation waves may have a role in reprogramming of the genetically inactive sperm and egg chromatin for somatic development. Disturbances in this highly coordinated process may contribute to developmental failures and defects in mammals. The frequency and severity of abnormal phenotypes increase after interfering with or bypassing essential steps of gametogenesis, early embryogenesis, or both. Nevertheless, it is plausible that normal fertilization, assisted reproduction, and embryo cloning are all susceptible to similar dysregulation of epigenetic components. Although the mouse may be an excellent model for early human development, species and strain differences in the molecular and cellular events shortly after fertilization may have important implications for the efficiency of epigenetic reprogramming and the incidence of reprogramming defects. Some species, i.e., rabbit and sheep, do not require drastic genome-wide demethylation for early development, most likely because the transition from maternal to embryonic control occurs relatively late during preimplantation development. A better understanding of key reprogramming factors—in particular the demethylase activity in the fertilized egg—is crucial for improving human infertility treatment and the efficiency of mammalian embryo cloning.

In mammals, both the paternal and the maternal genome are required for normal development (McGrath and Solter 1984; Surani et al. 1986). Genomic imprinting is an epigenetic mechanism by which the expression of a subset of genes becomes dependent on their parental origin (Bartolomei and Tilghman 1997). Following the establishment of imprinting in the male and female germ lines, respectively, the two parental genomes exhibit functional differences at fertilization. Methylation of 5′-cytosine residues in CpG dinucleotides is critical for regulating the temporal, spatial, and parent-specific gene expression patterns. DNA methylation establishes and maintains an inactive chromatin structure by posttranslational histone modifications (Wolffe and Matzke 1999; Jaenisch and Bird 2003). The sperm and egg genomes that are combined at fertilization are both highly methylated; however, there are important germ line-specific differences in the methylation patterns of genomic sequences (Reik et al. 2001; Haaf et al. 2004). Whereas the genomic methylation patterns and levels in somatic cells are generally stable and heritable, dramatic genome-wide changes occur in early embryos, where the two complementary parental genomes must be reprogrammed for somatic development. Methylation reprogramming may help to "revive" the inactive sperm and egg genomes and to restore a broad developmental potential in embryonic cells. This entire process appears to be maternally driven. However, the cellular machinery and factors in the fertilized egg that can reprogram the two very different gamete nuclei as well as a somatic cell nucleus that has been introduced into an oocyte during cloning remain to be elucidated. This chapter reviews recent cytological and molecular experiments that have addressed fundamental questions related to the reprogramming mechanisms and capabilities of mammalian oocytes.

1 Methylation Reprogramming in Early Mouse Embryos

Immunofluorescent staining with an antibody against 5-methylcytosine (mC) provides a valuable tool to directly visualize the genome-wide demethylation and remethylation waves in preimplantation mouse embryos (Rougier et al. 1998; Mayer et al. 2000a; Santos et al. 2002). The oocyte genome completes its meiotic maturation after sperm entry by extrusion of the second polar body. Very shortly after fertilization, the mouse zygote shows equally high methylation levels of sperm nucleus, maternal meiotic metaphase II chromosomes, and second polar body (Haaf et al. 2004). The activated oocyte then remodels the gamete chromatin into functional male and female pronuclei that oppose each other. In the normal diploid mouse zygote, the paternal genome is rapidly and drastically demethylated before onset of the first DNA replication. The maternal genome, although exposed to the same cytoplasm, is resistant to this active demethylation process. Bisulfite sequencing studies revealed that paternal zygotic demethylation affects widely different classes of repetitive and single-copy sequences (Oswald et al. 2000). Only the control regions of imprinted genes seem to be protected against the maternal demethylase activity.

The mouse oocyte efficiently demethylates multiple male pronuclei in polyspermic embryos (Santos et al. 2002), whereas it cannot demethylate the additional female genome in parthenogenetic, gynogenetic, and triploid digynic embryos (Barton et al. 2001). This suggests that active demethylation depends on a sperm-derived factor. Before the male pronucleus can be formed, the highly compacted sperm chromatin must be decondensed and the protamines be exchanged by histones (Perreault 1992; Kanka 2003). During this period, the paternal DNA is unusually loosely packaged and provides a unique opportunity for binding of a demethylating enzyme whose molecular nature remains unknown. Paternal zygotic demethylation may be facilitated by strikingly different histone modifications in paternal and maternal pronuclei. The paternal zygotic genome becomes transiently associated with hyperacetylated histone H4 (Adenot et al. 1997; Santos et al. 2002), whereas the maternal genome is preferentially associated with methylated histone H3 (Cowell et al. 2002).

The global methylation level of the maternal genome is maintained up to the two-cell embryo stage. Interestingly, even breakdown of the pronuclear envelopes and first mitosis do not lead to an intermingling of the two parental chromosome sets (Fig. 1A). Topological genome separation is preserved at least up to the two-cell stage, each (the methylated maternal genome and the demethylated paternal genome) occupying approximately half of the nucleus (Fig. 1B, left nucleus). The existence of separate nuclear compartments may facilitate parent-specific methylation reprogramming in the early embryo (Mayer et al. 2000b; Haaf 2001). In contrast to first metaphase, where both sister chromatids of the maternal chromosomes are equally methylated, in the second metaphase only one of the two sister chromatids remains methylated (Fig. 1B, right metaphase). This sister chromatid differentiation is consistent with a replication-dependent demethylation mechanism of the maternal genome (Fig. 1C).

Because mC cannot be incorporated directly into replicating DNA, maintenance of DNA methylation patterns requires DNA methyltransferase 1 (DNMT1), which has a high affinity for hemimethylated sites that are generated transiently during DNA replication. DNMT1 detects methylated CpG sites in the parental DNA strand and adds methyl groups to the corresponding sites in the newly synthesized strand (Bestor 2000). Gradual demethylation of

Fig. 1 A–D **A, B** Parent-specific genome methylation patterns in early mouse embryos. Nuclei were stained with fluorescein isothiocyanate (FITC)-conjugated anti-mC antibody (*green*) and counterstained with 4′-6′-diamidino-2-phenylindole (DAPI) (*blue*). **A** One-cell embryo during first mitosis. The maternal chromosomes are methylated in both sister chromatids and spatially separated from the fully demethylated paternal chromosomes. **B** Two-cell embryo in which one cell goes through second mitosis. The interphase nucleus exhibits a methylated maternal and a demethylated paternal compartment. During second mitosis, only one of the two sister chromatids of the maternal chromosomes remains methylated. **C** Differential demethylation of maternal (*red*) and paternal (*blue*) chromosomes during mouse preimplantation development. Both DNA strands of the paternal chromosome (*DNA double helix*) are already demethylated (*blue dotted lines*) in the zygote before onset of the first DNA replication. The maternal chromosome is protected from this active demethylation process. Following the first DNA replication cycle in the absence of maintenance DNA methyltransferase, the maternal chromosome consists of two hemimethylated sister chromatids (*DNA double helices*). After the first cell division and another round of DNA replication, the maternal chromosome consists of a hemimethylated and a fully demethylated sister chromatid, resulting in differential sister chromatid staining. **D** Methylation dynamics in early mouse embryos. The paternal (*blue*) genome undergoes active zygotic demethylation, whereas the maternal (*red*) genome is gradually demethylated after the two-cell stage. Both parental genomes are equally demethylated at the morula stage and then remethylated. The newly established somatic methylation patterns (*green line*) are identical on both parental alleles. Embryonic lineages derived from the inner cell mass (*dark green*) are more heavily methylated than the trophoblast (*light green*). Imprinted genes (*dotted lines*) escape this genome-wide methylation reprogramming after fertilization and maintain their germ-line methylation patterns. Methylated imprinted alleles do not become demethylated, and demethylated imprinted alleles are not remethylated

the maternal genome is achieved by sequestration of DNMT1 from the nucleus into the cytoplasm, which prevents binding to its target sites in hemimethylated DNA (Cardoso and Leonhardt 1999; Ratnam et al. 2002). When half of the methyl groups are lost with every round of replication, full double-stranded demethylation in one chromatid occurs after two cell cycles (Fig. 1C, red ideograms). Consequently, four-cell embryos have a much weaker mC density over the maternal half of the nucleus. After the eight-cell embryo stage, paternal and maternal chromosomes show equivalently low methylation levels (Mayer et al. 2000a). Later, in mouse blastocyst-stage embryos, genome-wide de novo methylation (Fig. 1D, green graph) occurs preferentially in the inner cell mass, establishing somatic methylation patterns in cells that give rise to the different embryonic lineages. Trophoblast cells that give rise to the extraembryonic lineages become less heavily methylated (Dean et al. 2001; Reik et al. 2001).

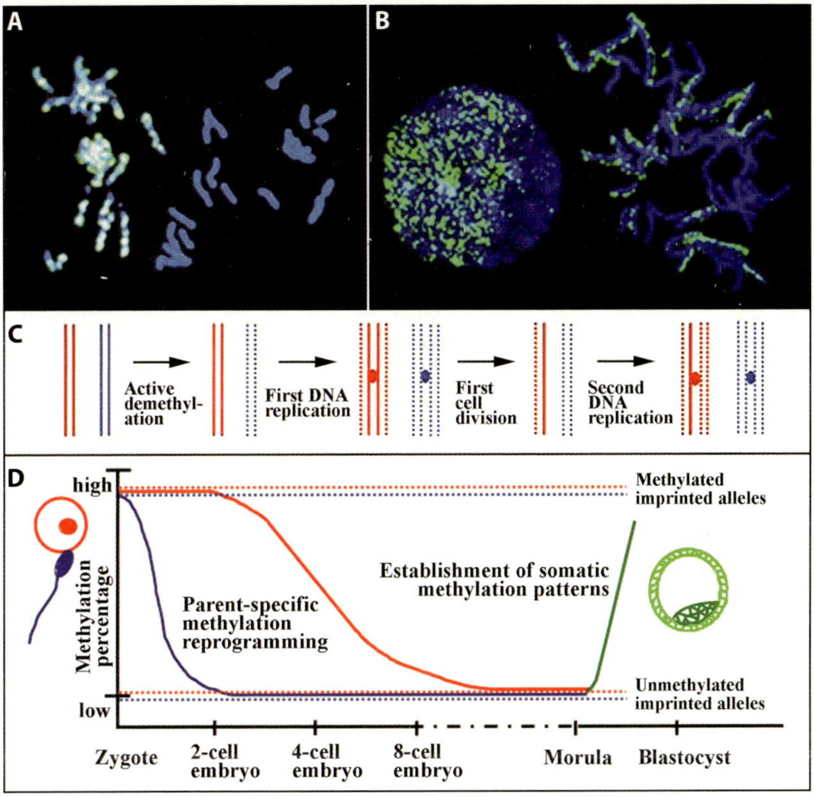

2
Species Differences in Methylation Reprogramming

If genome-wide methylation reprogramming in the early embryo is fundamental to the formation of totipotent embryonic cells, one would expect that the preimplantation methylation dynamics is conserved among mammalian species. Indeed, active demethylation of the paternal zygotic genome is observed in mouse, rat, pig, bovine, and human embryos (Mayer et al. 2000a; Dean et al. 2001; Beaujean et al. 2004a). However, the timing of remethylation already differs between species. In bovine embryos, considerable de novo methylation already occurs at the 8- to 16-cell stage, whereas in mouse, remethylation begins only in the blastocyst. In sheep and rabbit embryos anti-mC immunofluorescence revealed equally high methylation levels of the two parental genomes throughout preimplantation development (Beaujean et al. 2004a; Shi et al. 2004). The lack of detectable genome-wide methylation changes in these two species suggests that neither active nor passive demethylation is an obligatory requirement for epigenetic reprogramming after fertilization.

Interestingly, mouse sperm injected into sheep oocytes is significantly demethylated, although to a lesser extent than mouse sperm in murine oocytes. Ram sperm, which is not demethylated in sheep oocytes, can be partially demethylated in bovine oocytes (Beaujean et al. 2004b). Evidently, the demethylating activity of the ooplasm differs among species, being the highest in mouse, medium in bovine, and low in sheep (and rabbit) oocytes. However, the demethylation process must also involve a sperm-derived factor(s), i.e., differences in male pronuclear chromatin structure. The biological significance of the observed species differences in methylation reprogramming remains unclear. The timing and degree of demethylation are likely to play an important role for remodeling the two complementary germ line genomes into a diploid somatic genome (Haaf 2001; Haaf et al. 2004). The mouse embryonic genome, which is the most rapidly and drastically demethylated of all analyzed species, is already activated in the two-cell stage (Schulz 1993). Demethylation of the paternal genome in human, pig, and bovine zygotes is also associated with a relatively early transition from maternal to embryonic control of development (Memili and First 2000; Kanka 2003). By contrast, in rabbit and sheep embryos, which maintain high methylation levels after fertilization, maternal factors seem to control the preimplantation period, and transition to embryonic control of development occurs only at the 8- to16-cell stage (Manes 1973).

3 Methylation Reprogramming Defects

In mouse, cow, and most other mammalian species, the paternal and maternal genomes are demethylated by different mechanisms and at different times during preimplantation development (Fig. 1C, D). Disturbances in this spatially and temporally highly coordinated process provide one important explanation for the high rate of embryo loss after fertilization (Shi and Haaf 2002). Immunofluorescence staining demonstrated 20% abnormal methylation patterns in mouse two-cell embryos that were flushed from the oviducts of superovulated females, compared to 10% from non-superovulated females. Of the embryos, 14% from superovulated females, but only 5% from non-superovulated matings, failed to develop in culture to the blastocyst stage. This reflects an overall reduction in the reprogramming capability of the oocyte and embryo quality after hormone treatment. The dramatic differences in methylation reprogramming and development of in vitro fertilized mouse embryos that were cultured in different media may be due to a suboptimal environment at or shortly after fertilization. Since the preimplantation embryo is much less protected than the germ cells, this may be the time when environmental factors, i.e., nutrition and drugs, have the greatest impact on epigenetic reprogramming. Acetaldehyde, the toxic metabolic product of ethanol—and by extrapolation alcohol consumption—can cause methylation disturbances and developmental arrest in early mouse embryos. The efficiency of methylation reprogramming also depends on genetic factors (strain- and species-specific differences). Embryos from most inbred mouse strains or hybrids can efficiently develop in culture to the blastocyst stage, whereas embryos from Naval Medical Research Institute, USA (NMRI) and other so-called blocking strains show high incidences (20%–60%) of abnormal methylation patterns and arrest in in vitro development.

Disturbances in the establishment or maintenance of the appropriate parent-specific methylation patterns may also contribute to the medical problems of assisted reproductive technologies (ART). Genome-wide alterations cause early developmental failure and embryo loss (Barton et al. 2001; Shi and Haaf 2002), whereas methylation changes at specific gene loci have been associated with aberrant fetal growth and abnormal phenotypes. Initially it was shown in mouse and ruminants that isolation, manipulation, and culture of gametes and early embryos can affect the methylation and regulation of imprinted genes, leading to phenotypic defects (Koshla et al. 2001, Young et al. 2001). Recent studies in children conceived with ART also reported unexpectedly high incidences of certain rare human imprinting diseases, such as Beckwith-Wiedemann and Angelman syndromes, resulting from epigenetic DNA methylation defects, specifically an abnormal hypomethylation of the normally methylated maternal alleles (Maher et al. 2003; Ludwig et al. 2005). In light of growing concerns about epigenetic disturbances resulting from superovulation and embryo culture, there is clearly a need for both basic research on reproductive epigenetic events and long-term follow up studies of children born of ART.

Similar to ART, somatic cell nuclear transfer technologies interfere with essential reprogramming events in gametogenesis and early embryogenesis; however, the epigenetic insults in cloned embryos are much more frequent and pronounced. A somatic cell nucleus that has been introduced into an oocyte during cloning can be reprogrammed to some extent for somatic development, but the embryo cloning efficiency is generally low and somewhat variable between species (Solter 2000; Shi et al. 2003). Methylcytosine staining of cloned bovine embryos demonstrated incomplete or delayed demethylation of the donor genome (or both effects) (Bourc'his et al. 2001; Dean et al.

2001). The genome of cloned embryos is likely to be a mixture of normally and abnormally methylated sequences. Expression profiling of more than 10,000 genes showed frequent (4%) abnormal gene expression in placentas and livers of neonatal cloned mice (Humpherys et al. 2002). Reactivation of key embryonic genes that are necessary for the development of pluripotent cell lineages may be particularly inefficient in clones derived from somatic cell nuclei (Bortvin et al. 2003). Disruption of the allele-specific methylation and expression patterns of imprinted genes were observed in more than 95% of cloned mouse blastocysts (Mann et al. 2003). Collectively, these results suggest that the frequent developmental failures and defects resulting from cloning are largely due to epigenetic reprogramming defects.

References

Adenot PG, Mercier Y, Renard J-P, Thompson EM (1997) Differential H4 acetylation of paternal and maternal chromatin precedes DNA replication and differential transcriptional activity in pronuclei of 1-cell mouse embryos. Development 124:4615–4625

Bartolomei MS, Tilghman SM (1997) Genomic imprinting in mammals. Annu Rev Genet 31:493–525

Barton SC, Arney KL, Shi W, Fundele R, Surani MA, Haaf T (2001) Genome-wide methylation patterns in normal and uniparental early mouse embryos. Hum Mol Genet 10:2983–2987

Beaujean N, Hartshorne G, Cavilla J, Taylor JE, Gardner J, Wilmut I, Meehan R, Young L (2004a) Non-conservation of mammalian preimplantation methylation dynamics. Curr Biol 14:R266–R267

Beaujean N, Taylor JE, McGarry M, Gardner JO, Wilmut I, Loi P, Ptak G, Galli C, Lazzari G, Bird A, Young LE, Meehan RR (2004b) The effect of interspecific oocytes on demethylation of sperm DNA. Proc Natl Acad Sci USA 101:7636–7640

Bestor TH (2000) The DNA methyltransferases of mammals. Hum Mol Genet 9:2395–2402

Bortvin A, Eggan K, Skaletsky H, Akutsu H, Berry DL, Yanagimachi R, Page DC, Jaenisch R (2003) Incomplete reactivation of Oct4-related genes in mouse embryos cloned from somatic nuclei. Development 130:1673–1680

Bourc'his D, Le Bourhis D, Patin D, Niveleau A, Comizzoli P, Renard JP, Viegas-Pequignot E (2001) Delayed and incomplete reprogramming of chromosome methylation patterns in bovine cloned embryos. Curr Biol 11:1542–1546

Cardoso MC, Leonhardt H (1999) DNA methyltransferase is actively retained in the cytoplasm during early development. J Cell Biol 147:25–32

Cowell IG, Aucott R, Mahadevaiah SK, Burgoyne PS, Huskisson N, Bongiorni S, Prantera G, Fanti L, Pimpinelli S, Wu R, Gilbert DM, Shi W, Fundele R, Morrison H, Jeppesen P, Singh P (2002) Heterochromatin, HP1 and methylation at lysine 9 of histone H3 in animals. Chromosoma 111:22–36

Dean W, Santos F, Stojkovic M, Zakhartchenko V, Walter J, Wolf E, Reik W (2001) Conservation of methylation reprogramming in mammalian development: aberrant reprogramming in cloned embryos. Proc Natl Acad Sci USA 98:13734–13738
Haaf T (2001) The battle of the sexes after fertilization: behaviour of paternal and maternal chromosomes in the early mammalian embryo. Chromosome Res 9:263–271
Haaf T, Shi W, Fundele R, Arney KL, Surani MA, Barton SC (2004) Differential demethylation of paternal and maternal genomes in the preimplantation mouse embryo: implications for mammalian development. In: Schmid M, Nanda I (eds) Chromosomes today, vol. 14. Kluwer Academic Publishers, Dordrecht, Boston, London, pp 207–214
Humpherys D, Eggan K, Akutsu H, Friedman A, Hochedlinger K, Yanagimachi R, Lander ES, Golub TR, Jaenisch R (2002) Abnormal gene expression in cloned mice derived from embryonic stem cell and cumulus cell nuclei. Proc Natl Acad Sci USA 99:12889–12894
Jaenisch R, Bird A (2003) Epigenetic regulation of gene expression: how the genome integrates intrinsic and environmental signals. Nat Genet 33:245–254
Kanka J (2003) Gene expression and chromatin structure in the pre-implantation embryo. Theriogenology 59:3–19
Koshla S, Dean W, Brown D, Reik W, Feil R (2001) Culture of preimplantation mouse embryos affects fetal development and the expression of imprinted genes. Biol Reprod 64:918–926
Ludwig M, Katalinic A, Groß S, Sutcliffe A, Varon R, Horsthemke B (2005) Increased prevalence of imprinting defects in patients with Angelman syndrome born to subfertile couples. J Med Genet 42:289–291
Maher ER, Brueton LA, Bowdin SC, Luharia A, Cooper W, Cole TR, Macdonald F, Sampson JR, Barrat CL, Reik W, Hawkins MM (2003) Beckwith-Wiedemann syndrome and assisted reproductive technology (ART). J Med Genet 40:62–64
Manes C (1973) The participation of the embryonic genome during early cleavage in the rabbit. Dev Biol 32:453–459
Mann MR, Chung YG, Nolen LD, Verona RI, Latham KE, Bartolomei MS (2003) Disruption of imprinted gene methylation and expression in cloned preimplantation stage mouse embryos. Biol Reprod 69:902–914
Mayer W, Niveleau A, Walter J, Fundele R, Haaf T (2000a) Demethylation of the zygotic paternal genome. Nature 403:501–502
Mayer W, Smith A, Fundele R, Haaf T (2000b) Spatial separation of parental genomes in preimplantation mouse embryos. J Cell Biol 148:629–634
McGrath J, Solter D (1984) Completion of mouse embryogenesis requires both the maternal and paternal genomes. Cell 37:179–183
Memili E, First NL (2000) Zygotic and embryonic gene expression in cow: a review of timing and mechanisms of early gene expression as compared with other species. Zygote 8:87–96
Oswald J, Engemann S, Lane N, Mayer W, Olek A, Fundele R, Dean W, Reik W, Walter J (2000) Active demethylation of the paternal genome in the mouse zygote. Curr Biol 10:475–478
Perreault SD (1992) Chromatin remodeling in mammalian zygotes. Mutat Res 296:43–55

Ratnam S, Mertineit C, Ding F, Howell CY, Clarke HJ, Bestor TH, Chaillet JR, Trasler JM (2002) Dynamics of Dnmt1 methyltransferase expression and intracellular localization during oogenesis and preimplantation development. Dev Biol 235:304–314

Reik W, Dean W, Walter J (2001) Epigenetic reprogramming in mammalian development. Science 293:1089–1093

Rougier D, Bourc'his D, Gomes DM, Niveleau A, Plachot M, Pàldi A, Viegas-Péquignot E (1998) Chromosome methylation patterns during mammalian development. Genes Dev 12:2108–2113

Santos F, Hendrich B, Reik W, Dean W (2002) Dynamic reprogramming of DNA methylation in the early mouse embryo. Dev Biol 241:172–182

Schulz RM (1993) Regulation of zygotic gene activation in the mouse. BioEssays 8:531–538

Shi W, Haaf T (2002) Aberrant methylation patterns at the two-cell stage as an indicator of early developmental failure. Mol Reprod Dev 63:329–334

Shi W, Zakhartchenko V, Wolf E (2003) Epigenetic reprogramming in mammalian nuclear transfer. Differentiation 71:91–113

Shi W, Dirim F, Wolf E, Zakhartchenko V, Haaf T (2004) Methylation reprogramming and chromosomal aneuploidy in in vivo fertilized and cloned rabbit preimplantation embryos. Biol Reprod 71:340–347

Solter D (2000) Mammalian cloning: advances and limitations. Nat Rev Genet 1:199–207

Surani MA, Barton SC, Norris ML (1986) Nuclear transplantation in the mouse: heritable differences between parental genomes after activation of the embryonic genome. Cell 45:127–136

Wolffe AP, Matzke MA (1999) Epigenetics: regulation through repression. Science 286:481–486

Young LE, Fernandes K, McEvoy TG, Butterwith SC, Gutierrez CG, Carolan C, Broadbent PJ, Robinson JJ, Wilmut I, Sinclair KD (2001) Epigenetic change in IGF2R is associated with fetal overgrowth after sheep embryo culture. Nat Genet 27:153–154

CTMI (2006) 310:23–44

Epigenetic Regulation in *Drosophila*

F. Lyko[1] (✉) · C. Beisel[2] · J. Marhold[1] · R. Paro[2]

[1]Division of Epigenetics, Deutsches Krebsforschungszentrum,
Im Neuenheimer Feld 580, 69120 Heidelberg, Germany
f.lyko@dkfz.de

[2]ZMBH, University of Heidelberg,
Im Neuenheimer Feld 282, 69120 Heidelberg, Germany

Abstract Epigenetic regulation of gene transcription relies on molecular marks like DNA methylation or histone modifications. Here we review recent advances in our understanding of epigenetic regulation in the fruit fly *Drosophila melanogaster*. In the past, DNA methylation research has primarily utilized mammalian model systems. However, several recent landmark discoveries have been made in other organisms. For example, the interaction between DNA methylation and histone methylation was first described in the filamentous fungus *Neurospora crassa*. Another example is provided by the interaction between epigenetic modifications and the RNA interference (RNAi) machinery that was first reported in the fission yeast *Schizosaccharomyces pombe*. Another organism with great experimental power is the fruit fly *Drosophila*. Epigenetic regulation by chromatin has been extensively analyzed in the fly and several of the key components have been discovered in this organism. In this chapter, we will focus on three aspects that represent the complexity of epigenetic gene regulation. (1) We will discuss the available data about the DNA methylation system, (2) we will illuminate the interaction between DNA methylation and chromatin regulation, and (3) we will provide an overview over the Polycomb system of epigenetic chromatin modifiers that has proved to be an important paradigm for a chromatin system regulating epigenetic programming.

1 DNA Methylation in *Drosophila*

Until recently, it was assumed that *Drosophila* genomic DNA is completely unmethylated. This assumption was based on the fact that most researchers had failed to detect methylated bases in the fly genome. Thus, the fly appeared to belong to an exceptional group of organisms that lack an epigenetic DNA modification that is otherwise conserved from bacteria to humans. It was proposed that DNA methylation might be dispensable for less complex genomes (Bird 1995) or for genomes with non-canonical centrosome sequences (Dong et al. 2001). However, the *Drosophila* genome is not particularly simple (Adams et al. 2000) and the centrosome organization is conserved in a wide range of organisms, including humans (Blower et al. 2002). In addition, DNA methylation has been described in several other insect species (Field et al. 2004).

The sequencing of the *Drosophila* genome revealed the presence of a single candidate DNA methyltransferase gene, which belongs to the Dnmt2 family of eukaryotic DNA methyltransferases (Hung et al. 1999; Tweedie et al. 1999). When the *Drosophila Dnmt2* gene was first described, it was generally assumed that Dnmt2 proteins are not enzymatically active. This assumption was mainly based on the fact that the protein had failed to reveal DNA methyltransferase activity in vitro. However, Dnmt2 proteins are widely conserved in evolution and also show perfect conservation of the catalytic (cytosine-5) DNA methyltransferase motifs. Only the Dnmt2 homolog from fission yeast contains an insertional mutation in one of the essential catalytic motifs (Wilkinson et al. 1995). Intriguingly, the DNA methyltransferase activity of this protein was restored upon removal of the inserted amino acid (Pinarbasi et al. 1996). This finding strongly suggested that Dnmt2 proteins represent active DNA methyltransferases.

The presence of a putative DNA methyltransferase gene in the *Drosophila* genome raised the possibility that DNA methylation in this organism might have escaped detection in the past. Additional experiments showed that *Dnmt2* is developmentally regulated, with the highest messenger RNA (mRNA) expression levels during early development (Hung et al. 1999; Lyko et al. 2000b). This finding focused the subsequent DNA methylation analysis on genomic DNA from embryos, rather than the later developmental stages that were predominantly used in previous studies. Chromatographic analysis of genomic DNA from early stages of embryonic development indeed revealed a low but significant level of cytosine methylation (Gowher et al. 2000; Lyko et al. 2000a). Interestingly, most of the 5-methylcytosine was found in the context of CpT/A dinucleotides (Lyko et al. 2000a). This finding is of substantial interest because it implies that DNA methylation in *Drosophila* is not maintained by symmetrical methylation of CpG dinucleotides. In contrast to vertebrate cells that stably maintain their methylation patterns throughout development, DNA methylation in *Drosophila* appears to be a transient epigenetic signal during early developmental stages.

Importantly, the recent data on *Drosophila* methylation do not contradict the previous studies that claimed the absence of 5-methylcytosine from *Drosophila* DNA. For example, it has also been demonstrated that CpG methylation is not detectable in embryos (Urieli-Shoval et al. 1982). However, these experiments lacked the requisite sensitivity to uncover low levels of CpT/A methylation. Another landmark paper demonstrated that pupal DNA is unmethylated (Patel and Gopinathan 1987). Because DNA methylation in *Drosophila* is developmentally regulated, this finding cannot be extrapolated to other stages of development. The same restriction also applies to other studies that demonstrated the absence of methylation in adult flies (Bird and Taggart 1980; Rae and Steele 1979).

2
The Dnmt2 Methyltransferase

The Dnmt2 family of proteins is distinguished by several characteristic features (Dong et al. 2001): (1) They represent the most widely conserved family of eukaryotic DNA methyltransferases, and homologs have been described in various organisms ranging from protozoans to humans. (2) Dnmt2 proteins are characterized by a compact structure that resembles bacterial methylases. (3) Dnmt2 proteins contain only a catalytic DNA methyltransferase domain, and lack an extended regulatory domain, like other eukaryotic DNA methyltransferases. By now, catalytic activity of Dnmt2 proteins has been shown for the *Entamoeba histolytica* (Fisher et al. 2004), *Drosophila* (Kunert et al. 2003), mouse (Mund et al. 2004), and human (Hermann et al. 2003) homologs. Although it seems that Dnmt2 enzymes are very well capable of methylating non-CpG dinucleotides, a consensus target sequence has not been established yet.

In *Drosophila*, Dnmt2 is both necessary and sufficient for DNA methylation (Kunert et al. 2003). When Dnmt2 was knocked down by RNA interference

(RNAi), embryonic DNA methylation was completely lost. When Dnmt2 was overexpressed in adult flies, significant hypermethylation could be detected (Kunert et al. 2003). In light of these findings, it seems likely that Dnmt2 is the only DNA methyltransferase in *Drosophila*. The indications for a second enzyme are restricted to the detection of an unknown protein with a Dnmt1-specific antibody (Hung et al. 1999) and cannot be substantiated by the available genome sequence. In conclusion, the combined data thus indicate that *Drosophila* DNA is methylated at low levels by the Dnmt2 methyltransferase specifically during embryogenesis. These characteristics establish major differences to the mammalian DNA methylation systems that depend heavily on the Dnmt1 and Dnmt3 enzymes and are highly active during late stages of development. However, the exceptional conservation of Dnmt2 proteins makes it likely that Dnmt2 represents the ancestral eukaryotic DNA methyltransferase and that Dnmt1 and Dnmt3 proteins appeared later during evolution. In addition, there are more direct indications for an evolutionary conservation of the Dnmt2 methylation system: The genome sequences of *Drosophila pseudoobscura* and *Anopheles gambiae* also contain a single candidate DNA methyltransferase homolog, which belongs to the Dnmt2 family (Marhold et al. 2004c). Consistent with the observations made in *Drosophila melanogaster*, embryos from *Drosophila pseudoobscura* and *Anopheles gambiae* also revealed low but significant levels of 5-methylcytosine (Marhold et al. 2004c). Thus, the Dnmt2 methylation system appears to have been conserved over at least 250 million years separating *Drosophila melanogaster* from *Anopheles gambiae*.

In light of the wide evolutionary conservation of Dnmt2 proteins, it will be interesting to determine the function of Dnmt2-mediated DNA methylation. Mouse embryonic stem cells with a disrupted *Dnmt2* gene proliferate normally and show no apparent phenotype (Okano et al. 1998). Consistent with these findings, RNAi-mediated knockdown of Dnmt2 did not affect the viability of *Drosophila* embryos (Kunert et al. 2003). However, the presence of subtle phenotypes in post-embryonic stages (J. Marhold and F. Lyko, unpublished data) suggested that DNA methylation is required for an as-yet-unidentified cellular function. It will be important to establish and analyze *Dnmt2* mutant fly strains that will provide intriguing insights into the biological function of DNA methylation.

3
DNA Methylation-Dependent Chromatin Structures

Recent studies in the filamentous fungus *Neurospora crassa* and in *Arabidopsis thaliana* revealed a close interaction between histone methylation and

DNA methylation. In *Neurospora*, *dim-2*-mediated DNA methylation has been shown to be dependent on the activity of the histone H3 methyltransferase *dim-5* (Tamaru and Selker 2001). In *Arabidopsis*, CHROMOMETHYLASE3-mediated DNA methylation has been shown to be dependent on the activity of the histone H3 methyltransferase KRYPTONITE (Jackson et al. 2002). A similar interaction between DNA methylation and histone methylation can be also found in *Drosophila*. The Su(var)3-9 histone methyltransferase specifically methylates lysine 9 of histone H3 (H3K9) (Schotta et al. 2002). Because in *Su(var)3-9* null mutant larvae H3K9 methylation was strongly reduced at the chromocenters, the protein seems to be specific for the methylation of histones in centromeric heterochromatin (Schotta et al. 2002). Null mutant flies for *Su(var)3-9* are viable and fertile (Tschiersch et al. 1994), similar to *knockout* mice for the murine *Drosophila Su(var)3-9* homologs *Suv39h1* and *Suv39h2* (Peters et al. 2001). Immunofluorescence staining of *Su(var)3-9*-null mutant fly embryos revealed a dramatic reduction if not complete loss of DNA methylation (Kunert et al. 2003), demonstrating a conservation of the interaction between DNA methylation and histone methylation.

Moreover, histone methylation also plays an important role in the transmission of epigenetic information from DNA methylation to repressive chromatin structures. Ectopic expression of the mouse DNA methyltransferase DNMT3a in *Drosophila* leads to lethality, which is characterized by irregular chromosome condensation and dysregulation of histone modifications (Weissmann et al. 2003). This lethality could be partially rescued when the ectopic expression of DNMT3a was induced in a *Su(var)3-9* mutant background (Weissmann et al. 2003). These data together with additional results from *Arabidopsis* suggest a mutual and complex relationship between DNA methylation and H3K9 methylation (Weissmann and Lyko 2003).

4
The Methyl-DNA Binding Protein MBD2/3

Epigenetic information encoded by methylated DNA has to be translated into specific chromatin structures in order to repress genes or establish specialized genomic compartments (or both). In vertebrates, this process involves a family of proteins that specifically bind to methylated CpG dinucleotides via a methyl-CpG binding domain (MBD) (Hendrich and Tweedie 2003). The first methyl-DNA binding protein investigated was MeCP2. Point mutations within the MBD of MeCP2 cause Rett syndrome, a childhood neurodevelopmental disorder (Amir et al. 1999). It has been shown that MeCP2 interacts with the SIN3 corepressor complex (Jones et al. 1998; Nan et al. 1998). This complex contains the SIN3 corepressor and (among other proteins) the histone deacetylases HDAC1 and HDAC2. Similar interactions have also been described for MBD2, another member of the MBD protein family. MBD2 recruits the MI-2/NuRD (nucleosome remodeling and histone deacetylase) complex to methylated sites of DNA. This complex also contains histone deacetylase activity and additionally an ATP-dependent nucleosome-remodeling activity, which is mediated by MI-2 (Tong et al. 1998; Wade et al. 1999; Zhang et al. 1999).

The *Drosophila* genome contains homologs for all vertebrate NuRD subunits, many of which have been shown to be necessary for proper development of the fly. *Drosophila MI-2* mutants die as first or second instar larvae, indicating that MI-2 is essential for embryogenesis (Kehle et al. 1998). The histone deacetylase RPD3, which is also contained in the *Drosophila* NuRD complex, is important for embryonic segmentation (Mannervik and Levine 1999). The fly genome also contains a single gene encoding a methyl-DNA binding domain protein, which has been designated MBD2/3. This protein shares more than 70% amino acid similarity to the vertebrate methyl-CpG binding proteins MBD2 and MBD3 (Tweedie et al. 1999). Alternative splicing of *MBD2/3* generates two distinct isoforms of the protein: The long isoform contains the methyl-DNA binding domain, a *Drosophila*-specific domain, and a coiled-coil domain, whereas the shorter protein (MBD2/3Δ) is lacking the *Drosophila*-specific domain and parts of the methyl-DNA binding domain (Roder et al. 2000; Tweedie et al. 1999). The long isoform can be detected during early to mid-embryogenesis, while the shorter protein is expressed in mid- and late embryogenesis (Ballestar et al. 2001; Marhold et al. 2002). Post-embryonic stages did not appear to express the protein (Marhold et al. 2002). The expression of MBD2/3 thus closely coincides with the expression of Dnmt2 and the presence of methylated DNA in the *Drosophila* genome (Kunert et al. 2003; Lyko et al. 2000b).

MBD2/3 has been shown to function as a transcriptional corepressor. Using reporter assays of transfected cell lines, it was shown that both MBD2/3 isoforms repress transcription, probably through the recruitment of histone deacetylase and nucleosome-remodeling activities (Ballestar et al. 2001; Roder et al. 2000). An association of MBD2/3 with RPD3 and MI-2 has been suggested previously, based on results from co-immunoprecipitation experiments (Tweedie et al. 1999). Biochemical fractionation of protein extracts also suggested the presence of MBD2/3Δ in the *Drosophila* MI-2/NuRD complex (Ballestar et al. 2001; Marhold et al. 2004a).

MBD2/3 associates with DNA during embryogenesis. During cleavage and syncytial blastoderm stages, MBD2/3 is excluded from DNA. This pattern changes dramatically during cellular blastoderm, when the protein forms bright foci that precisely colocalize with DNA (Marhold et al. 2002). Moreover, MBD2/3 also associates with the activated Y-chromosome in primary spermatocytes (Marhold et al. 2002). Here, the protein formed bright nuclear foci, while no such signals were detectable in mutants lacking an active Y-chromosome. These data indicate that MBD2/3 might be involved in the epigenetic regulation of the *Drosophila* genome during large-scale genome activation.

Similar to MBD2 knockout mice (Hendrich et al. 2001), null mutants of *MBD2/3* are viable and fertile, but reveal chromosomal segregation defects, suggesting that MBD2/3 plays an important role in the stabilization of pericentric heterochromatin (Marhold et al. 2004b). Confocal analysis of MBD2/3 mutant embryos, which were stained for MI-2, showed an abnormal MI-2 staining, suggesting that a proportion of MI-2 complexes is targeted by MBD2/3 (Marhold et al. 2004b). This appeared similar to a subset of vertebrate MI-2 complexes that are recruited by MBD2 (Ng et al. 1999; Wade et al. 1999; Zhang et al. 1999). Thus, MBD2/3 shares important functional characteristics with the mammalian methyl-DNA binding protein MBD2.

The similarity between *Drosophila* MBD2/3 and mammalian MBD2 has been discussed and is particularly controversial with respect to the methyl-DNA binding properties of MBD2/3. One report described the binding of a CpG-methylated probe to the short isoform of MBD2/3, which lacks parts of the MBD (Roder et al. 2000). In other reports no binding could be detected (Ballestar et al. 2001; Tweedie et al. 1999). Remarkably, only probes with methylated CpGs were used, which does not reflect the endogenous methylation pattern of *Drosophila*, which is ^{m}CpA/T (Lyko et al. 2000a). Indeed, recent work with probes containing methylated cytosine residues in the context of CpA and CpT showed specific binding to the long isoform of MBD2/3, but not to the short isoform (Marhold et al. 2004b). Under these conditions, no binding to CpG-methylated probes could be observed. Moreover, experimental demethylation of the *Drosophila* genome by RNAi of the DNA methyltransferase Dnmt2 or treatment with the DNA methyltransferase inhibitor 5-azacytidine led to the delocalization of MBD2/3 from DNA (Marhold et al. 2004b). These results provided an additional confirmation for a functional conservation between MBD2/3 and mammalian MBD2.

5
A Functional DNA Methylation System in *Drosophila*

It has been proposed that *Drosophila melanogaster* belongs to an atypical group of animals with no detectable genomic DNA methylation (Bird 1995).

After the detection of *Dnmt2* and *MBD2/3* genes, it was subsequently suggested that both genes encode inactive proteins and represent evolutionary vestiges of a simple DNA methylation system (Tweedie et al. 1999). However, several lines of evidence now demonstrate that *Drosophila* DNA is methylated and that *Dnmt2* encodes an active DNA methyltransferase (Kunert et al. 2003; Tang et al. 2003). Importantly, this process seems to be conserved in other dipteran insects (Marhold et al. 2004c). Several characteristics that may have contributed to its longstanding elusiveness distinguish *Drosophila* DNA methylation from mammalian DNA methylation. Similar arguments also pertain to the functional characteristics of MBD2/3. Until recently, the ability of the protein to bind methylated DNA has only been analyzed in a very restricted experimental context. All in vitro assays were performed with probes that were methylated at CpGs, which does not reflect the endogenous pattern of DNA methylation in the fly. In addition, the binding of MBD2/3 to *Drosophila* DNA had not been investigated in vivo. Recent results demonstrate that MBD2/3 binds specifically to CpT/A-methylated probes in vitro and that the protein becomes mislocalized in embryos with reduced DNA methylation (Marhold et al. 2004b). In summary, it thus appears likely that *Drosophila* contains an active DNA methylation system (Fig. 1). This system might utilize both DNA methylation and chromatin-based mechanisms to establish and maintain epigenetic information during development.

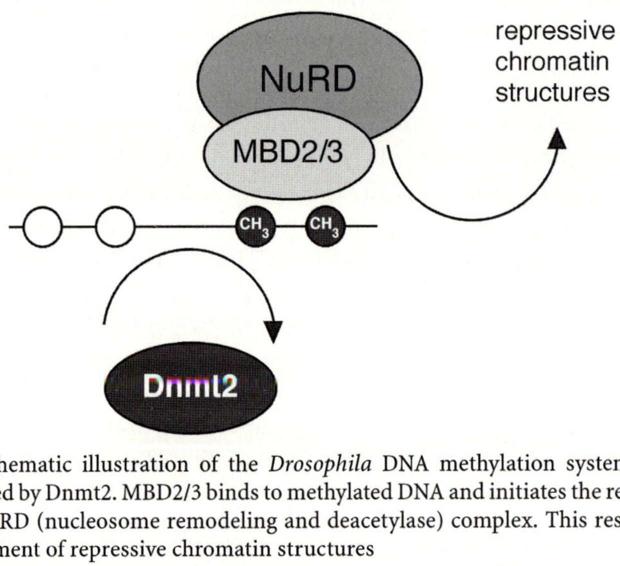

Fig. 1 Schematic illustration of the *Drosophila* DNA methylation system. DNA is methylated by Dnmt2. MBD2/3 binds to methylated DNA and initiates the recruitment of the NuRD (nucleosome remodeling and deacetylase) complex. This results in the establishment of repressive chromatin structures

6
Chromatin-Based Maintenance of Gene Expression

The Polycomb group (PcG) and Trithorax group (TrxG) proteins were initially identified as part of a regulatory system maintaining the expression state of the homeotic (*Hox*) genes of *Drosophila*. The products of the *Hox* genes are required to determine segmental identity. Initially, their expression level is set in the early embryo by transcriptional regulators encoded by the maternal and segmentation genes. With the downregulation of the segmentation gene products during mid-embryogenesis, the PcG and TrxG proteins take over and continue to maintain the silenced and the active state, respectively, through all following developmental stages. It is now well established that PcG and TrxG proteins regulate *Hox* gene expression in all multicellular organisms from *Caenorhabditis elegans* to mammals. Apart from their role in *Hox* gene regulation, additional target genes have been identified and their importance has been emphasized by their involvement in mammalian X-chromosome inactivation, cancer formation, and the maintenance of embryonic and adult stem cell fate (Muyrers-Chen and Paro 2001; Silva et al. 2003; Valk-Lingbeek et al. 2004).

Both the PcG and TrxG proteins act as large multisubunit protein complexes by modulating the chromatin structure in the vicinity of their target genes. Genetic studies in *Drosophila* identified several switchable *cis*-regulatory DNA elements, termed PcG response elements (PREs) or cellular memory modules (CMMs), that enable PcG proteins to bind and to maintain the status of transcriptional activity of the corresponding gene. In contrast to the low number of genetically identified target loci, PcG and TrxG proteins bind to about 200 chromosomal loci on polytene chromosomes. However, due to the low resolution of cytogenetic chromosome mapping, it is impossible to infer the precise localization of PREs and the corresponding PcG/TrxG responsive genes. A recent study used the knowledge gained from known PRE sequences to devise an algorithm that predicts PREs at a genome-wide level in *Drosophila* (Ringrose et al. 2003). This bioinformatic approach identified 167 candidate PREs and the genes that are possibly regulated by them (Table 1). Since PREs can be located tens of kilobases away from their responsive genes and only a subset of the predicted PREs overlap directly with gene sequences, the functional relationship between these genes and the new PREs has to be demonstrated. Nevertheless, the data provided new insights into the wide spectrum of pathways regulated by PcG and TrxG proteins and have the potential to uncover new information for the mammalian systems, in which no PRE sequence could be identified so far.

Table 1 Classes of predicted PcG/TrxG-regulated genes. Among the genes regulated by the 167 candidate PREs, about half can be related to specific developmental functions

Quantity	Molecular function/developmental process
26	Transcription factors (10 homeodomain proteins)
17	Embryonic patterning
10	Oogenesis
27	Late cell fate specification (Nervous system development, 13) (Eye development, 10)
2	Tumor suppressors (p53-like transcription factors)

7 Characterization of PcG/TrxG Complexes

Questions concerning the molecular mechanisms of PcG/TrxG-mediated gene regulation such as (1) the targeting of PcG and TrxG proteins, (2) their mechanism of conveying epigenetic inheritance through DNA replication and mitosis, and (3) the integration of their counteracting and cross-regulatory activities remain largely unsolved. In recent years the biochemical purification and characterization of distinct PcG and TrxG complexes has shed light on these problems. Especially the identification of accompanying enzymatic activities gave new insights into the opposing PcG and TrxG complexes and how they generate epigenetic signals that survive mitosis and DNA replication.

Two physically and functionally distinct classes of PcG complexes have been purified from *Drosophila*. The Polycomb repressive complex 1 (PRC1) and the related chromatin-associated silencing complex for homeotics (CHRASCH) contain the founding member of the PcG, the Polycomb (PC) protein. PcG proteins associated with these complexes are present in *Drosophila* and vertebrates but are absent in plants and *C. elegans*. The other complex (PRC2) contains the PcG proteins Extra sex combs (ESC), Enhancer of zeste [E(Z)] and Suppressor of zeste 12 [SU(Z)12] as core components. These proteins have been found in all multicellular model organisms from plants to mammals. In the following sections we focus on recent progress regarding molecular interactions and enzymatic activities of the PcG complexes and associated proteins and correlate them with the known antagonistic activities of TrxG proteins.

8
PC-Containing PcG Complexes and the Compaction of Chromatin

The core composition of PRC1 purified from *Drosophila* embryos and human HeLa cells has been conserved (Levine et al. 2002; Saurin et al. 2001; Shao et al. 1999), although the *Drosophila* complex seems to contain additional accessory proteins (see below). Consistent with the hypothesis that PcG complexes repress genes by affecting chromatin structure, purified PRC1 inhibited chromatin remodeling of in vitro assembled nucleosomal arrays by the human SWI/SNF (hSWI/SNF) complex. The hSWI/SNF complex is the homolog of the *Drosophila* Brahma (BRM) complex, a large ATP-dependent chromatin remodeling machine that contains the TrxG proteins BRM, Moira (MOR), OSA, and SNR1 (Crosby et al. 1999; Kal et al. 2000; Vazquez et al. 1999). A reconstituted recombinant PRC1 core complex (PCC) consisting of the four core components—PC, Polyhomeotic (PH), Posterior sex combs (PSC), and dRING/Sex combs extra (SCE)—also inhibited remodeling and repressed transcription on a chromatinized DNA template (Francis et al. 2001; King et al. 2002).

A prerequisite for the inhibition of chromatin remodeling was the preincubation of the chromatin template with PRC1, which speaks against the simple inactivation of SWI/SNF but rather for modulation of the chromatin substrate. The chromatin compaction of reconstituted nucleosomal arrays mediated by the PCC has been visualized by electron microscopy (Francis et al. 2004). Upon incubation of nucleosomal arrays with PCC, the relaxed beads-on-a-string conformation transformed to a compacted structure in which individual nucleosomes could not be detected. The observed chromatin compaction was independent of the histone tails and not mediated by the linker DNA, which does not exclude an important function of histone tails in PcG/TrxG mediated gene regulation in vivo (see Sect. 10). Previous work demonstrated a key role for PSC in the in vitro inhibition of chromatin remodeling and transcription (Francis et al. 2001; King et al. 2002) and showed that PH is dispensable for the PCC mediated in vitro functions (Lavigne et al. 2004). The underlying reasons for these observations could be visualized by electron microscopy. PSC alone and the PCC without PH were able to compact the chromatin templates. In addition, it could be shown that the C-terminus of PSC was required for chromatin compaction and inhibition of chromatin remodeling and transcription (Francis et al. 2004).

Another in vitro activity of mouse and *Drosophila* PCCs is their ability to recruit and repress a second chromatin template (Lavigne et al. 2004). PREs, the binding sites for PcG proteins, can be located tens of kilobases apart from their regulated genes but PcG proteins could be also localized at promoter regions (Orlando et al. 1998; Strutt and Paro 1997). In addition, the *Drosophila* PRE prediction proposed a PRE at or near the promoter of PRE-containing genes (Ringrose et al. 2003). Together, this indicates a mechanism of how PcG proteins could propagate their silencing effect, namely by interaction of PRE-bound PcG proteins with the promoter region of the corresponding gene. The PcG-promoter interaction was further confirmed by the identification of promoter associated factors like TBP (TATA-box binding protein) and several TBP-associated factors (TAFs) as accessory components of the PRC1 complex purified from *Drosophila* embryos (Saurin et al. 2001). TBP co-immunoprecipitates with PC, PH, and PSC from embryonic nuclear extracts, and chromatin immunoprecipitation assays detected the colocalization of PC, PH, and the general transcription factors TFIIB and TFIIF at PcG-regulated *Hox* gene promoters in *Drosophila* cells (Breiling et al. 2001; Saurin et al. 2001).

Two additional proteins that co-purified with *Drosophila* PRC1 have been identified as the PcG protein Sex comb on midleg (SCM) and Zeste (Saurin et al. 2001; Shao et al. 1999). Zeste is a sequence-specific DNA binding protein that is associated with both PcG-mediated gene silencing and TrxG-mediated gene activation. It colocalizes with many PC binding sites at polytene chromosomes and has been shown to be involved in the maintenance of repression (Hur et al. 2002; Rastelli et al. 1993). The incorporation of Zeste into recombinant PCC enhanced the inhibitory activity of this complex and mediated a preference for templates containing Zeste binding sites (Mulholland et al. 2003). In contrast, Zeste has also been shown to be recruited to an activated Fab-7 minimal PRE, which subsequently led to the binding of BRM (Dejardin and Cavalli 2004).

In an attempt to identify the protein complex that is responsible for histone H2A ubiquitination, another PcG complex has been purified from human HeLa cells that contains Ring1 and Ring2 (the human homologs of dRing/SCE) Bmi1 (a PSC homolog) and HPH2 (a PH homolog) (Wang et al. 2004). Because of its similarity to PRC1—PC is the only missing subunit from the four core components—the complex was termed human PRC1-like (hPRC1L). Ring2 and dRing/SCE have been identified as E3 ubiquitin ligases that ubiquitinate histone H2A at lysine 119. A point mutation within the RING domain of dRing/SCE that results in an inactive enzyme leads to *Ultrabithorax* (a *Drosophila Hox* gene) derepression in wing imaginal discs (Wang et al. 2004). Since dRing/SCE is a core component of PRC1, it will not be surprising if PRC1 also shows H2A ubiquitination activity.

Lastly, the CHRASCH complex that was purified from a *Drosophila* cell line is also similar to PRC1. The main difference has been found in the association of the PcG protein Pipsqueak (PSQ), a sequence-specific DNA binding protein that mediates the binding of the complex to PREs that contain the (GA)n motif

(Huang and Chang 2004; Huang et al. 2002), suggesting that the CHRASCH complex might be involved in the regulation of a specific subset of PC target genes.

9 ESC/E(Z)-Containing Complexes and Histone Deacetylation/Methylation

The PC-containing complexes are considered to be responsible for the direct repression of gene activity, whereas the ESC/E(Z)-containing complexes are thought to be required earlier, for the establishment of molecular marks that initiate PcG-mediated silencing. Consistent with this notion, ESC has a special place among the known PcG proteins. Its mRNA is only expressed during oogenesis and the first hours of embryonic development, and the function of the protein seems to be required for only a brief 3- to 4-h period of embryogenesis (Simon et al. 1995). E(Z) has been shown to be required for the binding of other PcG proteins to chromosomes and colocalizes with other PcG proteins at many sites (Carrington and Jones 1996; Rastelli et al. 1993).

In recent years, several groups have independently purified ESC/E(Z)-containing complexes from *Drosophila* and human cells (Cao et al. 2002; Czermin et al. 2002; Furuyama et al. 2003; Kuzmichev et al. 2002; Muller et al. 2002; Tie et al. 2001, 2003; van der Vlag and Otte 1999). All the complexes contain the same PcG proteins [ESC, E(Z) and SU(Z)12], the histone binding protein p55 that is also a component of the nucleosome remodeling complexes NuRD and NURF, and the chromatin assembly factor CAF-1 (Marhold et al. 2004a; Martinez-Balbas et al. 1998; Tyler et al. 1996).

The composition of ESC/E(Z) complexes undergoes dynamic changes during *Drosophila* embryonic and larval development. In embryonic extracts, a 600-kDa (termed PRC2) and a 1-MDa complex could be separated, with the latter disappearing at the end of embryogenesis and a new 3-MDa complex becoming detectable in larval extracts (Furuyama et al. 2003; Tie et al. 2003). In addition to the four core components, the 1-MDa complex also contained the RPD3 histone deacetylase and the PcG protein Polycomb-like (PCL), whereas the 3-MDa complex contained RPD3 and the SIR2 histone deacetylase (Furuyama et al. 2004; Tie et al. 2003). Although mutations in either *Rpd3* or *Sir2* do not result in PcG phenotypes, they enhance the phenotypes of PcG mutants (Chang et al. 2001; Furuyama et al. 2004; Mannervik and Levine 1999), which suggests a functional role in PcG regulation.

The presence of a SET [Su(var)3-9, E(Z), and Trithorax] histone methyltransferase (HMT) domain in E(Z) provided a strong indication for HMT activity in ESC/E(Z) complexes. Indeed, PRC2 complexes can methylate lysines 9 and 27 of histone H3 (H3K9me, H3K27me) (Cao et al. 2002; Czermin et al. 2002; Kuzmichev et al. 2002; Muller et al. 2002). In addition, mammalian E(Z) was shown to methylate lysine 26 in histone H1 (H1K26me) under certain conditions (Kuzmichev et al. 2004). This specificity of mammalian E(Z) is triggered by the interaction with specific isoforms of EED [the mammalian E(Z) homolog]. Complexes with shorter isoforms of EED methylate H3K27, whereas complexes with longer isoforms methylate histone H1. The additional domain that is present in the longer splice variant seems to be responsible for H1K26 methylation and is absent in *Drosophila* ESC (Levine et al. 2004). Together, these results provided convincing evidence for a functional role of histone methylation in epigenetic regulation by ESC/E(Z) complexes.

10 The Complexity of PcG/TrxG-Regulated Chromatin

During the last decade, several covalent histone modifications have been recognized to play a fundamental role in chromatin-influenced processes such as transcription (Strahl and Allis 2000; Turner 2002). Regulation of gene activity has been correlated with acetylation of histones by histone acetyltransferases (HATs) and deacetylation by deacetylases (HDACs) (Marmorstein and Roth 2001). In fission yeast, histone acetylation provides an epigenetic tag that is stably inherited through mitosis and meiosis (Ekwall et al. 1997). The role of histone acetylation in PcG/TrxG-dependent processes is supported by the observation that disruption of PRE-mediated silencing in *Drosophila* transgenes is accompanied by local accumulation of hyperacetylated histone H4 (Cavalli and Paro 1999). Thus, the active status of a PRE appears to involve histone acetylation. The direct interaction of TrxG proteins with histone acetyltransferase activity could be demonstrated by the purification of the TAC1 complex that contained TRX, dCBP (a member of the CBP/p300 HAT family), and the antiphosphatase Sbf1 (Petruk et al. 2001).

The discovery that E(Z) methylates H3K9 and H3K27 suggested an attractive hypothesis how E(Z) could set a chromatin mark that is responsible for establishing and maintaining the silent state. PC contains a chromodomain that binds a histone H3 tail peptide methylated at lysine 27 and with lower affinity also a peptide methylated at lysine 9 (Fischle et al. 2003). This interaction could potentially stabilize PRC1 complexes at their site of action. Indeed, differently methylated histones have been identified at different gene locations, suggesting that different combinations of H3K9me/H3K27me contribute to locus-specific stabilities of PRC1 complexes [see Ringrose et al. (2004) and Ringrose and Paro (2004) for a more thorough discussion of this topic]. As such, histone methylation seems to play a crucial role in PcG/TrxG-dependent processes. The TrxG proteins TRX and ASH1 (Absent Small or Homeotic discs 1) that are present in two separate complexes in *Drosophila* embryonic extracts were identified as antirepressors required throughout development to specifically counteract the silencing effects of the PcG (Klymenko and Muller 2004; Papoulas et al. 1998). Both proteins also contain a SET domain and exhibit HMT activity specific for lysine 4 of histone H3, which was previously connected to transcriptional activation (Beisel et al. 2002; Czermin et al. 2002; Strahl and Allis 2000).

To date, 18 *PcG* and 17 *trxG* genes have been genetically identified in *Drosophila* (Ringrose and Paro 2004) of which only 10 and 5 proteins, respectively, are described in this review based on the available information with respect to their molecular function. This suggests that additional complexes may exist. Indeed, there is strong evidence that the function and composition of the core complexes are modulated in a tissue- and target gene-specific way (Furuyama et al. 2003; Otte and Kwaks 2003; Rastelli et al. 1993; Strutt and Paro 1997). For example, co-immunoprecipitation experiments of PC with early *Drosophila* embryonic extracts showed an interaction with E(Z), ESC, and Pleiohomeotic (PHO), a DNA binding PcG protein (Poux et al. 2001). An interaction with PH and PSC could not be detected in these experiments. Five DNA binding proteins have been linked to PcG/TrxG-dependent regulatory mechanisms in *Drosophila*, namely Zeste, PSQ, PHO, PHO-Like, and GAF, but how they mediate PRE binding remains elusive (for a review of interactions see Ringrose and Paro 2004).

To fully understand the molecular mechanisms underlying PcG/TrxG-mediated epigenetic inheritance, we need to obtain more information about the identity and molecular interactions of these proteins and complexes. Histone-modifying activities have been tightly connected to the PcG/TrxG system, but their exact function remains to be uncovered. Regarding the regulation of these proteins, the investigation of enzymatic activities that modify PcG and TrxG proteins will also give new insights into this complex system.

Acknowledgements Work in F.L.'s laboratory is supported by grants from the Deutsche Forschungsgemeinschaft. Work in R.P.'s laboratory is supported by grants from the Deutsche Forschungsgemeinschaft and the EU-FP6 NoE "Epigenome."

References

Adams MD, Celniker SE, Holt RA, Evans CA, Gocayne JD, Amanatides PG, Scherer SE, Li PW, Hoskins RA, Galle RF, George RA, Lewis SE, Richards S, Ashburner M, Henderson SN, Sutton GG, Wortman JR, Yandell MD, Zhang Q, Chen LX, Brandon RC, Rogers YH, Blazej RG, Champe M, Pfeiffer BD, Wan KH, Doyle C, Baxter EG, Helt G, Nelson CR, Gabor GL, Abril JF, Agbayani A, An HJ, Andrews-Pfannkoch C, Baldwin D, Ballew RM, Basu A, Baxendale J, Bayraktaroglu L, Beasley EM, Beeson KY, Benos PV, Berman BP, Bhandari D, Bolshakov S, Borkova D, Botchan MR, Bouck J, Brokstein P, Brottier P, Burtis KC, Busam DA, Butler H, Cadieu E, Center A, Chandra I, Cherry JM, Cawley S, Dahlke C, Davenport LB, Davies P, de Pablos B, Delcher A, Deng Z, Mays AD, Dew I, Dietz SM, Dodson K, Doup LE, Downes M, Dugan-Rocha S, Dunkov BC, Dunn P, Durbin KJ, Evangelista CC, Ferraz C, Ferriera S, Fleischmann W, Fosler C, Gabrielian AE, Garg NS, Gelbart WM, Glasser K, Glodek A, Gong F, Gorrell JH, Gu Z, Guan P, Harris M, Harris NL, Harvey D, Heiman TJ, Hernandez JR, Houck J, Hostin D, Houston KA, Howland TJ, Wei MH, Ibegwam C, et al (2000) The genome sequence of Drosophila melanogaster. Science 287:2185–2195

Amir RE, Van den Veyver IB, Wan M, Tran CQ, Francke U, Zoghbi HY (1999) Rett syndrome is caused by mutations in X-linked MECP2, encoding methyl-CpG-binding protein 2. Nat Genet 23:185–188

Ballestar E, Pile LA, Wassarman DA, Wolffe AP, Wade PA (2001) A Drosophila MBD family member is a transcriptional corepressor associated with specific genes. Eur J Biochem 268:5397–5406

Beisel C, Imhof A, Greene J, Kremmer E, Sauer F (2002) Histone methylation by the Drosophila epigenetic transcriptional regulator Ash1. Nature 419:857–862

Bird AP (1995) Gene number, noise reduction and biological complexity. Trends Genet 11:94–100

Bird AP, Taggart MH (1980) Variable patterns of total DNA and rDNA methylation in animals. Nucleic Acids Res 8:1485–1497

Blower MD, Sullivan BA, Karpen GH (2002) Conserved organization of centromeric chromatin in flies and humans. Dev Cell 2:319–330

Breiling A, Turner BM, Bianchi ME, Orlando V (2001) General transcription factors bind promoters repressed by Polycomb group proteins. Nature 412:651–655

Cao R, Wang L, Wang H, Xia L, Erdjument-Bromage H, Tempst P, Jones RS, Zhang Y (2002) Role of histone H3 lysine 27 methylation in Polycomb-group silencing. Science 298:1039–1043

Carrington EA, Jones RS (1996) The Drosophila Enhancer of zeste gene encodes a chromosomal protein: examination of wild-type and mutant protein distribution. Development 122:4073–4083

Cavalli G, Paro R (1999) Epigenetic inheritance of active chromatin after removal of the main transactivator. Science 286:955–958

Chang YL, Peng YH, Pan IC, Sun DS, King B, Huang DH (2001) Essential role of Drosophila Hdac1 in homeotic gene silencing. Proc Natl Acad Sci U S A 98:9730–9735

Crosby MA, Miller C, Alon T, Watson KL, Verrijzer CP, Goldman-Levi R, Zak NB (1999) The trithorax group gene moira encodes a brahma-associated putative chromatin-remodeling factor in Drosophila melanogaster. Mol Cell Biol 19:1159–1170

Czermin B, Melfi R, McCabe D, Seitz V, Imhof A, Pirrotta V (2002) Drosophila enhancer of Zeste/ESC complexes have a histone H3 methyltransferase activity that marks chromosomal Polycomb sites. Cell 111:185–196

Dejardin J, Cavalli G (2004) Chromatin inheritance upon Zeste-mediated Brahma recruitment at a minimal cellular memory module. EMBO J 23:857–868

Dong A, Yoder JA, Zhang X, Zhou L, Bestor TH, Cheng X (2001) Structure of human DNMT2, an enigmatic DNA methyltransferase homolog that displays denaturant-resistant binding to DNA. Nucleic Acids Res 29:439–448

Ekwall K, Olsson T, Turner BM, Cranston G, Allshire RC (1997) Transient inhibition of histone deacetylation alters the structural and functional imprint at fission yeast centromeres. Cell 91:1021–1032

Field LM, Lyko F, Mandrioli M, Prantera G (2004) DNA methylation in insects. Insect Mol Biol 13:109–115

Fischle W, Wang Y, Jacobs SA, Kim Y, Allis CD, Khorasanizadeh S (2003) Molecular basis for the discrimination of repressive methyl-lysine marks in histone H3 by Polycomb and HP1 chromodomains. Genes Dev 17:1870–1881

Fisher O, Siman-Tov R, Ankri S (2004) Characterization of cytosine methylated regions and 5-cytosine DNA methyltransferase (Ehmeth) in the protozoan parasite Entamoeba histolytica. Nucleic Acids Res 32:287–297

Francis NJ, Saurin AJ, Shao Z, Kingston RE (2001) Reconstitution of a functional core polycomb repressive complex. Mol Cell 8:545–556

Francis NJ, Kingston RE, Woodcock CL (2004) Chromatin compaction by a polycomb group protein complex. Science 306:1574–1577

Furuyama T, Tie F, Harte PJ (2003) Polycomb group proteins ESC and E(Z) are present in multiple distinct complexes that undergo dynamic changes during development. Genesis 35:114–124

Furuyama T, Banerjee R, Breen TR, Harte PJ (2004) SIR2 is required for polycomb silencing and is associated with an E(Z) histone methyltransferase complex. Curr Biol 14:1812–1821

Gowher H, Leismann O, Jeltsch A (2000) DNA of Drosophila melanogaster contains 5-methylcytosine. EMBO J 19:6918–6923

Hendrich B, Tweedie S (2003) The methyl-CpG binding domain and the evolving role of DNA methylation in animals. Trends Genet 19:269–277

Hendrich B, Guy J, Ramsahoye B, Wilson VA, Bird A (2001) Closely related proteins MBD2 and MBD3 play distinctive but interacting roles in mouse development. Genes Dev 15:710–723

Hermann A, Schmitt S, Jeltsch A (2003) The human Dnmt2 has residual DNA-(cytosine-C5) methyltransferase activity. J Biol Chem 278:31717–31721

Huang DH, Chang YL (2004) Isolation and characterization of CHRASCH, a polycomb-containing silencing complex. Methods Enzymol 377:267–282

Huang DH, Chang YL, Yang CC, Pan IC, King B (2002) pipsqueak encodes a factor essential for sequence-specific targeting of a polycomb group protein complex. Mol Cell Biol 22:6261–6271

Hung MS, Karthikeyan N, Huang B, Koo HC, Kiger J, Shen CJ (1999) Drosophila proteins related to vertebrate DNA (5-cytosine) methyltransferases. Proc Natl Acad Sci USA 96:11940–11945

Hur MW, Laney JD, Jeon SH, Ali J, Biggin MD (2002) Zeste maintains repression of Ubx transgenes: support for a new model of Polycomb repression. Development 129:1339–1343

Jackson JP, Lindroth AM, Cao X, Jacobsen SE (2002) Control of CpNpG DNA methylation by the KRYPTONITE histone H3 methyltransferase. Nature 416:556–560

Jones PL, Veenstra GJ, Wade PA, Vermaak D, Kass SU, Landsberger N, Strouboulis J, Wolffe AP (1998) Methylated DNA and MeCP2 recruit histone deacetylase to repress transcription. Nat Genet 19:187–191

Kal AJ, Mahmoudi T, Zak NB, Verrijzer CP (2000) The Drosophila brahma complex is an essential coactivator for the trithorax group protein zeste. Genes Dev 14:1058–1071

Kehle J, Beuchle D, Treuheit S, Christen B, Kennison JA, Bienz M, Muller J (1998) dMi-2, a hunchback-interacting protein that functions in polycomb repression. Science 282:1897–1900

King IF, Francis NJ, Kingston RE (2002) Native and recombinant polycomb group complexes establish a selective block to template accessibility to repress transcription in vitro. Mol Cell Biol 22:7919–7928

Klymenko T, Muller J (2004) The histone methyltransferases Trithorax and Ash1 prevent transcriptional silencing by Polycomb group proteins. EMBO Rep 5:373–377

Kunert N, Marhold J, Stanke J, Stach D, Lyko F (2003) A Dnmt2-like protein mediates DNA methylation in Drosophila. Development 130:5083–5090

Kuzmichev A, Nishioka K, Erdjument-Bromage H, Tempst P, Reinberg D (2002) Histone methyltransferase activity associated with a human multiprotein complex containing the Enhancer of Zeste protein. Genes Dev 16:2893–2905

Kuzmichev A, Jenuwein T, Tempst P, Reinberg D (2004) Different EZH2-containing complexes target methylation of histone H1 or nucleosomal histone H3. Mol Cell 14:183–193

Lavigne M, Francis NJ, King IF, Kingston RE (2004) Propagation of silencing; recruitment and repression of naive chromatin in trans by polycomb repressed chromatin. Mol Cell 13:415–425

Levine SS, Weiss A, Erdjument-Bromage H, Shao Z, Tempst P, Kingston RE (2002) The core of the polycomb repressive complex is compositionally and functionally conserved in flies and humans. Mol Cell Biol 22:6070–6078

Levine SS, King IF, Kingston RE (2004) Division of labor in polycomb group repression. Trends Biochem Sci 29:478–485

Lyko F, Ramsahoye BH, Jaenisch R (2000a) DNA methylation in Drosophila melanogaster. Nature 408:538–540

Lyko F, Whittaker AJ, Orr-Weaver TL, Jaenisch R (2000b) The putative Drosophila methyltransferase gene dDnmt2 is contained in a transposon-like element and is expressed specifically in ovaries. Mech Dev 95:215–217

Mannervik M, Levine M (1999) The Rpd3 histone deacetylase is required for segmentation of the Drosophila embryo. Proc Natl Acad Sci U S A 96:6797–6801

Marhold J, Zbylut M, Lankenau DH, Li M, Gerlich D, Ballestar E, Mechler BM, Lyko F (2002) Stage-specific chromosomal association of Drosophila dMBD2/3 during genome activation. Chromosoma 111:13–21
Marhold J, Brehm A, Kramer K (2004a) The Drosophila methyl-DNA binding protein MBD2/3 interacts with the NuRD complex via p55 and MI-2. BMC Mol Biol 5:20
Marhold J, Kramer K, Kremmer E, Lyko F (2004b) The Drosophila MBD2/3 protein mediates interactions between the MI-2 chromatin complex and CpT/A-methylated DNA. Development 131:6033–6039
Marhold J, Rothe N, Pauli A, Mund C, Kuehle K, Brueckner B, Lyko F (2004c) Conservation of DNA methylation in dipteran insects. Insect Mol Biol 13:117–123
Marmorstein R, Roth SY (2001) Histone acetyltransferases: function, structure, and catalysis. Curr Opin Genet Dev 11:155–161
Martinez-Balbas MA, Tsukiyama T, Gdula D, Wu C (1998) Drosophila NURF-55, a WD repeat protein involved in histone metabolism. Proc Natl Acad Sci U S A 95:132–137
Mulholland NM, King IF, Kingston RE (2003) Regulation of Polycomb group complexes by the sequence-specific DNA binding proteins Zeste and GAGA. Genes Dev 17:2741–2746
Muller J, Hart CM, Francis NJ, Vargas ML, Sengupta A, Wild B, Miller EL, O'Connor MB, Kingston RE, Simon JA (2002) Histone methyltransferase activity of a Drosophila Polycomb group repressor complex. Cell 111:197–208
Mund C, Musch T, Strodicke M, Assmann B, Li E, Lyko F (2004) Comparative analysis of DNA methylation patterns in transgenic Drosophila overexpressing mouse DNA methyltransferases. Biochem J 378:763–768
Muyrers-Chen I, Paro R (2001) Epigenetics: unforeseen regulators in cancer. Biochim Biophys Acta 1552:15–26
Nan X, Ng HH, Johnson CA, Laherty CD, Turner BM, Eisenman RN, Bird A (1998) Transcriptional repression by the methyl-CpG-binding protein MeCP2 involves a histone deacetylase complex. Nature 393:386–389
Ng HH, Zhang Y, Hendrich B, Johnson CA, Turner BM, Erdjument-Bromage H, Tempst P, Reinberg D, Bird A (1999) MBD2 is a transcriptional repressor belonging to the MeCP1 histone deacetylase complex. Nat Genet 23:58–61
Okano M, Xie S, Li E (1998) Dnmt2 is not required for de novo and maintenance methylation of viral DNA in embryonic stem cells. Nucleic Acids Res 26:2536–2540
Orlando V, Jane EP, Chinwalla V, Harte PJ, Paro R (1998) Binding of trithorax and Polycomb proteins to the bithorax complex: dynamic changes during early Drosophila embryogenesis. EMBO J 17:5141–5150
Otte AP, Kwaks TH (2003) Gene repression by Polycomb group protein complexes: a distinct complex for every occasion? Curr Opin Genet Dev 13:448–454
Papoulas O, Beek SJ, Moseley SL, McCallum CM, Sarte M, Shearn A, Tamkun JW (1998) The Drosophila trithorax group proteins BRM, ASH1 and ASH2 are subunits of distinct protein complexes. Development 125:3955–3966
Patel CV, Gopinathan KP (1987) Determination of trace amounts of 5-methylcytosine in DNA by reverse-phase high-performance liquid chromatography. Anal Biochem 164:164–169

Peters AH, O'Carroll D, Scherthan H, Mechtler K, Sauer S, Schofer C, Weipoltshammer K, Pagani M, Lachner M, Kohlmaier A, Opravil S, Doyle M, Sibilia M, Jenuwein T (2001) Loss of the Suv39h histone methyltransferases impairs mammalian heterochromatin and genome stability. Cell 107:323–337

Petruk S, Sedkov Y, Smith S, Tillib S, Kraevski V, Nakamura T, Canaani E, Croce CM, Mazo A (2001) Trithorax and dCBP acting in a complex to maintain expression of a homeotic gene. Science 294:1331–1334

Pinarbasi E, Elliott J, Hornby DP (1996) Activation of a yeast pseudo DNA methyltransferase by deletion of a single amino acid. J Mol Biol 257:804–813

Poux S, Melfi R, Pirrotta V (2001) Establishment of Polycomb silencing requires a transient interaction between PC and ESC. Genes Dev 15:2509–2514

Rae PM, Steele RE (1979) Absence of cytosine methylation at C-C-G-G and G-C-G-C sites in the rDNA coding regions and intervening sequences of Drosophila and the rDNA of other insects. Nucleic Acids Res 6:2987–2995

Rastelli L, Chan CS, Pirrotta V (1993) Related chromosome binding sites for zeste, suppressors of zeste and Polycomb group proteins in Drosophila and their dependence on Enhancer of zeste function. EMBO J 12:1513–1522

Ringrose L, Paro R (2004) Epigenetic regulation of cellular memory by the polycomb and trithorax group proteins. Annu Rev Genet 38:413–443

Ringrose L, Rehmsmeier M, Dura JM, Paro R (2003) Genome-wide prediction of Polycomb/Trithorax response elements in Drosophila melanogaster. Dev Cell 5:759–771

Ringrose L, Ehret H, Paro R (2004) Distinct contributions of histone H3 lysine 9 and 27 methylation to locus-specific stability of polycomb complexes. Mol Cell 16:641–653

Roder K, Hung MS, Lee TL, Lin TY, Xiao H, Isobe KI, Juang JL, Shen CJ (2000) Transcriptional repression by Drosophila methyl-CpG-binding proteins. Mol Cell Biol 20:7401–7409

Saurin AJ, Shao Z, Erdjument-Bromage H, Tempst P, Kingston RE (2001) A Drosophila Polycomb group complex includes Zeste and dTAFII proteins. Nature 412:655–660

Schotta G, Ebert A, Krauss V, Fischer A, Hoffmann J, Rea S, Jenuwein T, Dorn R, Reuter G (2002) Central role of Drosophila SU(VAR)3-9 in histone H3-K9 methylation and heterochromatic gene silencing. EMBO J 21:1121–1131

Shao Z, Raible F, Mollaaghababa R, Guyon JR, Wu CT, Bender W, Kingston RE (1999) Stabilization of chromatin structure by PRC1, a Polycomb complex. Cell 98:37–46

Silva J, Mak W, Zvetkova I, Appanah R, Nesterova TB, Webster Z, Peters AH, Jenuwein T, Otte AP, Brockdorff N (2003) Establishment of histone h3 methylation on the inactive X chromosome requires transient recruitment of Eed-Enx1 polycomb group complexes. Dev Cell 4:481–495

Simon J, Bornemann D, Lunde K, Schwartz C (1995) The extra sex combs product contains WD40 repeats and its time of action implies a role distinct from other Polycomb group products. Mech Dev 53:197–208

Strahl BD, Allis CD (2000) The language of covalent histone modifications. Nature 403:41–45

Strutt H, Paro R (1997) The polycomb group protein complex of Drosophila melanogaster has different compositions at different target genes. Mol Cell Biol 17:6773–6783

Tamaru H, Selker EU (2001) A histone H3 methyltransferase controls DNA methylation in Neurospora crassa. Nature 414:277–283
Tang LY, Reddy MN, Rasheva V, Lee TL, Lin MJ, Hung MS, Shen CK (2003) The eukaryotic DNMT2 genes encode a new class of cytosine-5 DNA methyltransferases. J Biol Chem 278:33613–33616
Tie F, Furuyama T, Prasad-Sinha J, Jane E, Harte PJ (2001) The Drosophila Polycomb group proteins ESC and E(Z) are present in a complex containing the histone-binding protein p55 and the histone deacetylase RPD3. Development 128:275–286
Tie F, Prasad-Sinha J, Birve A, Rasmuson-Lestander A, Harte PJ (2003) A 1-megadalton ESC/E(Z) complex from Drosophila that contains polycomblike and RPD3. Mol Cell Biol 23:3352–3362
Tong JK, Hassig CA, Schnitzler GR, Kingston RE, Schreiber SL (1998) Chromatin deacetylation by an ATP-dependent nucleosome remodelling complex. Nature 395:917–921
Tschiersch B, Hofmann A, Krauss V, Dorn R, Korge G, Reuter G (1994) The protein encoded by the Drosophila position-effect variegation suppressor gene Su(var)3-9 combines domains of antagonistic regulators of homeotic gene complexes. EMBO J 13:3822–3831
Turner BM (2002) Cellular memory and the histone code. Cell 111:285–291
Tweedie S, Ng HH, Barlow AL, Turner BM, Hendrich B, Bird A (1999) Vestiges of a DNA methylation system in Drosophila melanogaster? Nat Genet 23:389–390
Tyler JK, Bulger M, Kamakaka RT, Kobayashi R, Kadonaga JT (1996) The p55 subunit of Drosophila chromatin assembly factor 1 is homologous to a histone deacetylase-associated protein. Mol Cell Biol 16:6149–6159
Urieli-Shoval S, Gruenbaum Y, Sedat J, Razin A (1982) The absence of detectable methylated bases in Drosophila melanogaster DNA. FEBS Lett 146:148–152
Valk-Lingbeek ME, Bruggeman SW, van Lohuizen M (2004) Stem cells and cancer; the polycomb connection. Cell 118:409–418
van der Vlag J, Otte AP (1999) Transcriptional repression mediated by the human polycomb-group protein EED involves histone deacetylation. Nat Genet 23:474–478
Vazquez M, Moore L, Kennison JA (1999) The trithorax group gene osa encodes an ARID-domain protein that genetically interacts with the brahma chromatin-remodeling factor to regulate transcription. Development 126:733–742
Wade PA, Gegonne A, Jones PL, Ballestar E, Aubry F, Wolffe AP (1999) Mi-2 complex couples DNA methylation to chromatin remodelling and histone deacetylation. Nat Genet 23:62–66
Wang H, Wang L, Erdjument-Bromage H, Vidal M, Tempst P, Jones RS, Zhang Y (2004) Role of histone H2A ubiquitination in Polycomb silencing. Nature 431:873–878
Weissmann F, Lyko F (2003) Cooperative interactions between epigenetic modifications and their function in the regulation of chromosome architecture. Bioessays 25:792–797
Weissmann F, Muyrers-Chen I, Musch T, Stach D, Wiessler M, Paro R, Lyko F (2003) DNA hypermethylation in Drosophila melanogaster causes irregular chromosome condensation and dysregulation of epigenetic histone modifications. Mol Cell Biol 23:2577–2586

Wilkinson CR, Bartlett R, Nurse P, Bird AP (1995) The fission yeast gene pmt1+ encodes a DNA methyltransferase homologue. Nucleic Acids Res 23:203–210
Zhang Y, Ng HH, Erdjument-Bromage H, Tempst P, Bird A, Reinberg D (1999) Analysis of the NuRD subunits reveals a histone deacetylase core complex and a connection with DNA methylation. Genes Dev 13:1924–1935

CTMI (2006) 310:45–59

Epimutations in Human Disease

B. Horsthemke (✉)

Institut für Humangenetik, Universitätsklimikum Essen,
Hufelandstrasse 55, 45122 Essen, Germany
b.horsthemke@uni-essen.de

Abstract Epigenetics is the study of genes during development. Gene expression states are set by transcriptional activators and repressors and locked in by cell-heritable chromatin states. Inappropriate expression or repression of genes can change developmental trajectories and result in disease. Aberrant chromatin states leading to aberrant gene expression patterns (epimutations) have been detected in several recognizable syndromes as well as in cancer. They can occur secondary to a DNA mutation in a *cis*- or *trans*-acting factor, or as a "true" or primary epimutation in the absence of any DNA sequence change. Primary epimutations often occur after fertilization and lead to somatic mosaicism. It has been estimated that the rate of primary epimutations is one or two orders of magnitude greater than somatic DNA mutation. Therefore, the contribution of epimutations to human disease is probably underestimated.

1
Epigenetic Inheritance

Variations in the DNA sequence account for most of the heritable diversity of a species. DNA sequence changes that are not repaired before mitosis are transmitted to daughter cells and, if deleterious, can impair cellular function. Mutations that are present in the germline will be transmitted to the next generation. Mutation research is therefore an integral component of genetics and has made important contributions to the understanding of human disease.

Over the past 30 years, it has been increasingly recognized that genetic disease can be caused not only by chromosomal aberrations and DNA mutations, but also by aberrant gene activity states that occur in the absence of a DNA sequence change. During the development of an organism, different sets of genes are switched on or off by transcription factors. The activity states of these genes are stably transmitted through several rounds of cell division, but they may also be reversed. The transmission of cell-heritable, potentially reversible gene activity states is the basis of epigenetic inheritance. Epigenetic inheritance is essential for the normal development and function of an organism. Therefore, it is not unreasonable to assume that epigenetic variation contributes significantly to phenotypic variance. It is also possible that variation of the epigenetic system, which buffers the developmental process against genetic and environmental perturbations (Waddington 1959), allows hidden genetic variation to be phenotypically manifested (Sollars et al. 2003). In turn, some of the epigenetic variation has a genetic basis.

While it is well established that epigenetic states are transmitted from cells to daughter cells, it is a matter of debate whether epigenetic states can be transmitted from one generation to another. It is generally believed that the genome is reprogrammed during gametogenesis and early embryonic development. Nevertheless, transgenerational epigenetic inheritance has been demonstrated in experimental model systems. However, evidence for transgenerational inheritance in normal populations is lacking, although there are some suggestive data (Chong and Whitelaw 2004).

Epigenetic states are not encoded in the DNA sequence, but in the configuration of the chromatin. Roughly speaking, chromatin exists in a transcriptionally competent or a transcriptionally silent state. Distinguishing features of different epigenetic states are the modification of the DNA and histones, the presence or absence of different non-histone proteins and the position of the nucleosomes. Acetylation of histone 3 at lysine 9 (H3K9), for example, is typically found in transcriptionally competent chromatin, whereas H3K9 methylation is a hallmark of the transcriptionally silent state. In mammals, the silent state of certain regions is also marked by the methylation of cytosine residues located in CpG dinucleotides at the 5′ end of genes. Mitotic transmission of these states implies that the DNA methylation patterns are copied onto the newly synthesized daughter strand and that the appropriate histone and non-histone proteins are (re)assembled on the two daughter helices. The DNA methylation patterns are replicated by the maintenance DNA methyltransferase DNMT1, which recognizes hemimethylated DNA. The assembly of the appropriate proteins is less well understood.

2
Classification of Epimutations

It is obvious that errors in replicating the DNA methylation pattern and/or the histone pattern can affect the epigenetic state of a gene in the daughter cell. Such epimutations (Holliday 1987) can lead to inappropriate activation of a gene that should be silent, or inactivation of a gene that should be active. Epimutations can also result from mutations in *cis*-regulatory elements or *trans*-acting factors. In the following, we will refer to epimutations that occur without any DNA sequence change as primary or "true" epimutations, and to epimutations that result from a DNA mutation as secondary epimutations.

2.1
Secondary Epimutations

Secondary epimutations are most often the result of a hereditary DNA sequence change and present in all cells of a patient. The underlying genetic defect can be close to the affected gene (*cis*) or impair the function of an epigenetic protein encoded somewhere else in the genome (*trans*).

2.1.1
Secondary Epimutations Resulting from a *cis*-Acting DNA Mutation

There are at least two genetic diseases in which a secondary epimutation represents the major pathogenetic mechanism. These are the fragile X mental retardation syndrome (FMR1) and the facioscapulohumeral muscular dystrophy (FSHD). FMR1 is an X-linked dominant disease caused by the expansion of an unstable trinucleotide repeat (CGG) within exon 1 of the *FMR1* gene. It is one of the most common causes of mental retardation. The number of repeats varies in the human population. Repeats with more than 58 copies are unstable and can expand to several hundred copies during the proliferation of the diploid oogonia in the fetal ovary. After fertilization of an oocyte carrying an expanded *FMR1* allele, the CGG repeat and *FMR1* promoter are methylated. DNA methylation, histone deacetylation and the establishment of repressive chromatin in this region silence the *FMR1* gene.

FSHD is an autosomal dominant disorder that has been linked to a 3.3-kb tandemly repeated sequence (*D4Z4*) in the subtelomeric region of the long arm of chromosome 4. In normal individuals the number of *D4Z4* repeats varies between 11 and 150 units, whereas FSHD patients have fewer than 11 repeats. Gabellini et al. have shown that a sequence element within *D4Z4* specifically binds a multiprotein complex consisting of the transcriptional repressor YY1, the architectural protein HMGB2 and nucleolin, and that this multiprotein complex mediates transcriptional repression of adjacent genes in 4q35 genes (Gabellini et al. 2002), probably by establishing a repressive chromatin structure over a very large distance. Based upon these results, the authors propose that deletion of *D4Z4* is associated with an open chromatin structure in 4q35 and the inappropriate expression of several genes within this region. However, these findings have remained controversial.

Imprinting defects resulting from mutations in a *cis*-acting imprinting control element are another example of secondary epimutations. Imprinting is an epigenetic process by which the male and the female germline mark specific chromosome regions so that only the maternal or paternal allele of certain genes is active. Imprint establishment and imprint maintenance are under the control of imprinting centres (IC). The IC on human chromosome 15 contains two critical elements, which are defined by the shortest region of deletion overlap (SRO) in Angelman syndrome (AS) and Prader–Willi syndrome (PWS) patients with an imprinting defect (AS-SRO and PWS-SRO, respectively) (Buiting et al. 1995). The AS-SRO element is necessary for the establishment of the maternal imprint in the female germline. A deletion of this element prevents maternal imprinting of the mutated chromosome. A child inheriting this chromosome will develop AS, which is a neurogenetic syndrome characterized by severe mental retardation, lack of speech, jerky movements and a happy disposition (estimated prevalence 1/15,000 newborns). It is caused by the loss of function of the *UBE3A* gene, which encodes an enzyme involved in targeted protein degradation. In the brain, the gene is active on the maternal chromosome only. In contrast to many other imprinted genes, mono-allelic expression of *UBE3A* is not associated with differential DNA methylation of the promoter/exon 1 region. There is some tentative evidence that the paternal allele is silenced by an antisense RNA which originates at the neighbouring *SNRPN* locus (Rougeulle et al. 1998; Runte et al. 2001). In normal individuals, *SNRPN* is methylated on the maternal chromosome and expressed from the paternal chromosome (Ozcelik et al. 1992; Zeschnigk et al. 1997). In AS patients with an imprinting defect (which accounts for approximately 3% of cases), the maternal *SNRPN* allele is unmethylated and expressed, and the maternal *UBE3A* allele is silenced. Of these patients, 10% have an AS-SRO deletion, whereas 90% have a primary epimutation (see Sect. 2.2). Most of the other AS patients have a large maternally derived chromosomal deletion, a maternal *UBE3A* mutation or paternal uniparental disomy 15.

The PWS-SRO of the chromosome 15 IC is necessary for the postzygotic maintenance of the paternal imprint (Bielinska et al. 2000). A paternally derived deletion of this element leads to an epigenetic state that resembles the maternal imprint. A child with such a chromosome will develop PWS, which is characterized by neonatal muscular hypotonia, hypogonadism, hyperphagia and obesity, short stature, small hands and feet, sleep apnoea, behavioural problems and mild to moderate mental retardation (estimated prevalence, 1/25,000 newborns). PWS is caused by the loss of function of imprinted genes which are active on the paternal chromosome only. Although all of the genes in the critical region are known, it is unclear which are the "PWS genes". In patients with an imprinting defect, which is found in approximately 1% of cases, all paternally expressed genes are silent. Of these patients, 10% have a PWS-SRO deletion, whereas 90% have a primary epimutation (see Sect. 2.2). Almost all of the other PWS patients have a large paternally derived chromosomal deletion, or maternal uniparental disomy.

Most of the IC deletions are familial deletions. Since deletions of the AS-SRO affect maternal imprinting only, they are silently transmitted through the paternal germline. Likewise, deletions of the PWS-SRO, which affect the paternal imprint only, are silently transmitted through the maternal germline. This explains why in some families only a few and distantly related individuals are affected. In some cases, the IC deletion has occurred de novo or is the result of germline mosaicism. There is only one case in which the deletion occurred postzygotically (Bielinska et al. 2000).

In contrast to AS and PWS, more than 50% of patients with transient neonatal diabetes mellitus or Beckwith–Wiedemann syndrome (BWS) have an imprinting defect. BWS is an overgrowth syndrome characterized by high birth weight, hypoglycaemia, macroglossia, exomphalos and increased risk of Wilms' tumour (estimated prevalence, 1/25,000 newborns). It is caused by overexpression of the paternally active *IGF2* gene and silencing of the maternally expressed *H19* gene or by silencing of the maternally active *CDKN1C* gene. These genes map to the short arm of chromosome 11, but are controlled by two different ICs, the *IGF2/H19* IC (IC1) and *LIT1/KCNQ1OT1* (IC2), which controls imprinting of *CDKN1C*. Similar to imprinting defects in AS and PWS, secondary epimutations in BWS are very rare. Sparago et al. have recently identified two families segregating a microdeletion in the *IGF2/H19* IC. Maternal transmission of the deletions resulted in hypermethylation of the *IGF2/H19* IC, biallelic *IGF2* expression, *H19* silencing and BWS (Sparago et al. 2004). Prawitt et al. (2005) have identified a family with a similar deletion. A deletion of IC2 has been described by Niemitz and colleagues (2004). When inherited maternally, the deletion caused BWS with silencing of *CDKN1C*. When inherited paternally, there is no phenotype, suggesting that the *LIT1/KCNQ1OT1* RNA itself is not necessary for normal development in humans.

A unique epimutation affecting the α-globin gene *HBA2* has recently been described by Tufarelli and colleagues (2003). The authors studies an individual with an inherited form of α-thalassaemia who has a deletion that results in a truncated, widely expressed gene (*LUC7L*) becoming juxtaposed to the structurally normal α-globin gene *HBA2*. Although it retains all of its local and remote *cis*-regulatory elements, expression of *HBA2* is silenced. *LUC7L* is transcribed from the opposite strand to the α-globin genes. In the patient, RNA transcripts from the truncated copy of *LUC7L* (missing the last three exons) extend into the *HBA2* CpG island, thus generating antisense transcripts with respect to *HBA2*. Antisense RNA transcription appears to mediate methylation of the *HBA2* CpG island during early development and silencing of *HBA2* expression.

There are also several examples of chromosomal translocations affecting the epigenetic state of genes adjacent to the breakpoints. This is in particular the case in translocations involving the X chromosome. A very instructive case was published by Jones and colleagues (1997). The authors studied a male patient with an unbalanced X;13 translocation [46,XY,der(13)t(X;13)(q10q10)] and bilateral retinoblastoma. DNA replication and methylation studies suggested that the extra copy of Xq, which is attached to the long arm of one chromosome 13, was inactivated and that inactivation had spread to chromosome 13 and silenced the *RB1* gene in 13q14. This epimutation is equivalent to a constitutional *RB1* mutation and explains the development of bilateral tumours in this patient.

2.1.2 Secondary Epimutations Resulting from *trans*-Acting DNA Mutations

In the last few years, many epigenetic players have been identified. They include DNA methyltransferases, methyl-CpG binding proteins, histone modifying enzymes, chromatin-remodelling factors and others. Loss of function of these proteins has a major impact on the epigenetic control of gene expression. In contrast to epimutations caused by a *cis*-acting DNA mutation, epimutations caused by *trans*-acting DNA mutations can affect many different genes on different chromosomes. In humans, several recognizable syndromes have been linked to a mutation in one of the epigenetic players. Mutations in the de novo DNA methyltransferase *DNMT3a*, for example, cause autosomal-recessive ICF syndrome (immunodeficiency, centromere instability and facial anomalies). The patients die of severe recurrent infections. Chromosome instability correlates with severe hypomethylation of the satellite DNA.

X-linked α-thalassaemia mental retardation (ATRX) syndrome is a developmental disorder characterized by mental retardation, facial dysmorphism, abnormal genitalia and anaemia resulting from reduced expression of the α-globin genes. It results from mutations in the *ATRX* gene, which encodes a member of the SWI/SNF family of chromatin remodelling factors.

Rett syndrome (RTT) is an X-linked dominant neurodevelopmental disorder. Girls with RTT have apparently normal development throughout the first 6 months of life; they subsequently begin to loose previously acquired skills and develop microcephaly, hand stereotypies, autistic features, seizures and gait apraxia. RTT is caused by mutations in the *MeCP2* gene, which encodes the methyl-CpG-binding protein 2. One function of MeCP2 is to recruit the Sin3A corepressor complex, which contains histone deacetylase, and to set up repressive chromatin. Initially it was believed that the loss of MeCP2 leads to widespread loss of gene repression. This, however, does not appear to be the case. To date, only two genes (*BDNF* and *DLX5*) have been identified as target genes (Chen et al. 2003; Horike et al. 2005).

2.2 Primary or "True" Epimutations

Compared to secondary epimutations, primary epimutations appear to be more frequent. As mentioned above, only 10% of imprinting defects in AS and PWS are caused by an IC mutation; in 90% of cases the imprinting defect is a primary epimutation. In BWS almost all imprinting defects are primary epimutations. Epimutations affecting genomic imprints can occur during imprint erasure in primordial germ cells, imprint establishment during later stages of gametogenesis or imprint maintenance after fertilization. If it occurs in the germline, all cells of the patient are affected. If it occurs after fertilization, it often results in somatic mosaicism.

Buiting et al. have found that in PWS patients with an imprinting defect not caused by an IC mutation the affected chromosome is always derived from the paternal grandmother (Buiting et al. 2003). This finding suggests that the (grand)maternal imprint was not erased in the paternal germline. Thus, the child inherited an epigenetic state from the grandmother. This is the best example of transgenerational epigenetic inheritance in man.

Wey et al. have recently described a PWS patient with a mosaic imprinting defect (Wey et al. 2004). In this patient, the epimutation most likely occurred after fertilization, although we cannot exclude the possibility that somatic mosaicism results from the postzygotic correction of an inherited epimutation.

In contrast to PWS, mosaic imprinting defects in AS are relatively common. Nazlican et al. have estimated that at least 30% of AS patients with a primary imprinting defect are mosaics (Nazlican et al. 2004). In two patients studied, somatic mosaicism was proved by molecular and cellular cloning, respectively. X inactivation studies of cloned fibroblasts from one patient suggest that the imprinting defect occurred before the blastocyst stage. To quantify the degree of mosaicism, the authors developed a quantitative methylation assay based on real-time PCR. In 24 patients tested, the percentage of normal cells ranged from less than 1% to 40%. Regression analysis suggested that patients with a higher percentage of normally methylated cells tend to have milder clinical symptoms than patients with a lower percentage. Some mosaic patients have "atypical Angelman syndrome" characterized by obesity, muscular hypotonia and ability to speak (Gillessen-Kaesbach et al. 1999). We might assume that the role of mosaic imprinting defects on chromosome 15 in mental retardation is underestimated.

Primary epimutations have not only been recognized in "imprinting disorders", but also in cancer. In 1983, A. Feinberg and B. Vogelstein discovered altered DNA methylation in cancer cells (Feinberg and Vogelstein 1983). Subsequently, these and other authors demonstrated that hypomethylation can lead to inappropriate activation of oncogenes. In 1986, S. Baylin and colleagues identified hypermethylation of the calcitonin gene in human lung cancers and lymphomas (Baylin et al. 1986), but the role of these changes in tumour development were unknown. Soon after the discovery of the first tumour suppressor gene (the retinoblastoma gene *RB1*), our own group found that the *RB1* promoter is methylated in a significant subset of retinoblastomas (Greger et al. 1989, 1994), suggesting that tumour-suppressor silencing can also occur by an epigenetic pathway. Subsequently, methylation of tumour-suppressor genes has been found in virtually all tumours, and the field of cancer epigenetics is rapidly growing (Feinberg and Tycko 2004).

In general, tumour-associated epimutations are found only in premalignant or malignant cells. There is only one case in which an inherited cancer epimutation has been described. Suter et al. have reported two individuals with soma-wide, allele-specific and mosaic hypermethylation of the DNA mismatch repair gene *MLH1* (Suter et al. 2004). Both individuals lacked evidence of DNA sequence mutation in any mismatch repair gene, but had multiple primary tumours that show mismatch repair deficiency. The epimutation was also present in spermatozoa of one of the individuals, indicating a germline defect and the potential for transmission to offspring.

Primary epimutations appear to play a role in cardiovascular disease also. Similar to tumours, atherosclerotic lesions are characterized by global DNA hypomethylation and local DNA hypermethylation. These similarities should not be surprising, because a key step of the atherogenetic process is the proliferation and migration of smooth muscle cells. Once within the intima, the phenotype of the smooth muscle cells switches from contractile to "dedifferentiated". It has been suggested that methylation of oestrogen receptor-α gene (*ESR1*) could contribute to these processes (Ying et al. 2000).

3
Causes of Primary Epimutations

Whereas secondary epimutations result from DNA mutations, it is less clear what triggers the occurrence of primary epimutations. Primary epimutations probably represent stochastic errors in the establishment or maintenance of an epigenetic state. An interesting model for tumour-suppressor methylation has been proposed by S. Clark (Stirzaker et al. 2004). According to this model, a combination of transient gene silencing and methylation seeding leads to the recruitment of the methyl-CpG-binding protein MBD2, histone deacetylase and DNA methyltransferase. This then leads to the spreading of DNA and histone methylation and consequently the establishment of silent chromatin.

The spontaneous epimutation rate can be modified by genetic and environmental factors.

3.1
Genetic Factors

It is tempting to speculate that certain DNA sequence variants are more susceptible to epimutations than others. Murrell et al. have recently obtained tentative evidence for a genetic predisposition to epimutations in the BWS region (Murrell et al. 2004). Four single nucleotide polymorphisms (SNPs) were found in a differentially methylation region of the *IGF2* gene (T123C, G358A, T382G and A402G), which occurred in three out of 16 possible haplotypes: TGTA, CATG and CAGA. There was a significant increase in the frequency of the CAGA haplotype and a significant decrease in the frequency of the CATG haplotype in BWS patients compared to controls. Our group has found preferential maternal transmission of a particular AS-SRO haplotype in families with an AS imprinting defect patient (Zogel et al. 2006).

Epigenetic states may also be affected by sequence variations in genes that encode epigenetic players. It has been shown, for example, that sequence variants of the *MTHFR* gene, which encodes the 5,10-methylenetetrahydrofolate reductase, are associated with variation in DNA methylation (Paz et al. 2002; Castro et al. 2004). The reductase is a key regulatory enzyme of the one-carbon metabolism. Changes in MTHFR activity affect the levels of *S*-adenosylmethionine (SAM), which is the methyl donor of both DNA methyltransferases and histone methyltransferases. As described by Zogel et al. (2006), women who are homozygous for the *MTHFR* 677C>T variant might have an increased risk of conceiving a child with an AS imprinting defect.

3.2 Environmental Factors

SAM levels are also dependent on folic acid, and the enzymes involved in one-carbon metabolism use vitamin B cofactors. As folic acid and vitamin B are provided by nutrition, it should not be surprising that epigenetic states can be influenced by the diet. Changes in DNA methylation by folate have been observed in various types of cancers as well as in animal models (Garfinkel and Ruden 2004).

In 2002, Cox et al. suggested that assisted reproduction (ART) might be associated with an increased risk of imprinting defects (Cox et al. 2002). The authors described two children with Angelman syndrome and an imprinting defect who were conceived by intracytosplasmic sperm injection (ICSI). In both cases, an IC deletion was excluded. Thus, the imprinting defects were primary epimutations that had occurred spontaneously. In 2003, Orstavik reported another ICSI child with AS and an imprinting defect (Orstavik et al. 2003), and three groups reported a 3- to 6-fold increased prevalence of ART in children with BWS (DeBaun et al. 2003; Gicquel et al. 2003; Maher et al. 2003). In a large case-control study, Halliday et al. have recently found that the risk of BWS in children conceived by ART is nine times greater than in the general population (Halliday et al. 2004). Despite this highly increased risk, the absolute risk of conceiving a child with BWS or AS after ART remains very low. It is of interest to note that imprinting defects in AS and BWS are characterized by hypomethylation of imprinting control regions on the maternal chromosome.

In view of experimental data in animals and the tentative epidemiological evidence in humans, it is possible, but far from being proved, that ART is associated with an increased risk of imprinting disorders. Assuming the association was genuine, it is unclear whether the risk can be attributed to infertility itself and/or the technique. To shed more light on these questions, Ludwig et al. have recently conducted a cohort study on patients with Angelman syndrome (Ludwig et al. 2005). These authors found an increased prevalence of imprinting defects in patients with Angelman syndrome born to subfertile couples (defined as having had a time to pregnancy >2 years and/or infertility treatment; relative risk, 6.25; 95% CI 1.68–16.00). Interestingly, the relative risk was the same in untreated couples with time to pregnancy exceeding 2 years and in couples treated by ICSI or hormonal stimulation alone, although the increase did not reach statistical significance, possibly due to the small sample size. The relative risk was highest in couples with time to pregnancy exceeding 2 years and infertility treatment (relative risk, 12.50; 95 CI 1.40–45.13). The findings suggest that imprinting defects and subfertility can have a common, possibly genetic cause, and that superovulation rather than ICSI may further increase the risk of conceiving a child with an imprinting defect. Superovulation may lead to the maturation of epigenetically imperfect oocytes that would not have been ovulated without treatment, or may disturb the process of DNA methylation in the oocyte. As shown by animal studies, another risk factor may be the culture of gametes and the early embryo (Khosla et al. 2001).

4
Epigenetic Candidate Diseases

Are primary epimutations restricted to "imprinting disorders" and cancer? This author thinks they are not. The apparent restriction of primary epimutations to "imprinting disorders" and cancer is most probably due to an ascertainment bias. As soon as it had emerged that Prader–Willi and Angelman syndromes involved imprinted genes, it was a reasonable hypothesis that a subset of patients should have PWS or AS because of an error in the imprinting process (Glenn et al. 1993; Reis et al. 1994). Likewise, after the identification of the first tumour suppressor gene (*RB1*), it was tempting to speculate that in some tumours it might be inactivated by DNA methylation (Greger et al. 1989). These investigations were facilitated by the fact that tissues enriched for the epimutation (tumour specimens) were available for analysis. Primary epimutations, which often occur in a mosaic form, are difficult to detect, unless they confer selective growth advantage onto the affected cell. Hyperplastic tissues and tumours are often subjected to a biopsy or removed, so that a relatively pure sample of the affected tissue can be studied. If, however, the epimutation leads to selective growth disadvantage, it is very difficult, if not impossible, to obtain material from the affected cell lineage. Thus, epigenetic defects leading to dysgenesis/dysplasia or agenesis/aplasia are likely to escape detection. It is not unreasonable to assume that a developmental master gene is silenced by an epimutation, similar to the epigenetic silencing of a tumour suppressor gene. Of course, the same reasoning holds true for somatic DNA mutations, but it has been estimated that the rate of primary epimutations may be one or two orders of magnitude greater than somatic DNA mutation (Bennett-Baker et al. 2003). This suggests that the contribution of epimutations to human disease is probably underestimated.

Epigenetic candidate diseases defy simple Mendelian inheritance. Thus, the study of diseases that meet one or more of the following criteria may prove fruitful:

- The disease occurs mainly sporadically, with only a few familial cases.
- There is a broad phenotypic spectrum with mainly unilateral manifestation.
- There are discordant monozygotic twins.

These criteria are also met by the so-called complex diseases, which are indeed good candidates for epigenetic involvement (Petronis 2001). Furthermore, ageing may be regarded as an epigenetic disease.

Acknowledgements Work performed in the author's laboratory was supported by the Deutsche Forschungsgemeinschaft. I thank all former and present members of my epigenetic research group for collaboration and M. Zeschnigk and K. Buiting for critical comments on the manuscript.

References

Baylin SB, Hoppener JW, de Bustros A, Steenbergh PH, Lips CJ, Nelkin BD (1986) DNA methylation patterns of the calcitonin gene in human lung cancers and lymphomas. Cancer Res 46:2917–2922

Bennett-Baker PE, Wilkowski J, Burke DT (2003) Age-associated activation of epigenetically repressed genes in the mouse. Genetics 165:2055–2062

Bielinska B, Blaydes SM, Buiting K, Yang T, Krajewska-Walasek M, Horsthemke B, Brannan CI (2000) De novo deletions of SNRPN exon 1 in early human and mouse embryos result in a paternal to maternal imprint switch. Nat Genet 25:74–78

Buiting K, Saitoh S, Gross S, Dittrich B, Schwartz S, Nicholls RD, Horsthemke B (1995) Inherited microdeletions in the Angelman and Prader-Willi syndromes define an imprinting centre on human chromosome 15. Nat Genet 9:395–400

Buiting K, Gross S, Lich C, Gillessen-Kaesbach G, el-Maarri O, Horsthemke B (2003) Epimutations in Prader-Willi and Angelman syndromes: a molecular study of 136 patients with an imprinting defect. Am J Hum Genet 72:571–577

Castro R, Rivera I, Ravasco P, Camilo ME, Jakobs C, Blom HJ, de Almeida IT (2004) 5,10-methylenetetrahydrofolate reductase (MTHFR) 677C→T and 1298A→C mutations are associated with DNA hypomethylation. J Med Genet 41:454–458

Chen WG, Chang Q, Lin Y, Meissner A, West AE, Griffith EC, Jaenisch R, Greenberg ME (2003) Derepression of BDNF transcription involves calcium-dependent phosphorylation of MeCP2. Science 302:885–889

Chong S, Whitelaw E (2004) Epigenetic germline inheritance. Curr Opin Genet Dev 14:692–696

Cox GF, Burger J, Lip V, Mau UA, Sperling K, Wu BL, Horsthemke B (2002) Intracytoplasmic sperm injection may increase the risk of imprinting defects. Am J Hum Genet 71:162–164

DeBaun MR, Niemitz EL, Feinberg AP (2003) Association of in vitro fertilization with Beckwith-Wiedemann syndrome and epigenetic alterations of LIT1 and H19. Am J Hum Genet 72:156–160

Feinberg AP, Tycko B (2004) The history of cancer epigenetics. Nat Rev Cancer 4:143–153

Feinberg AP, Vogelstein B (1983) Hypomethylation distinguishes genes of some human cancers from their normal counterparts. Nature 301:89–92

Gabellini D, Green MR, Tupler R (2002) Inappropriate gene activation in FSHD: a repressor complex binds a chromosomal repeat deleted in dystrophic muscle. Cell 110:339–348

Garfinkel MD, Ruden DM (2004) Chromatin effects in nutrition, cancer, and obesity. Nutrition 20:56–62

Gicquel C, Gaston V, Mandelbaum J, Siffroi JP, Flahault A, Le Bouc Y (2003) In vitro fertilization may increase the risk of Beckwith-Wiedemann syndrome related to the abnormal imprinting of the KCN1OT gene. Am J Hum Genet 72:1338–1341

Gillessen-Kaesbach G, Demuth S, Thiele H, Theile U, Lich C, Horsthemke B (1999) A previously unrecognised phenotype characterised by obesity, muscular hypotonia, and ability to speak in patients with Angelman syndrome caused by an imprinting defect. Eur J Hum Genet 7:638–644

Glenn CC, Nicholls RD, Robinson WP, Saitoh S, Niikawa N, Schinzel A, Horsthemke B, Driscoll DJ (1993) Modification of 15q11-q13 DNA methylation imprints in unique Angelman and Prader-Willi patients. Hum Mol Genet 2:1377–1382

Greger V, Passarge E, Hopping W, Messmer E, Horsthemke B (1989) Epigenetic changes may contribute to the formation and spontaneous regression of retinoblastoma. Hum Genet 83:155–158

Greger V, Debus N, Lohmann D, Hopping W, Passarge E, Horsthemke B (1994) Frequency and parental origin of hypermethylated RB1 alleles in retinoblastoma. Hum Genet 94:491–496

Halliday J, Oke K, Breheny S, Algar E, D JA (2004) Beckwith-Wiedemann syndrome and IVF: a case-control study. Am J Hum Genet 75:526–528

Holliday R (1987) The inheritance of epigenetic defects. Science 238:163–170

Horike SI, Cai S, Miyano M, Cheng JF, Kohwi-Shigematsu T (2005) Loss of silent-chromatin looping and impaired imprinting of DLX5 in Rett syndrome. Nat Genet 37:31–40

Jones C, Booth C, Rita D, Jazmines L, Brandt B, Newlan A, Horsthemke B (1997) Bilateral retinoblastoma in a male patient with an X; 13 translocation: evidence for silencing of the RB1 gene by the spreading of X inactivation. Am J Hum Genet 60:1558–1562

Khosla S, Dean W, Reik W, Feil R (2001) Culture of preimplantation embryos and its long-term effects on gene expression and phenotype. Hum Reprod Update 7:419–427

Ludwig M, Katalinic A, Groß S, Sutcliffe A, Varon R, Horsthemke B (2005) Increased prevalence of imprinting defects in patients with Angelman syndrome born to subfertile couples. J Med Genet 42:289–291

Maher ER, Brueton LA, Bowdin SC, Luharia A, Cooper W, Cole TR, Macdonald F, Sampson JR, Barratt CL, Reik W, Hawkins MM (2003) Beckwith-Wiedemann syndrome and assisted reproduction technology (ART). J Med Genet 40:62–64

Murrell A, Heeson S, Cooper WN, Douglas E, Apostolidou S, Moore GE, Maher ER, Reik W (2004) An association between variants in the IGF2 gene and Beckwith-Wiedemann syndrome: interaction between genotype and epigenotype. Hum Mol Genet 13:247–255

Nazlican H, Zeschnigk M, Claussen U, Michel S, Boehringer S, Gillessen-Kaesbach G, Buiting K, Horsthemke B (2004) Somatic mosaicism in patients with Angelman syndrome and an imprinting defect. Hum Mol Genet 13:2547–2555

Niemitz EL, DeBaun MR, Fallon J, Murakami K, Kugoh H, Oshimura M, Feinberg AP (2004) Microdeletion of LIT1 in familial Beckwith-Wiedemann syndrome. Am J Hum Genet 75:844–849

Orstavik KH, Eiklid K, van der Hagen CB, Spetalen S, Kierulf K, Skjeldal O, Buiting K (2003) Another case of imprinting defect in a girl with Angelman syndrome who was conceived by intracytoplasmic semen injection. Am J Hum Genet 72:218–219

Ozcelik T, Leff S, Robinson W, Donlon T, Lalande M, Sanjines E, Schinzel A, Francke U (1992) Small nuclear ribonucleoprotein polypeptide N (SNRPN), an expressed gene in the Prader-Willi syndrome critical region. Nat Genet 2:265–269

Paz MF, Avila S, Fraga MF, Pollan M, Capella G, Peinado MA, Sanchez-Cespedes M, Herman JG, Esteller M (2002) Germ-line variants in methyl-group metabolism genes and susceptibility to DNA methylation in normal tissues and human primary tumors. Cancer Res 62:4519–4524

Petronis A (2001) Human morbid genetics revisited: relevance of epigenetics. Trends Genet 17:142–146

Prawitt D, Enklaar T, Gaertner-Rupprecht B, Spangenberg C, Oswald M, Lausch E, Schmidtke P, Reutzel D, Fees S, Lucito R, Korzon M, Brozek I, Limon J, Housman D, Pelletier J, Zabel B (2005) Microdeletion of target sites for insulator protein CTCF in a chromosome 11p15 imprinting center in Beckwith-Wiedemann syndrome and Wilms' tumor. Proc Natl Acad Sci USA 102:4085–4090

Reis A, Dittrich B, Greger V, Buiting K, Lalande M, Gillessen-Kaesbach G, Anvret M, Horsthemke B (1994) Imprinting mutations suggested by abnormal DNA methylation patterns in familial Angelman and Prader-Willi syndromes. Am J Hum Genet 54:741–747

Rougeulle C, Cardoso C, Fontes M, Colleaux L, Lalande M (1998) An imprinted antisense RNA overlaps UBE3A and a second maternally expressed transcript. Nat Genet 19:15–16

Runte M, Huttenhofer A, Gross S, Kiefmann M, Horsthemke B, Buiting K (2001) The IC-SNURF-SNRPN transcript serves as a host for multiple small nucleolar RNA species and as an antisense RNA for UBE3A. Hum Mol Genet 10:2687–2700

Sollars V, Lu X, Xiao L, Wang X, Garfinkel MD, Ruden DM (2003) Evidence for an epigenetic mechanism by which Hsp90 acts as a capacitor for morphological evolution. Nat Genet 33:70–74

Sparago A, Cerrato F, Vernucci M, Ferrero GB, Silengo MC, Riccio A (2004) Microdeletions in the human H19 DMR result in loss of IGF2 imprinting and Beckwith-Wiedemann syndrome. Nat Genet 36:958–960

Stirzaker C, Song JZ, Davidson B, Clark SJ (2004) Transcriptional gene silencing promotes DNA hypermethylation through a sequential change in chromatin modifications in cancer cells. Cancer Res 64:3871–3877

Suter CM, Martin DI, Ward RL (2004) Germline epimutation of MLH1 in individuals with multiple cancers. Nat Genet 36:497–501

Tufarelli C, Stanley JA, Garrick D, Sharpe JA, Ayyub H, Wood WG, Higgs DR (2003) Transcription of antisense RNA leading to gene silencing and methylation as a novel cause of human genetic disease. Nat Genet 34:157–165

Waddington CH (1959) Canalization of development and genetic assimilation of acquired characters. Nature 183:1654–1655

Wey E, Bartholdi D, Riegel M, Nazlican H, Horsthemke B, Schinzel A, Baumer A (2004) Mosaic imprinting defect in a patient with an almost typical expression of the Prader-Willi syndrome. Eur J Hum Genet 13:273–277

Ying AK, Hassanain HH, Roos CM, Smiraglia DJ, Issa JJ, Michler RE, Caligiuri M, Plass C, Goldschmidt-Clermont PJ (2000) Methylation of the estrogen receptor-alpha gene promoter is selectively increased in proliferating human aortic smooth muscle cells. Cardiovasc Res 46:172–179

Zeschnigk M, Schmitz B, Dittrich B, Buiting K, Horsthemke B, Doerfler W (1997) Imprinted segments in the human genome: different DNA methylation patterns in the Prader-Willi/Angelman syndrome region as determined by the genomic sequencing method. Hum Mol Genet 6:387–395

Zogel C, Böhringer S, Groß S, Varon R, Buiting K, Horsthemke B (2006) Identification of cis- and trans-acting factors possibly modifying the risk of epimutations on chromosome 15. Eur J Hum Genet (in press)

CTMI (2006) 310:61–80

Epigenotypes of Latent Herpesvirus Genomes

J. Minarovits (✉)

Microbiological Research Group, National Center for Epidemiology,
Pihenö u 1, 1529 Budapest,Hungary
mini@microbi.hu

Abstract Epigenotypes are modified cellular or viral genotypes which differ in transcriptional activity in spite of having an identical (or nearly identical) DNA sequence. Restricted expression of latent, episomal herpesvirus genomes is also due to epigenetic modifications. There is no virus production (lytic viral replication, associated with the expression of all viral genes) in tight latency. *In vitro* experiments demonstrated that DNA methylation could influence the activity of latent (and/or crucial lytic) promoters of prototype strains belonging to the three herpesvirus subfamilies (α-, β-, and γ-herpesviruses). *In vivo*, however, DNA methylation is not a major regulator of herpes simplex virus type 1 (HSV-1, a human α-herpesvirus) latent gene expression in neurons of infected mice. In these cells, the promoter/enhancer region of latency-associated transcripts (LATs) is enriched with acetyl histone H3, suggesting that histone modifications may control HSV-1 latency in terminally differentiated, quiescent neurons. Epstein–Barr virus (EBV, a human γ-herpesvirus) is associated with a series of neoplasms. Latent, episomal EBV genomes are subject to host cell-dependent epigenetic modifications (DNA methylation, binding of proteins and protein complexes, histone modifications). The distinct viral epigenotypes are associated with distinct EBV latency types, i.e., cell type-specific usage of latent EBV promoters controlling the expression of latent, growth transformation-associated EBV genes. The contribution of major epigenetic mechanisms to the regulation of latent EBV promoters is variable. DNA methylation contributes to silencing of Wp and Cp (alternative promoters for transcripts coding for the nuclear antigens EBNA 1–6) and LMP1p, LMP2Ap, and LMP2Bp (promoters for transcripts encoding transmembrane proteins). DNA methylation does not control, however, Qp (a promoter for EBNA1 transcripts only) in lymphoblastoid cell lines (LCLs), although *in vitro* methylated Qp-reporter gene constructs are silenced. The invariably unmethylated Qp is probably switched off by binding of a repressor protein in LCLs.

Herpesviruses are "large genome" double-stranded DNA viruses defined by the morphology of the virion. They are widespread in nature and are associated with a broad range of diseases in their host species. In many cases they efficiently infect a significant proportion of the host population, and infected individuals carry the viral genomes lifelong in their cells. A "core" gene set or complement (around 40 conserved genes) is present in most herpesvirus genomes and codes for structural proteins of the virion and viral enzymes. In addition, the viral DNA encompasses less conserved "non-core" genes as well which encode control or modulator proteins and virion tegument or surface components. Highly divergent "non-core" genes code for proteins that interfere with the immunological defense mechanisms of the host or induce malignant transformation of certain host cells (tumorigenesis). The genomes of extant herpesvirus strains (members of the subfamilies designated as α-, β-, and γ-herpesviruses) are products of a more than 200 million years of co-evolution that affected the genomes of the host species as well (cospeciation) (McGeoch and Davison 1999). Latency is a remarkable property of herpesviruses, which ensures maintenance of their genetic information in their hosts for an extended period in the absence of productive (lytic) replication. Members of all three herpesvirus subfamilies are capable of establishing latent infection, which is associated with a restricted expression of the viral genome.

1
Cellular and Viral Epigenotypes

Epigenotypes are modified genotypes that differ in transcriptional activity in spite of having an identical (or nearly identical) DNA sequence (Wolffe and Matzke 1999; Whitelaw and Martin 2001). Thus, in diploid cells gene expression patterns of nearly sequence-identical alleles may differ even within the same cell, cell type, or tissue due to differentially imprinted (modified) domains carried by the maternal chromosome (maternal epigenotype) and the paternal chromosome (paternal epigenotype) (Obata and Kono 2002; Adam et al. 1996; Mitsuya et al. 1997; Forejt et al. 1999; Su et al. 2004; Saitoh et al. 1996).

Epigenetic modifications (DNA methylation, DNA associated proteins and protein complexes, histone modifications) regulate cell type- or tissue-specific and developmental stage-specific gene expression as well (reviewed by Ehrlich 2003; Levine et al. 2004; Egger et al. 2004). Thus, in principle, one could define cell type-specific, tissue-specific, and developmental stage-specific epigenotypes—which determine gene expression probabilities in mammalian genomes or in important domains of such genomes—by high-resolution mapping of CpG methylation, *in vivo* DNA–protein interactions, and analysis of histone modifications region by region.

The association of certain genetic loci with transcriptionally repressive nuclear subcompartments (such as the constitutive heterochromatin and the nuclear periphery) or subcompartments favoring transcription (euchromatin) provides a tool for coregulation of gene batteries in a heritable, cell-type specific manner (Kosak et al. 2002; Alcobia et al. 2000; Brown et al. 1997). Thus, subnuclear compartmentalization and the possibility for relocation of certain chromosomal regions (subcompartment switch), which can occur at a specific stage of development or cellular differentiation, can also be considered as an epigenetic mechanism and an important feature of the epigenotype.

Developmental and lineage-specific coregulation of gene sets can be mediated by complex *cis*-acting elements called locus control regions (LCRs) as well (reviewed by Li et al. 2002). LCRs can provide accessibility for ubiquitous and tissue- or cell type-specific transcription factors to an insulated chromatin domain. It is interesting to note that the mouse T cell receptor α/δ locus LCR can contribute, when active, to the alteration of the epigenotype by inducing localized demethylation (Santoso et al. 2000).

The genomes of DNA viruses and proviral (DNA) genomes of retroviruses can integrate into the host cell nuclei; they are targets, therefore, to epigenetic modifications as well (Sutter and Doerfler 1980; Vardimon et al. 1980; Toth et al. 1990; Wettstein and Stevens 1983; Jaenisch et al. 1985; Bednarik 1996). Latent herpesvirus genomes can persist as unintegrated episomes that co-replicate with the cellular genome (Nonoyama and Pagano 1972; Adams and Lindahl 1975). Episomal genomes of Epstein–Barr virus (EBV, a human γ-herpesvirus associated with a series of malignant tumors) are attached to the nuclear matrix (Jankelevich et al. 1992) and are subject to host cell-dependent epigenetic modifications (DNA methylation, protein–DNA interactions, and histone modifications) resulting in cell type-specific promoter usage and viral gene expression (Minarovits et al. 1991; Salamon et al. 2001;

Li and Minarovits 2003; Chau and Lieberman 2004). Thus, there are distinct viral epigenotypes associated with well-defined transcriptional patterns of latent, growth transformation-associated viral genes (latency types).

2 Epigenetic Modifications of α-Herpesvirus Genomes

α-Herpesviruses have a relatively short reproductive cycle and their productive (lytic) replication results in destruction of the infected cells. *In vitro* they spread rapidly after infection in sensitive cell types. *In vivo* they establish latency first in sensory ganglia and in cells of the central nervous system, although equine herpesvirus 1 and equine herpesvirus 4 target the lymphoreticular system (Welch et al. 1992) and latent Marek's disease virus genomes are associated with lymphomas in chicken (reviewed by Morimura et al. 1998). T lymphoblastoid cell lines derived from such lymphomas carry methylated viral genomes (Kanamori et al. 1987). The latent genomes of herpes simplex virus are associated with nucleosomes (Deshamne and Fraser 1989) and exist in an extrachromosomal (episomal) state (Mellerick and Fraser 1987). Depending on the host cell, both DNA methylation and modification of histone tails may affect expression of the best-characterized α-herpesvirus genomes.

2.1 Methylation Patterns of Latent HSV-1 Genomes *In Vitro* and *In Vivo*

In a pioneering study, Youssoufian et al. investigated a major epigenetic regulatory mechanism, cytosine methylation (a heritable form of DNA modification) using cells either carrying latent herpes simplex virus type 1 (HSV-1) genomes or undergoing productive infection (Youssoufian et al. 1982). Digestion with cytosine methylation-sensitive and cytosine methylation-resistant restriction enzyme pairs followed by Southern blotting revealed that the latent HSV-1 genomes were highly methylated in cells of a persistently infected lymphoblastoid T cell line (CEM) treated with concanavalin A. In contrast, only unmethylated viral genomes could be detected during productive infection in this reversible *in vitro* model of viral latency. This study suggested that DNA methylation might play a role in the maintenance of HSV-1 latency.

In contrast, using a similar assay, HSV-1 DNA isolated from the brains of latently infected mice in an *in vivo* model of HSV-1 latency were found to be predominantly unmethylated (Dressler et al. 1987). Acutely infected brains (productive HSV-1 replication) and purified virion DNA yielded identical restriction fragments after digestion with *Sma*I (methylation sensitive) and

*Xma*I (methylation insensitive) enzymes, indicating that the viral genomes were unmethylated. Thus, extensive methylation of latent HSV-1 genomes is dispensable for *in vivo* latency and HSV-1 latency is not associated with *de novo* methylation of the viral genomes *in vivo*.

In another *in vivo* model of HSV-1 latency, latent viral genomes isolated from mouse dorsal root ganglia were analyzed by sequencing bisulfite-treated DNA samples (Kubat et al. 2004a). This method permits positive identification of all 5-methylcytosine residues in individual DNA strands (Frommer et al. 1992). The promoter of ICP4, an immediate–early (IE) gene—which is not expressed in latent HSV-1 infection but expressed during acute (productive) infection—was found to be predominantly unmethylated in DNA samples isolated from latently infected ganglia, acutely infected ganglia, and HSV-1 virions (Kubat et al. 2004a). During latency the episomal HSV-1 genomes express latency-associated transcripts (LATs) from a single transcription unit (Stevens et al. 1987 Dobson et al. 1989). Kubat et al. observed that similarly to the ICP4 promoter, the LAT promoter was also predominantly unmethylated in latency, acute infection, and in virion DNA (Kubat et al. 2004a). These results confirmed the conclusions of Dressler et al. and also suggested that DNA methylation is not a major regulator of HSV-1 latent gene expression in neurons.

Thus, CpG methylation is apparently not used for silencing of crucial lytic HSV-1 promoters during *in vivo* latency, in spite of the fact that a well-characterized lytic HSV-1 promoter (the thymidine kinase promoter) can be silenced by DNA methylation *in vitro* (Christy and Scangos 1982; Buschhausen et al. 1985, 1987; Tasseron-de Jong et al. 1989).

2.2
Histone Modifications Associated with Latent HSV-1 Genomes

Modification of histone tails influences chromatin structure and might regulate transcription by affecting chromatin configuration (Jenuwein and Allis 2001; Egger et al. 2004). Acetylated histones mark transcriptionally active domains of chromatin while histone deacetylation is associated with silencing of promoter activity.

Arthur et al. used a neonatal rat dorsal root ganglion-derived neuronal culture system to study HSV-1 latency and reactivation *in vitro*. They constructed recombinant viruses carrying reporter genes under the control of viral promoters and observed that, in latently infected cultures, inhibition of histone deacetylases by trichostatin A (TSA) switched on the activity of HSV-1 IE110 promoter (Arthur et al. 2001). They speculated that this could be an indirect effect mimicking nerve growth factor (NGF) withdrawal (which also switches on the lytic cycle) because TSA can induce expression of certain cellular factors blocking NGF action (Sano and Kitayama 1996).

Kubat et al., using the same *in vivo* model system as in their DNA methylation analysis, determined the level of histone H3 acetylation (at lysines 9 and 14) using a chromatin immunoprecipitation assay, at a region located within the LAT promoter, a region corresponding to the HSV-1 DNA polymerase gene (expressed in the early phase of the lytic cycle) and promoters of two IE genes (UL54/ICP27 promoter and ICP4 promoter). They found that the LAT promoter is enriched with acetyl histone H3 (K9, K14) compared to the viral polymerase gene and the IE promoters analyzed. They speculated that in latently infected, terminally differentiated, quiescent neurons, HSV-1 employs a relatively dynamic epigenetic mechanism—i.e., histone acetylation and deacetylation—to activate the LAT promoter and repress lytic viral promoters, respectively (Kubat et al. 2004a). In a follow-up study, they observed that the LAT enhancer (including *rcr*, a region critical for induced reactivation of HSV-1) is hyperacetylated even in the absence of LAT transcription and suggested that this *cis*-acting regulator region maintains a transcriptionally permissive chromatin domain (Kubat et al. 2004b).

Histone deacetylation might explain the highly efficient silencing of heterologous promoters incorporated into recombinant HSV-1 genomes that occurs after establishment of the latent state (a phenomenon observed by Lokensgard et al. 1994). The mouse phosphoglycerate kinase (PGK) promoter is highly active and the murine metallothionein promoter (MT1) shows a moderate activity after acute (productive) infection of dorsal root ganglia by the recombinant viruses; they do not drive, however, the expression of the lacZ reporter gene in the latent phase of infection (Lokensgard et al. 1994).

3 Epigenetic Modifications Regulate the Activity of a β-Herpesvirus Promoter

β-Herpesviruses are highly species-specific viruses that grow slowly in cultured cells. The site(s) of residence for β-herpesviruses (monocytes and their bone marrow progenitors, endothelial cells) is still a subject of intensive research. Infection with human cytomegalovirus (HCMV, human herpesvirus 5, the best-characterized β-herpesvirus) and similar viruses isolated from other primates, domestic animals, and rodents causes typical cytopathologic changes involving cell enlargement.

Honess et al. analyzed dinucleotide frequencies in herpesvirus DNA and observed a local deficiency of CpG dinucleotides in the major IE genes of human, murine, and simian cytomegalovirus genomes (Honess et al. 1989). The complete genome of human and murine cytomegalovirus is not CpG deficient; as a matter of fact, the observed CpG frequency is higher than the expected one (Takacs et al. 2001b). Because a relative CpG deficiency and a surplus of TpG+CpA dinucleotides is thought to be a consequence of DNA methylation in vertebrate genomes (methylcytosine is relatively unstable), Honess et al. suggested that IE regions of latent cytomegalovirus genomes undergo local methylation. Unfortunately, this idea has not been tested yet, probably due to the difficulties in identifying the sites of cytomegalovirus latency.

The HCMV IE1 promoter is considered to be a "strong" promoter and it is frequently used, therefore, to drive the expression of reporter genes or other genes to be expressed *in vitro* or *in vivo*. Methylation of the IE enhancer/promoter at CpG dinucleotides by the *Spiroplasma* methyltransferase SssI completely repressed the activity of the promoter, probably via silencing by a methylcytosine binding protein (Prosch et al. 1996).

When incorporated into an adenovirus vector, the cytomegalovirus promoter and enhancer could drive expression of the human fibroblast growth factor 4 gene after intramuscular injection of rats (peak activity at 6 h) (Brooks et al. 2004). The promoter was practically silenced, however, at 24 h after injection and remained silent at day 3 and 28. In parallel, the promoter-enhancer sequences became methylated. Thus, in principle, DNA methylation could modulate (silence) the activity of HCMV IE1 promoter in latent human cytomegalovirus genomes too.

When incorporated into a heterologous viral genome (recombinant HSV-1), the silent HCMV IE1 promoter could be activated in latently infected neuronal cultures by NGF withdrawal or through inhibition of histone deacetylases (trichostatin A treatment) (Arthur et al. 2001). Thus, histone acetylation/deacetylation could potentially also regulate HCMV IE1 promoter activity and influence HCMV latency and reactivation.

4
Epigenetic Modifications of Latent γ-Herpesvirus Genomes

γ-Herpesviruses replicate in epithelial cells (lytic infection) and establish latency in lymphoid cells *in vivo*. In γ-herpesvirus-infected lymphocytes and lymphoblastoid cell lines cultured *in vitro*, both virus production and tight latency can be observed. Latent γ-herpesvirus transcripts frequently encode oncoproteins and protein non-coding small RNAs that may contribute to immortalization and malignant transformation of host cells. Although γ-herpesviruses are favorite subjects of epigenetic studies, there are no data yet on the association of latent, episomal γ-herpesvirus genomes with different nuclear subcompartments (euchromatin or heterochromatin). Association of EBV genomes with the nuclear matrix (Jankelevich et al. 1992) and a colinear arrangement of the functional units of a 30-kb region the EBV genome with the rearranged human immunoglobulin gene loci (Niller et al. 2004a) suggest, however, that a viral LCR may control latent EBV gene expression in a host cell-dependent manner (Niller et al. 2004b). The putative EBV LCR consists of a c-Myc binding site (Niller et al. 2003), transcription units of two protein non-coding viral RNAs (EBER 1 and 2), and *oriP*, the latent origin of EBV replication (which acts as a long range enhancer as well). How such a structure regulates host cell-dependent activity of latent EBV promoters remains to be studied. Low- and high-resolution DNA methylation maps have been established in several regions of certain latent γ-herpesvirus genomes (herpesvirus saimiri, EBV, Kaposi's sarcoma-associated herpes virus), and recent studies suggest a role for histone modifications and protein–DNA interactions in the control of latent γ-herpesvirus promoters, and latent origins of DNA replication too (see Sect. 4.4).

4.1 Methylation Patterns of Herpesvirus Saimiri DNA *In Vitro* and *In Vivo*

Herpesvirus saimiri causes lymphoma or leukemia in certain New World primates. The virion DNA was found to be unmethylated (Desrosiers 1982), but methylated viral DNA molecules were detected in lymphoid tumor cell lines (Desrosiers et al. 1979) and in DNA samples isolated from peripheral blood of one owl monkey and three white-lipped marmosets with leukemia (Desrosiers 1982). This suggests that *de novo* methylation of the viral genomes occurred *in vivo* and was not a result of prolonged *in vitro* cultivation of the tumor cells. Using CpG methylation-sensitive and resistant isoschizomers, an unmethylated region was also mapped in viral genomes carried by a cell line established from a marmoset tumor (Desrosiers 1982).

4.2 Host Cell-Dependent Epigenotypes of Latent Epstein–Barr Virus Genomes

EBV replicates in epithelial cells of the oropharynx and establishes latency in B cells. The virus is associated with a series of human neoplasms (endemic Burkitt's lymphoma, posttransplant lymphoproliferative disease, AIDS-associated B cell lymphoma, peripheral T cell lymphoma, Hodgkin's disease, nasopharyngeal carcinoma, midline granuloma, etc.; reviewed by Klein 1996; Liebowitz 1998; Li and Minarovits 2003). EBV immortalizes human B cells *in vitro* very efficiently, resulting in the generation of lymphoblastoid cell lines (LCLs). Depending on the activity of the latent viral promoters, different sets of latent gene products are expressed in various host cells carrying EBV episomes. These products define the major latency types of EBV (type I, II, and III latency, also called viral latency program; reviewed by Li and Minarovits 2003).

In Burkitt's lymphomas (BLs) and BL-derived cell lines maintaining the BL biopsy phenotype (group I BL lines, type I latency), only transcripts for a single nuclear antigen (EBNA 1) are expressed from a promoter called Qp (located to the *Bam*HI Q fragment of the viral genome). EBNA 1 binds to the latent origin of viral DNA replication (*oriP*) and it is necessary for the maintenance of viral episomes. Two protein non-coding viral RNAs (EBER 1 and 2) are also expressed and might contribute to the pathogenesis of BL (reviewed by Niller et al. 2004c). The significance of *Bam*HI A rightward transcripts (BARTs, BARF0; Chen et al. 1992) which can also be detected in group I BL lines is unknown at present (see references in de Jesus et al. 2003). There are no data on methylation of the BART promoter, the adjacent, inactive BARF1 promoter, and TRp (a promoter localized in the first terminal repeat of the EBV genome, see below) in BL lines. With these exceptions, one could conclude that the promoters that are active in type I latency are unmethylated (Tao et al. 1988; Minarovits et al. 1992; Salamon et al. 2001), while all of the silent promoters studied (Wp and Cp for transcription of EBNA 1–6 RNAs; LMP promoters for RNAs encoding latent membrane proteins) are highly methylated (Jansson et al. 1992; Altiok et al. 1992; Robertson et al. 1995; Schaefer et al. 1997; Takacs et al. 1998, 2001a; Salamon et al. 2001, 2003; see Table 1). Type I latency with methylated Wp, Cp, and LMP1p was also described in gastric carcinomas carrying EBV genomes (Imai et al. 1994).

Chronic active EBV infection (CAEBV), a non-neoplastic T cell-lymphoproliferative disease, is a unique form of EBV latency. The expression pattern resembles type I latency in most cases, but the promoter usage (and probably the splicing of viral transcripts) is different since Cp (not Qp) seems to be used for EBNA 1 transcription only (Yoshioka et al. 2003a). In addition, the active Cp is highly methylated. This suggests that an unknown regulatory mechanism may overrule the inhibitory effect of Cp methylation in proliferating non-neoplastic T cells. Monoclonal or oligoclonal EBV-infected T cells proliferate in EBV-associated hemophagocytic syndrome as well, but the latency is less restricted (type III, see below) and EBNA 1 and EBNA 1–6 messages are generated from unmethylated Qp and methylated Wp/Cp, respectively (Yoshioka et al. 2003b).

Type II latency is best characterized in nasopharyngeal carcinomas (NPCs) and midline granulomas (see Table 2). Similarly to group I BL lines, NPCs also

Table 1 Activity and methylation patterns of latent Epstein–Barr virus promoters in type I latency

Designation	Activity	CpG methylation
EBNA 1–6/EBNA 1 promoters		
Wp	Off	+
Cp	Off	+
Qp	On	–
LMP promoters		
LMP1p	Off	+
TRp	Off	?
LMP2Ap	Off	+
LMP2Bp	Off	+
EBER promoters		
EBER1p	On	–
EBER2p	On	–
Promoters for *Bam*HI A transcripts		
BARTp	On	?
BARF1p	Off	?

initiate EBNA 1 transcripts at the unmethylated Qp and express EBER 1 and 2 from unmethylated EBER transcription units. They express BART transcripts too. In addition, a series of promoters that are silent in type I latency are switched on in NPC cells. These include one of the alternative promoters for LMP1 messages, TRp (Sadler and Raab-Traub 1995), or LMP1p (which is active in LCLs, see the following paragraph) and the LMP2A and LMP2B promoters. LMP1p and LMP2Bp share common regulatory elements, and the LMP1 and LMP2B genes are expressed co-ordinately in NPC biopsies (Chen et al. 1995). Latent membrane proteins modulate signal transduction pathways and contribute to immortalization and malignant transformation of host cells (reviewed by Longnecker 1998). A gene coding for a putative transforming protein (BARF1; Strockbine et al. 1998) is also expressed in NPCs (Decaussin et al. 2000). BARF1p is active during productive (lytic) EBV infection as well, and it was initially described as a lytic promoter. All of the active latent promoters (Qp, EBER1p, EBER2p, BARTp) studied in type II latency are unmethylated (Tao et al. 1998; Minarovits et al. 1992; de Jesus et al. 2003) while the silent Wp (in NPCs and midline granulomas) and Cp (studied in midline granulomas) is methylated (Hu et al. 1991; Minarovits et al. 1994). LMP1p is also unmethylated in NPCs (Hu et al. 1991; Falk et al. 1998) and midline granulomas (Minarovits et al. 1994) expressing LMP1 protein

Table 2 Activity and methylation patterns of latent Epstein–Barr virus promoters in type II latency

Designation	Activity	CpG Methylation
EBNA1–6/EBNA1 promoters		
Wp	Off	?
Cp	Off	+
Qp	On	–
LMP promoters		
LMP1p	?	–
TRp	On	?
LMP2Ap	On	?
LMP2Bp	On	–
EBER promoters		
EBER1p	On	–
EBER2p	On	–
Promoters for *Bam*HI A transcripts		
BARTp	On	–
BARF1p	On	?

or LMP1 mRNA, respectively, but the activity of this promoter (versus the alternative promoter, TRp) in type II latency, remains to be established.

Type III latency is characterized by expression of six nuclear antigens (EBNA 1–6) encoded by transcripts originating at Cp, a lymphoid-specific promoter (see Table 3; Sung et al. 1991; Jin and Speck 1992; Contreras-Brodin et al. 1991). Cp is active only in B lymphoblastoid cells (LCLs) and group III BL lines. EBNA 2 is the major transactivator protein of EBV, which is involved in immortalization and malignant transformation of B cells. It binds to a cellular protein (CBF1, C promoter binding factor 1) at regulatory sequences of viral and cellular promoters (Ling et al. 1994). EBNA 3, 4, 6, and 5 (also called EBNA 3a, b, c, and LP, respectively) modulate the activity of EBNA 2-regulated promoters. It is interesting to note that at the beginning of *in vitro* EBV infection and immortalization of human B cells, Cp is inactive while an alternative promoter, Wp, is active (Woisetschlaeger et al. 1990). Wp is methylated and switched off, however, at a later stage when the unmethylated Cp is activated (promoter switch) (Tierney et al. 2000). Cp may also replace Qp during the type I to type III latency switch accompanying a phenotypic change (drift) in cultivated BL lines (Kerr et al. 1992; Altiok et al. 1992). Wp methylation is variable, however, in LCLs and group III BL lines, suggesting that other epigenetic regulators may also contribute to the suppression of Wp activity

Table 3 Activity and methylation patterns of latent Epstein–Barr virus promoters in type III latency

Designation	Activity	CpG methylation
EBNA1–6/EBNA1 promoters		
Wp	Off	+/–
Cp	On	–
Qp	Off	–
LMP promoters		
LMP1p	On	–
TRp	?	?
LMP2Ap	On	–
LMP2Bp	On	–
EBER promoters		
EBER1p	On	–
EBER2p	On	–
Promoters for *Bam*HI A transcripts		
BARTp	On	?
BARF1p	Off	?

(Minarovits et al. 1991; Elliott et al. 2004). Qp is silent (in spite of the fact that it is unmethylated) in LCLs; it is probably repressed by binding of a cellular protein (Salamon et al. 2001). The active EBER transcription units and the active LMP promoters studied are unmethylated in LCLs (Minarovits et al. 1992; Salamon et al. 2001, 2003; Takacs et al. 2001a; see Table 3).

4.3 Reactivation of γ-Herpesviruses: A Dual Role for DNA Methylation?

Productive (lytic) replication of herpesviruses can be induced in cells carrying latent viral genomes by activation of IE genes. IE gene products activate thereafter transcription of lytic viral genes in a cascade-like manner. The promoter for the ORF50 gene encoding Lyta (KSHV/Rta), an IE protein of Kaposi's sarcoma-associated herpesvirus (KSHV; human herpesvirus 8), was heavily methylated in primary effusion lymphoma (PEL)-derived cell lines during latency, but induction of the lytic cycle by 12-*O*-tetradecanoylphorbol-13-acetate (TPA) caused Lyta promoter demethylation (Chen et al. 2001). This suggests that KSHV maintains latency by controlling (repressing) the activity of a key promoter, Lyta, via DNA methylation. In contrast, demethylation impairs activation of Rp, the IE promoter for the BRLF1 gene of EBV, by the IE protein BZLF1 (Z) (Bhende et al. 2004). Bhende et al. suggest that preferential binding of BZLF1 to the methylated form of its downstream target gene permits reactivation of highly methylated EBV genomes, probably by inviting a histone acetylase to the region. Rp methylation (which is due in large part to the action of DNA methyltransferase 3B), is not necessarily high, however, in latent EBV genomes; as a matter of fact, Rp is only moderately methylated ("mixed pattern") in certain LCLs (Tao et al. 2002).

4.4 Activator/Repressor Complexes and Histone Modifications in Regulatory Regions of γ-Herpesvirus Genomes

A series of cell type-specific or general transcription factors bind to regulatory regions of γ-herpesviruses. Most of them were identified in *in vitro* experiments. *In vivo* footprinting studies partly supported and partly extended these results. A consequent finding at the EBNA 2-activated Cp and LMP2Ap of EBV was the presence of a CBF1 footprint(s) in cells actively using these promoters and the absence of a typical CBF1 footprint in cells with silent Cp and LMP2Ap (Salamon et al. 2001, 2003). This implies that although CBF1 is present in cells with silent Cp and LMP2Ap, in the absence of EBNA 2 it does not bind to its recognition sites.

A multiprotein complex of Max, Mad1, and mSin3A binds to a transcriptional silencer and recruits histone deacetylases, inactivating the LMP1 promoter by altering the structure of chromatin (Sjöblom-Hallén et al. 1999). Inhibition of histone deacetylases induces LMP1 transcription in two BL lines (Daudi, P3HR-1) but not in Rael. LMP1p is highly methylated in Rael cells (Salamon et al. 2001), and 5-azacytidine, an inhibitor of DNA methyltransferases, is capable of switching on LMP1p in this BL line (Masucci et al. 1989; Sjöblom-Hallén et al. 1999). These results imply that more than one epigenetic mechanism contributes to transcriptional repression of LMP1p.

Histone modifications may extend to larger regions, not only to the vicinity of latent EBV promoters (Chau and Lieberman 2004). Methylation of histone H3 on lysine 4 (H3mK4) is a characteristic feature of a region encompassing LMP1p, EBER 1 and 2, *oriP*, Cp, and the W repeats in an LCL (type III latency), but it is confined to the EBER transcription units and *oriP* in a BL line (type I latency). Outside of EBER genes and *oriP*, it is replaced with methylated histone H3 at lysine 9 (H3mK9), which correlates with transcriptional repression (rep*, a highly methylated sequence is located in this area, 3′ of *oriP*; Niller et al. 2001). The heterochromatic region reaches LMP1p and Cp, which are silent in type I latency. In addition, Chau and Lieberman charted a region 5′ from Cp, at the boundary of the euchromatic and heterochromatic domain, where CTCF (CCCTC-binding factor) was bound. They suggest that CTCF blocks expansion of histone H3mK4 from the *oriP* enhancer region (which is hypomethylated or unmethylated in BLs and LCLs; Falk et al. 1998; Salamon et al. 2000) to Cp, interfering thereby with the activity of the enhancer.

Acknowledgements I'm grateful to Agnes Bakos and Daniel Salamon for critical reading of the manuscript. I also acknowledge the support of the National Science Foundation (OTKA), Hungary (grant T 042727).

References

Adam GIR, Cui H, Miller SJ, Flam F, Ohlsson R (1996) Allele-specific in situ hybridization (ASISH) analysis: a novel technique which resolves differential usage of H19 within the same cell lineage during human placental development. Development 122:839–847

Adams A, Lindahl T (1975) Epstein–Barr virus genomes with properties of circular DNA molecules in carrier cells. Proc Natl Acad Sci USA 72:1477–1481

Alcobia I, Dilao R, Parreira L (2000) Spatial associations of centromers in the nuclei of hematopoietic cells: evidence for cell-type-specific organizational patterns. Blood 95:1608–1615

Altiok E, Minarovits J, Hu LF, Contreras-Brodin B, Klein G, Ernberg I (1992) Host-cell-phenotype-dependent control of BCR2/BWR1 promoter complex regulates the expression of Epstein–Barr virus antigens 2–6. Proc Natl Acad Sci USA 89:905–909

Arthur JL, Scarpini CG, Connor V, Lachmann RH, Tolkowsky AM, Efstathiou S (2001) Herpes simplex virus type 1 promoter activity during latency establishment, maintenance, and reactivation in primary dorsal root neurons in vitro. J Virol 75:3885–3895

Bednarik DP (1996) The silencing of human immunodeficiency virus. In: Russo EA, Martienssen RA, Riggs AD (eds) Epigenetic mechanisms of gene regulation. Cold Spring Harbor Laboratory Press, Cold Spring Harbor, pp 547–559

Bhende P, Seaman W, Delecluse HJ, Kenney SC (2004) The EBV lytic switch protein Z, preferentially binds to and activates the methylated viral genome. Nat Genet 36:1099–1104

Brooks AR, Harkins RN, Wang P, Qian HS, Liu P, Rubanyi GM (2004) Transcriptional silencing is associated with extensive methylation of the CMV promoter following adenoviral gene delivery to muscle. J Gene Med 6:395–404

Brown KE, Guest SS, Smale ST, Hahm K, Merkenschlager M, Fischer A (1997) Association of transcriptionally silent genes with Ikaros complexes at centromeric heterochromatin. Cell 91:845–854

Buschhausen G, Graessmann M, Graessmann A (1985) Inhibition of herpes simplex thymidine kinase gene expression by DNA methylation is an indirect effect. Nucleic Acids Res 13:5503–5513

Buschhausen G, Wittig B, Graessmann M, Graessmann A (1987) Chromatin structure is required to block transcription of the methylated herpes simplex thymidine kinase gene. Proc Natl Acad Sci USA 84:1177–1181

Chau CM, Lieberman PM (2004) Dynamic chromatin boundaries delineate a latency control region of Epstein–Barr virus. J Virol 78:12308–12319

Chen F, Hu LF, Ernberg I, Klein G, Winberg G (1995) Coupled transcription of Epstein–Barr virus latent membrane protein (LMP)-1 and LMP2-B genes in nasopharyngeal carcinomas. J Gen Virol 76:131–138

Chen HL, Lung MM, Sham JS, Choy DT, Griffin BE, Ng MH (1992) Transcription of BamHI-A region of the EBV genome in NPC tissues and B cells. Virology 191:193–201

Chen J, Ueda K, Sakakibara S, Okuno T, Parravicini C, Corbellino M, Yamanishi K (2001) Activation of latent Kaposi's sarcoma-associated herpesvirus by demethylation of the promoter of the lytic transactivator. Proc Natl Acad Sci USA 98:4119–4124

Christy B, Scangos G (1982) Expression of transfected thymidine kinase genes is controlled by methylation. Proc Natl Acad Sci USA 79:6299–6303

Contreras-Brodin BA, Anvret M, Imreh S, Altiok E, Klein G, Masucci MG (1991) B cell phenotype-dependent expression of the Epstein–Barr virus nuclear antigens EBNA2-to EBNA-6: studies with somatic cell hybrids. J Gen Virol 72:3025–3033

de Jesus O, Smith PR, Spender LC, Karstegl CE, Niller HH, Huang D, Farrell PJ (2003) Updated Epstein–Barr virus (EBV) DNA sequence and analysis of a promoter for the BART (CST, BARF0) RNAs of EBV. J Gen Virol 84:1443–1450

Decaussin G, Sbih-Lammali F, de Turenne-Tessier M, Bouguermouh A, Ooka T (2000) Expression of BARF 1 gene encoded by Epstein–Barr virus in nasopharyngeal carcinoma biopsies. Cancer Res 60:5584–5588

Deshamne SL, Fraser NW (1989) During latency, herpes simplex virus type 1 DNA is associated with nucleosomes in a chromatin structure. J Virol 63:943–947

Desrosiers RC (1982) Specifically unmethylated cytidylic-guanylate sites in herpesvirus saimiri DNA in tumor cells. J Virol 43:427–435

Desrosiers RC, Mulder C, Fleckenstein B (1979) Methylation of herpesvirus saimiri DNA in lymphoid tumor cell lines. Proc Natl Acad Sci USA 76:3839–3843

Dobson AT, Sedarati F, Devi-Rao G, Flanagan WM, Farrell MJ, Stevens JG, Wagner EK, Feldman LT (1989) Identification of the latency-associated transcript promoter by expression of beta-globin mRNA in mouse sensory nerve ganglia latently infected with a recombinant herpes simplex virus. J Virol 63:3844–3851

Dressler GR, Rock DL, Fraser NW (1987) Latent herpes simplex virus type I DNA is not extensively methylated in vivo. J Gen Virol 68:1761–1765

Egger G, Liang G, Jones P (2004) Epigenetics in human disease and prospects for epigenetic therapy. Nature 429:457–463

Ehrlich M (2003) Expression of various genes is controlled by DNA methylation during mammalian development. J Cell Biochem 88:899–910

Elliott J, Goodhew EB, Krug LT, Shakhnovsky N, Yoo L, Speck SH (2004) Variable methylation of the Epstein–Barr virus Wp EBNA gene promoter in B-lymphoblastoid cell lines. J Virol 78:14602–14605

Falk KI, Szekely L, Aleman A, Ernberg I (1998) Specific methylation patterns in two control regions of Epstein–Barr virus latency: the LMP1-coding upstream regulatory region and an origin of DNA replication (oriP). J Virol 72:2969–2974

Forejt J, Saam JR, Gregorova S, Tilgham SM (1999) Monoallelic expression of reactivated imprinted genes in embryonal carcinoma cell hybrids. Exp Cell Res 252:416–422

Frommer M, McDonald LE, Millar DS, Collis CM, Watt F, Grigg GW, Molloy PL, Paul CL (1992) A genomic sequencing protocol that yields a positive display of 5-methylcytosine residues in individual DNA strands. Proc Natl Acad Sci USA 89:1827–1831

Honess RW, Gompels UA, Barrell BG, Craxton M, Cameron KR, Staden R, Chang YN, Hayward GS (1989) Deviations from expected frequencies of CpG dinucleotides in herpesvirus DNAs may be diagnostic of differences in the states of their latent genomes. J Gen Virol 70:837–855

Hu LF, Minarovits J, Contreras-Salazar B, Rymo L, Falk K, Klein G, Ernberg I (1991) Variable expression of latent membrane protein in nasopharyngeal carcinoma can be related to methylation status of the Epstein–Barr virus BNLF-1 5′ flanking region. J Virol 65:1558–1567

Imai S, Koizumi S, Sugiura M, Tokunaga M, Uemura Y, Yamamoto N, Tanaka S, Sato E, Osato T (1994) Gastric carcinoma: monoclonal epithelial malignant cells expressing Epstein–Barr virus latent infection protein. Proc Natl Acad Sci USA 91:9131–9135

Jaenisch R, Schnieke A, Harbers K (1985) Treatment of mice with 5-azacytidine activates silent retroviral genomes in different tissues. Proc Natl Acad Sci USA 82:1451–1455

Jankelevich S, Kolman JL, Bodnar JW, Miller G (1992) A nuclear attachment region organizes the Epstein–Barr viral plasmid in Raji cells into a single DNA domain. EMBO J 11:1165–1176

Jansson A, Masucci M, Rymo L (1992) Methylation at discrete sites within the enhancer region regulate the activity of the Epstein–Barr virus BamJI W promoter in Burkitt lymphoma lines. J Virol 66:62–69

Jenuwein T, Allis CD (2001) Translating the histone code. Science 293:2889–2897

Jin XW, Speck SH (1992) Identification of critical cis elements involved in mediating Epstein–Barr virus nuclear antigen 2-dependent activity of an enhancer located upstream of the viral BamHI C promoter. J Virol 66:2846–2852

Kanamori A, Ikuta K, Shigeharu U, Kato S, Hirai K (1987) Methylation of Marek's disease virus DNA in chicken T-lymphoblastoid cell lines. J Gen Virol 68:1485–1490

Kerr BM, Lear AL, Rowe M, Croom-Carter D, Young LS, Rookes S, Gallimore PH, Rickinson AB (1992) Three transcriptionally distinct forms of Epstein–Barr virus latency in somatic cell hybrids: cell phenotype dependence of promoter usage. Virology 187:189–201

Klein G (1996) EBV-B cell interactions: immortalization, rescue from apoptosis, tumorigenicity. Acta Microbiol Immunol Hung 43:97–105

Kosak ST, Skok JA, Medina KL, Riblet R, Le Beau MM, Fischer AG, Singh H (2002) Subnuclear compartmentalization of immunoglobulin loci during lymphocyte development. Science 296:158–162

Kubat NJ, Tran RK, McAnany P, Bloom DC (2004a) Specific histone tail modification and not DNA methylation is a determinant of herpes simplex virus type 1 latent gene expression. J Virol 78:1139–1149

Kubat NJ, Amelio AL, Giordani NV, Bloom DC (2004b) The herpes simplex virus type 1 latency-associated transcript (LAT) enhancer/rcr is hyperacetylated during latency independently of LAT transcription. J Virol 78:12508–12518

Levine SS, King IF, Kingston RE (2004) Division of labour in Polycomb group repression. Trends Biochem Sci 29:479–485

Li H, Minarovits J (2003) Host cell-dependent expression of latent Epstein–Barr virus genomes: regulation by DNA methylation. Adv Cancer Res 89:133–156

Li Q, Peterson KR, Fang X, Stamatoyannopoulos G (2002) Locus control regions. Blood 100:3077–3086

Liebowitz D (1998) Pathogenesis of Epstein–Barr virus. In: McCance DJ (ed) Human tumor viruses. ASM Press, Washington, pp 175–199

Ling PD, Hsieh JD, Ruf IK, Rawlins DE, Hayward SD (1994) EBNA-2 upregulation of Epstein–Barr virus promoters and the cellular CD23 promoter utilizes a common targeting intermediate, CBF1. J Virol 68:5375–5383

Lokensgard JR, Bloom DC, Dobson AT, Feldman LT (1994) Long-term promoter activity during herpes simplex virus latency. J Virol 68:7148–7158

Longnecker R (1998) Molecular biology of Epstein–Barr virus. In: McCance DJ (ed) Human tumor viruses. ASM Press, Washington, pp 135–174

Masucci MG, Contreras-Salazar B, Ragnar E, Falk K, Minarovits J, Ernberg I, Klein G (1989) 5-Azacytidine up regulates the expression of Epstein–Barr virus nuclear antigen 2 (EBNA-2) through EBNA-6 and latent membrane protein in the Burkitt's lymphoma line Rael. J Virol 63:3135–3141

McGeoch DJ, Davison AJ (1999) The molecular history of herpesviruses. In: Domingo E, Webster R, Holland J (eds) Origin and evolution of viruses. Academic Press, New York, pp 441–465

Mellerick DM, Fraser NW (1987) Physical state of the latent herpes simplex virus genome in a mouse model system: evidence suggesting an episomal state. Virology 158:265–275

Minarovits J, Minarovits-Kormuta S, Ehlin-Henriksson B, Falk K, Klein G, Ernberg I (1991) Host cell phenotype-dependent methylation patterns of Epstein–Barr virus DNA. J Gen Virol 72:1591–1599

Minarovits J, Hu LF, Marcsek Z, Minarovits-Kormuta S, Klein G, Ernberg I (1992) RNA polymerase III-transcribed EBER 1 and 2 transcription units are expressed and hypomethylated in the major Epstein–Barr virus carrying cell types. J Gen Virol 73:1687–1692

Minarovits J, Hu LF, Imai S, Harabuchi Y, Kataura A, Minarovits-Kormuta S, Osato T, Klein G (1994) Clonality, expression and methylation patterns of the Epstein–Barr virus genomes in lethal midline granulomas classified as peripheral angiocentric T cell lymphomas. J Gen Virol 75:77–84

Mitsuya K, Sui H, Meguro M, Kugoh H, Jinno Y, Niikawa N, Oshimura M (1997) Paternal expression of WT1 in human fibroblasts and lymphocytes. Hum Mol Genet 6:2243–2246

Morimura T, Ohashi K, Sugimoto C, Onuma M (1998) Pathogenesis of Marek's disease (MD) and possible mechanism of immunity induced by MD vaccine. J Vet Med Sci 60:1–8

Niller HH, Salamon D, Takacs M, Uhlig J, Wolf H, Minarovits J (2001) Protein–DNA interaction and CpG methylation at rep*/vIL-10p of latent Epstein–Barr virus genomes in lymphoid cell lines. Biol Chem 382:1411–1419

Niller HH, Salamon D, Ilg K, Koroknai A, Banati F, Bäml G, Rücker OL, Schwarzmann F, Wolf H, Minarovits J (2003) The in vivo binding site for oncoprotein c-Myc in the promoter for Epstein–Barr virus (EBV) encoding RNA (EBER) 1 suggests a specific role for EBV in lymphomagenesis. Med Sci Monit 9:HY1–HY9

Niller HH, Salamon D, Rahmann S, Ilg K, Koroknai A, Bánáti F, Schwarzmann F, Wolf H, Minarovits J (2004a) A 30 kb region of the Epstein–Barr virus genome is colinear with the rearranged human immunoglobulin gene loci: implications for a "ping-pong evolution" model for persisting viruses and their hosts. Acta Microbiol Immunol Hung 51:469–484

Niller HH, Salamon D, Banati F, Schwarzmann F, Wolf H, Minarovits J (2004b) The LCR of EBV makes Burkitt's lymphoma endemic. Trends Microbiol 12:495–499

Niller HH, Salamon D, Ilg K, Koroknai A, Banati F, Schwarzmann F, Wolf H, Minarovits J (2004c) EBV-associated neoplasms: alternative pathogenetic pathways. Med Hypotheses 62:387–391

Nonoyama M, Pagano JS (1972) Separation of Epstein–Barr virus DNA from large chromosomal DNA in non-virus producing cells. Nature 238:169–171

Obata Y, Kono T (2002) Maternal primary imprinting is established at a specific time for each gene throughout oocyte growth. J Biol Chem 277:5285–5289

Prosch S, Stein J, Staak K, Liebenthal C, Volk HD, Kruger DH (1996) Inactivation of the very strong HCMV immediate early promoter by CpG methylation in vitro. Biol Chem Hoppe Seyler 377:195–201

Robertson KD, Hayward SD, Ling PD, Samid D, Ambinder R (1995) Transcriptional activation of the Epstein–Barr virus latency C promoter after 5-azacytidine treatment: evidence that demethylation at a single CpG site is crucial. Mol Cell Biol 15:6150–6159

Sadler RH, Raab-Traub N (1995) The Epstein–Barr virus 3.5-kilobase latent membrane protein mRNA initiates from a TATA-less promoter within the first terminal repeat. J Virol 69:4577–4581

Saitoh S, Buiting K, Rogan PK, Buxton JL, Driscoll DJ, Arnemann J, König R, Malcolm S, Horsthemke B (1996) Minimal definition of the imprinting center and fixation of a chromosome 15q11–13 epigenotype by imprinting mutations. Proc Natl Acad Sci USA 93:7811–7815

Salamon D, Takacs M, Myöhänen S, Marcsek Z, Berencsi G, Minarovits J (2000) De novo methylation at nonrandom founder sites 5′ from an unmethylated minimal origin of DNA replication in latent Epstein–Barr virus genomes. Biol Chem 381:95–105

Salamon D, Takacs M, Ujvari D Uhlig J, Wolf H, Minarovits J, Niller HH (2001) Protein–DNA binding and CpG methylation at nucleotide resolution of latency-associated promoters Qp, Cp and LMP1p of Epstein–Barr virus. J Virol 75:2584–2596

Salamon D, Takacs M, Schwarzmann F, Wolf H, Minarovits J, Niller HH (2003) High-resolution methylation analysis and in vivo protein–DNA binding at the promoter of the viral oncogene LMP2A in B cell lines carrying latent Epstein–Barr virus genomes. Virus Genes 27:57–66

Sano M, Kitayama S (1996) Inhibition of nerve growth factor-induced outgrowth of neurites by trichostatin A requires de novo protein synthesis in PC12D cells. Brain Res 742:195–202

Santoso B, Ortiz BD, Winoto A (2000) Control of organ-specific demethylation by an element of the T-cell receptor-α locus control region. J Biol Chem 275:1952–1958

Schaefer BC, Strominger JL, Speck SH (1997) Host-cell-determined methylation of specific Epstein–Barr virus promoters regulates the choice between distinct viral latency programs. Mol Cell Biol 17:364–377

Sjöblom-Hallén A, Yang W, Jansson A, Rymo L (1999) Silencing the Epstein–Barr virus latent membrane protein 1 gene by the Max-Mad1-mSin3A modulator of chromatin structure. J Virol 73:2983–2993

Stevens JG, Wagner EK, Devi-Rao GB, Cook ML, Feldman LT (1987) RNA complementary to a herpesvirus alpha gene mRNA prominent in latently infected neurons. Science 235:1056–1059

Strockbine LD, Cohen JI, Farrah T, Lyman SD, Wagener F, DuBose R, Armitage SJ, Spriggs MK (1998) The Epstein–Barr virus BARF1 gene encodes a novel, soluble colony-stimulating factor-1 receptor. J Virol 72:4015–4021

Su A, Wiltshire T, Batalov S, Lapp H, Ching K, Block D, Zhang J, Soden R, Hayakawa M, Kreiman G, Cooke M, Walker JR, Hogenesch JB (2004) A gene atlas of the mouse and human protein-encoding transcriptomes. Proc Natl Acad Sci USA 101:6062–6067

Sung NS, Kenney S, Gutsch D, Pagano JS (1991) EBNA-2 transactivates a lymphoid specific enhancer in the BamHI-Cpromoter of Epstein–Barr virus. J Virol 65:2164–2169

Sutter D, Doerfler W (1980) Methylation of integrated adenovirus type 12 DNA sequences in transformed cells is inversely correlated with gene expression. Proc Natl Acad Sci USA 77:253–256

Takacs M, Myöhänen S, Altiok E, Minarovits J (1998) Analysis of methylation patterns in the regulatory region of the latent Epstein–Barr virus promoter BCR2 by automated fluorescent genomic sequencing. Biol Chem 379:417–422

Takacs M, Salamon D, Myöhänen S, Li H, Segesdi J, Ujvari D, Uhlig J, Niller HH, Wolf H, Berencsi G, Minarovits J (2001a) Epigenetics of latent Epstein–Barr virus genomes: high resolution methylation analysis of the bidirectional promoter region of latent membrane protein 1 and 2B genes. Biol Chem 382:699–705

Takacs M, Segesdi J, Balog K, Mezei M, Tóth G, Minárovits J (2001b) Relative deficiency in CpG dinucleotides is a widespread but not unique feature of Gammaherpesvirinae genomes. Acta Microbiol Immunol Hung 48:349–357

Tao Q, Robertson KD, Manns A, Hildesheim A, Ambinder RF (1998) The Epstein–Barr virus latent promoter Q is constitutively active, hypomethylated and methylation sensitive. J Virol 72:7075–7083

Tao Q, Huang H, Geiman TM, Lim CY, Fu L, Qiu GH, Robertson KD (2002) Defective de novo methylation of viral and cellular DNA sequences in ICF syndrome cells. Hum Mol Genet 11:2091–9102

Tasseron-de Jong JG, den Dulk H, van de Putte P, Giphart-Gassler M (1989) De novo methylation is a major event in the inactivation of transfected herpesvirus thymidine kinase genes in human cells. Biochim Biophys Acta 1007:215–223

Tierney RJ, Kirby HE, Nagra JK, Desmond J, Bell AI, Rickinson AB (2000) Methylation of transcription factor binding sites in the Epstein–Barr virus latent cycle promoter Wp coincides with promoter downregulation during virus-induced B-cell transformation. J Virol 74:10468–10479

Toth M, Muller U, Doerfler W (1990) Establishment of de novo DNA methylation patterns. Transcription factor binding and non-CpG sequences in an integrated adenovirus promoter. J Mol Biol 214:673–683

Vardimon L, Newmann R, Kuhlmann I, Sutter D, Doerfler W (1980) DNA methylation and viral gene expression in adenovirus transformed and infected cells. Nucleic Acids Res 8:2461–2473

Welch HM, Bridges CG, Lyon AM, Griffiths L, Edington N (1992) Latent equid herpesviruses 1 and 4: detection and distinction using the polymerase chain reaction and co-cultivation from lymphoid tissues. J Gen Virol 73:261–268

Wettstein FO, Stevens JG (1983) Shope papilloma virus DNA is extensively methylated in non-virus-producing neoplasms. Virology 126:493–504

Whitelaw E, Martin DIK (2001) Retrotransposons as epigenetic mediators of phenotypic variation in mammals. Nat Genet 27:361–365

Woisetschlaeger M, Yandava C, Fumarsky L, Strominger J, Speck S (1990) Promoter-switching in Epstein–Barr virus during the initial infection in B lymphocytes. Proc Natl Acad Sci USA 87:1725–1729

Wolffe AP, Matzke MA (1999) Epigenetics: regulation through repression. Science 286:481–486

Yoshioka M, Kikuta H, Ishiguro N, Ma X, Kobayashi K (2003a) Unique Epstein–Barr virus (EBV) latent gene expression, EBNA promoter usage and EBNA promoter methylation status in chronic active EBV infection. J Gen Virol 84:1133–1140

Yoshioka M, Kikuta H, Ishiguro N, Endo R, Kobayashi K (2003b) Latency pattern of Epstein–Barr virus and methylation status in Epstein–Barr virus-associated hemophagocytic syndrome. J Med Virol 70:410–419

Youssoufian H, Hammer SM, Hirsch MS, Mulder M (1982) Methylation of the viral genome in an in vitro model of herpes simplex virus latency. Proc Natl Acad Sci USA 79:2207–2210

CTMI (2006) 310:81–115

Epigenetics of Complex Diseases: From General Theory to Laboratory Experiments

A. Schumacher · A. Petronis (✉)

The Krembil Family Epigenetics Laboratory, Centre for Addiction and Mental Health, Rm 28, 250 College St., M4T 1R8 ON, Toronto, Canada
Arturas_Petronis@camh.net

Abstract Despite significant effort, understanding the causes and mechanisms of complex non-Mendelian diseases remains a key challenge. Although numerous molecular genetic linkage and association studies have been conducted in order to explain the heritable predisposition to complex diseases, the resulting data are quite often inconsistent and even controversial. In a similar way, identification of environmental factors causal to a disease is difficult. In this article, a new interpretation of the paradigm of "genes plus environment" is presented in which the emphasis is shifted to epigenetic misregulation as a major etiopathogenic factor. Epigenetic mechanisms are consistent with various non-Mendelian irregularities of complex diseases, such as the existence of clinically indistinguishable sporadic and familial cases, sexual dimorphism, relatively late age of onset and peaks of susceptibility to some diseases, discordance of monozygotic twins and major fluctuations on the course of disease severity. It is also suggested that a substantial portion of phenotypic variance that traditionally has been attributed to environmental effects may result from stochastic epigenetic events in the cell. It is argued that epigenetic strategies, when applied in parallel with the traditional genetic ones, may significantly advance the discovery of etiopathogenic mechanisms of complex diseases. The second part of this chapter is dedicated to a review of laboratory methods for DNA methylation analysis, which may be useful in the study of complex diseases. In this context, epigenetic microarray technologies are emphasized, as it is evident that such technologies will significantly advance epigenetic analyses in complex diseases.

1 Introduction

The possibility of understanding the molecular basis of human diseases is one of the most exciting perspectives of contemporary biomedical research. Since most (if not all) diseases exhibit inherited predisposition, significant research effort has been dedicated to identification of heritable risk factors. Despite major technological and computational developments, the progress in elucidation of aetiological DNA sequence variants in the overwhelming majority of human disease, primarily complex non-Mendelian disease, has been slow. The problems in understanding the molecular origin of complex diseases could be due to limitations in the current research strategy, which is focused primarily on DNA sequence variation (e.g. mutations, polymorphisms). As a rule, such DNA sequence variants are thought to be located in the coding or regulatory part of a gene, and this expectation originates from a series of discoveries in simple Mendelian disorders such as sickle cell anaemia, thalassaemia, phenylketonuria, Duchenne muscular dystrophy and cystic fibrosis. The idea of the essential role of DNA sequence variation has been generalized and extrapolated to the "fundamentally different group of diseases" (Risch 1990), namely complex non-Mendelian diseases. Complex diseases, unlike simple ones, exhibit irregular (non-Mendelian) mode of inheritance, discordance of monozygotic (MZ) twins, possible role of environmental factors, sexual dimorphism, a fluctuating course of disease and parental origin effects, among other features. Some methodological changes have been required in order to fit complex diseases into the already developed schemes of analyses. Such modifications basically consist of treating the genes as "predisposing" factors instead of "causative" factors and put some emphasis on environmental effects. It has to be admitted that to some extent the current paradigm has been successful in some complex diseases, especially familial cases, and a series of gene mutations has been identified in colon cancer, breast cancer and Alzheimer's disease, to name a few. The overwhelming proportion of non-Mendelian pathology, however, remains unexplained. In this context, epigenetics—with its multifaceted role in regulation of various genomic functions—arrives as a new frontier of molecular studies of complex disease. By definition, epigenetics refers to regulation of various genomic functions that are brought about by heritable, but potentially reversible changes in DNA modification (more specifically, methylation of cytosines) and chromatin structure (modifications of chromatin proteins such as histone acetylation, methylation or phosphorylation; Henikoff and Matzke 1997). Genes, even the ones that carry no mutations or disease predisposing polymorphisms, may be useless or even harmful if not expressed in the appropriate amount, at the right time of the cell cycle or in the right compartment of the nucleus. There is increasing evidence that cells can operate normally only if both DNA sequence and epigenetic components of the genome function properly. Thus far, the role of epigenetic factors has been primarily investigated in rare paediatric syndromes, such as Prader–Willi, Angelman (Nicholls 2000; Nicholls and Knepper 2001), Beckwith–Wiedemann (Feinberg 1999; Walter and Paulsen 2003; Weksberg et al. 2003) and Rett syndrome (Amir et al. 1999), and also the malignant transformation of cells in cancer (Baylin and Herman 2000; Jones and Laird 1999). This study describes the advantages of the epigenetic interpretation of common non-Mendelian complexities as well as epigenetic re-interpretation of a series of clinical and molecular findings in complex disorders. In the second part of the chapter, laboratory methods for DNA methylation analysis, which may be useful in the study of complex diseases, are reviewed and recommendations for their applications are provided.

2 Epigenetics and Complex Disease

The epigenetic theory of complex disease is based on three premises:

1. The epigenetic status of genes and genomes is far more dynamic in comparison to the DNA sequence and is subject to changes under the influence of developmental programs and the internal and external environment of the organism (Cooney et al. 2002; Sutherland and Costa 2003; Waterland and Jirtle 2003; Weaver et al. 2004). After mitotic division, the daughter chromosomes do not necessarily carry identical epigenetic patterns in comparison to the parental chromosomes. Over time, substantial epigenetic differences may be accumulated across the cells of the same cell line or the same tissue. In addition, quite significant epigenetic changes may occur even in the absence of evident environmental differences, i.e. due to stochastic reasons. In tissue culture, fidelity of maintenance methylation

in mammalian cells was detected to be between 97% and 99.9% and de novo methylation activity was as high as 3%–5% per mitosis (Riggs et al. 1998). It is important to note that epigenetic patterns are not established chaotically; there is a significant continuity of epigenetic patterns during mitotic divisions.

2. Some epigenetic signals can be transmitted along with DNA sequence across the germline generations, i.e. such signals exhibit partial meiotic stability. Although it has been generally accepted that, during the maturation of the germline, gamete's epigenetic signals are erased and new epigenetic profiles are established (Li 2002), it is now becoming clear that not all epigenetic signals are removed during gametogenesis, and epigenetically determined traits can be transmitted from one generation to another (Rakyan and Whitelaw 2003; Rakyan et al. 2001, 2002).
3. Epigenetic signals are critically important for the regulation of various genomic functions (Henikoff and Matzke 1997), and epigenetic misregulation may be as detrimental to an organism as are mutant genes. In addition to regulation of gene activity (Constancia et al. 1998; Ehrlich and Ehrlich 1993; Jones et al. 1998; Nan et al. 1998; Razin and Shemer 1999; Riggs et al. 1998; Siegfried et al. 1999), epigenetic factors play an important role in numerous other genomic functions (Bestor and al. 1994; Riggs and Porter 1996) including genetic recombination (Petronis 1996) and DNA mutability (Yang 1996).

The scientific value of the epigenetic model of complex disease lies in the possibility of integrating a variety of apparently unrelated data into a new theoretical framework, which provides the basis for new hypothesis and experimental approaches. The below overview is primarily based on the epigenetic re-analysis of various non-Mendelian irregularities in three major psychiatric diseases: bipolar disorder, schizophrenia and major depression.

2.1 Discordance of Monozygotic Twins, Environmental Impact, Stochasticity

Common phenotypic differences (discordance) in identical twins have been one of the hallmarks of complex non-Mendelian disease. For example, proband-wise, MZ concordance for major depression is 31% for male and 48% for female MZ twins (Kendler and Prescott 1999), 62%–79% in bipolar disorder (Bertelsen et al. 1977), and 41%–65% in schizophrenia (Cardno and Gottesman 2000) [for concordance rates in other diseases, in both MZ and dizygotic (DZ) twins, see Fig. 1]. The phenomenon of differential susceptibility to disease in genetically identical twins was identified decades ago; however, the

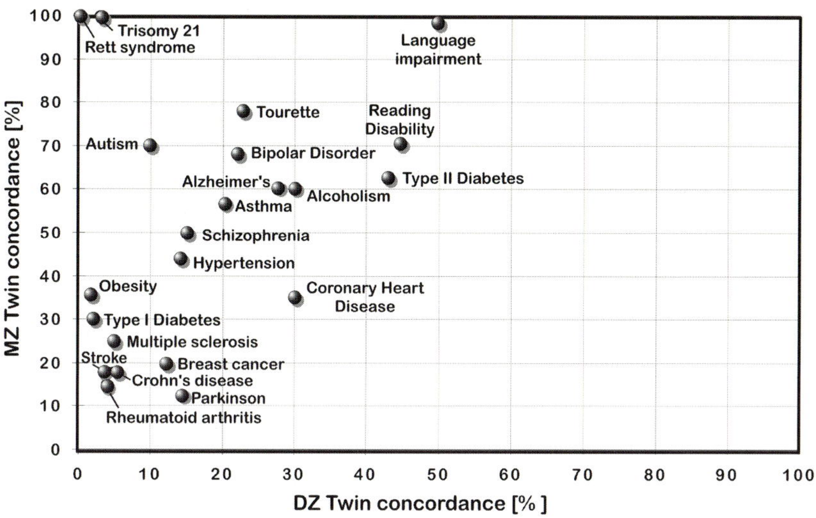

Fig. 1 Concordance of MZ and DZ twins for different disorders. As a rule, the degree of concordance in MZ twins is lower than 100% for nearly all complex diseases but substantially higher in comparison to the concordance rate in DZ twins

causes of such differences remain unknown. The traditional explanation for only one MZ twin having a clinical disease consists of the so-called "non-shared" environmental effects (Reiss et al. 1991), which supposedly produce disease in one of the two genetically predisposed co-twins. Several attempts to identify DNA sequence differences in MZ twins discordant for psychiatric diseases were carried out, but they did not detect any systemic DNA sequence differences in the tested twins (Deb-Rinker et al. 1999, 2002; Lavrentieva et al. 1999; Polymeropoulos et al. 1993; Tsujita et al. 1998; Vincent et al. 1998). Following the epigenetic model of complex disease (see Sect. 2.5 below), phenotypic differences in MZ twins result from their epigenetic differences. Due to the partial stability of epigenetic signals, a substantial degree of epigenetic dissimilarity can be accumulated over millions of mitotic divisions of cells in genetically identical organisms. This is well illustrated in inbred animals (Rakyan et al. 2002) and Beckwith–Wiedemann MZ twins (Weksberg et al. 2002), as well as in schizophrenia MZ twins (Petronis et al. 2003).

Epigenetic differences in identical twins may reflect differential exposure to a wide variety of environmental factors (that are very difficult to investigate directly) (Taubes 1995). There is an increasing list of such environmental factors that may have an impact on the epigenetic status of the genomes and individual genes (Jablonka and Lamb 1995; Ross 2003; Sutherland and Costa

2003). Epigenetic changes induced by diet have been of particular interest. For example, intake of affects both the global methylation level in the genome and regulation of imprinted genes (Ingrosso et al. 2003; Wolff et al. 1998). "Street" may also modify epigenetic regulation: Recent studies showed that methamphetamine that causes psychosis in humans alters DNA methylation as well as expression of genes in brain regions that are thought to be involved in schizophrenia (Numachi et al. 2004). This effect may be mediated via misregulation of DNA methylation enzymes, such as DNA methyltransferase, DNMT1, which was detected to be upregulated in the brain of schizophrenia patients (Veldic et al. 2004). During pregnancy, maternal dietary methyl supplements increase DNA methylation and change methylation-dependent epigenetic phenotypes in mammalian offspring (Cooney et al. 2002; Waterland and Jirtle 2003). Particularly interesting was the finding that pup licking and grooming and arched-back nursing by rat mothers altered the offspring's DNA methylation and histone modifications at a glucocorticoid receptor gene promoter in the hippocampus (Weaver et al. 2004).

In addition, quite significant epigenetic changes may occur even in the absence of evident environmental differences, i.e. due to stochastic reasons (Fig. 2). After mitotic division, the daughter chromosomes do not necessarily carry identical epigenetic patterns in comparison to the parental chromosomes, and this takes place without any specific environmental input. Over time, substantial epigenetic differences may be accumulated across the cells of the same cell line or the same tissue. As has been mentioned already, ex-

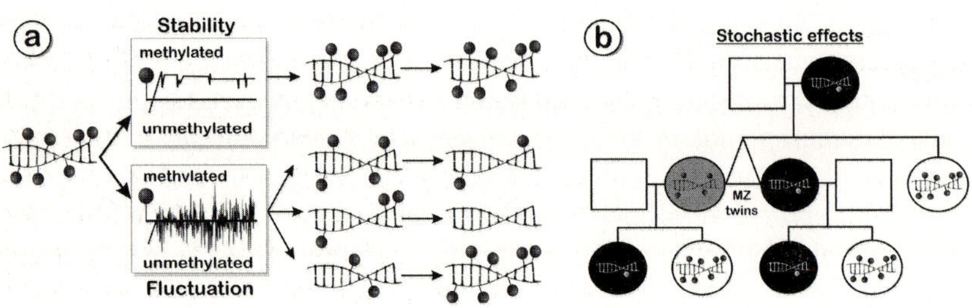

Fig. 2a, b Stochastic fluctuations in the methylation content of a genomic DNA fragment. **a** *Top*: Fully methylated regions may be particularly stable. The methylation level barely varies over time (*line*). *Bottom*: Epigenetic metastability of a short genomic region, indicated by fluctuations in the methylation level over time. The loss or gain of methyl groups at specific CpG dinucleotides results in different epigenotypes in the cells with identical genotypes. **b** An epimutation transmitted to identical twins causes the disease in only one co-twin (due to epigenetic differences in the brains), but the disease risk to the offspring of such discordant MZ twins is similar (due to the epigenetic similarities in their germline; Petronis 2004)

periments performed in tissue culture (where the genomes were identical and environment was fully controlled), fidelity of maintenance methylation in mammalian cells was detected to be a few percentage points short of 100; plus, evidence for some degree of de novo methylation activity was identified (Riggs et al. 1998). Such stochastic events may add up over the numerous mitotic divisions in two identical twins and result in quite substantial epigenetic differences is some genes and genomic regions, which results in phenotypic discordance.

2.2
Sex Effects and Critical Age

One of the important peculiarities of complex disease is sexual dimorphism—differential susceptibility to a disease in males and females—that cannot be explained by genetic risk factors on the sex chromosomes. For example, women experience a lifetime episode of major depression about twice as often as men (Piccinelli and Wilkinson 2000). Additionally, data of longitudinal studies suggest that depressed women have longer episodes of depression than men and a lower rate of spontaneous remission (Weissman and Olfson 1995). The onset of the gender gap in depression occurs between the ages of 11 years and 13 years, when a precipitous rise in depression rates for adolescent girls far exceeds that in adolescent boys, and by 15 years of age females are twice as likely as males to have experienced a major depressive episode (Cyranowski et al. 2000). These findings argue that changes in the endocrine milieu, as females progress from pre-puberty to post-puberty, might explain gender-linked increases in major depression (Warren and Brooks-Gunn 1989). In schizophrenia, the first episode usually occurs in one of three critical ages: adolescence/early adulthood, in the late 40s in women and in the sixth decade in both males and females (Howard et al. 2000), which evidently coincides with the periods of major hormonal rearrangements in the organism. Generally, intracellular effects of hormones are very consistent with the basic idea of epigenetic misregulation. Various hormones, including sex hormones, have an impact on epigenetic regulation. This is achieved by changing chromatin conformation (Csordas et al. 1986; Jantzen et al. 1987; Pasqualini et al. 1989; Truss et al. 1992), the local pattern of gene methylation (Saluz et al. 1986; Yokomori et al. 1995), or both. Hormone- induced epigenetic changes in critical genes may precipitate the onset of illness, and may also contribute to the differential susceptibility of the two sexes to complex diseases and the multiple age peaks seen in the onset of major psychosis. The findings of sex-specific effects in major depression, such as male-only linkage on 12q22-q23.2 (Abkevich et al. 2003) and female-only linkage on 2q33-35 (Zubenko et al.

2003), may be mediated via sex hormone-specific epigenetic modifications of the genes in these regions. Such findings argue that in some cases, to become a disease risk factor a gene (haplotype) must be epigenetically modified by oestrogens or androgens.

2.3 Parent-of-Origin Effects

In some psychiatric diseases, risk to offspring depends on the sex of the affected parent. For example, the risk of developing bipolar disorder is higher in the offspring whose mother is affected rather than the father (McMahon et al. 1995). Parent of origin-dependent clinical differences were also detected in schizophrenia (Crow et al. 1989; Ohara et al. 1997). Genetic linkage studies, although rarely performed in sex-specific fashion, also reveal parental origin effects in major psychosis (McMahon et al. 1997; Petronis et al. 2002; Schulze et al. 2003). One of the most common mechanisms of parent-of-origin effects is genomic imprinting (Hall 1990). The essence of genomic imprinting consists of the differential epigenetic modification of genes depending upon their parental origin (Barlow 1995). Disruption of the normal imprinting pattern often causes diseases that affect cell growth, development and behaviour (Pfeifer 2000). Animal studies investigating biological effects of genomic imprinting shed some new light on the impact of disrupted imprinting patterns on the development of the brain. Chimeric mice containing normal and uniparental cells have shown that parthenogenetic (Pg, complete maternal disomy) and androgenetic (Ag, complete paternal disomy) cells contribute differentially to specific regions of the brain (Allen et al. 1995; Schumacher 2001). In early development, Ag cells contribute substantially to those parts of the brain that are important for primary motivated behaviour (e.g. hypothalamus, septum and preoptic area) and proliferate extensively in the medio-basal forebrain. By contrast, Pg cells accumulate in the developing neocortex, striatum and telencephalic structures, where Ag cells are excluded (Keverne 1997). These results suggest an important role of epigenetic processes such as parent-of-origin effects or genomic imprinting in brain development. Such epigenetic events result in known aberrations of brain development. For example, Angelman syndrome—which presents with paroxysms of laughter, seizures, attention deficit, hyperactivity and aggressive behaviour—frequently shows anomalous cortical growth, resulting in cortical atrophy, microencephaly and ventricular dilation (Leonard et al. 1993; Schumacher 2001; Williams et al. 1989).

2.4 Familiality and Sporadicity in Complex Disease

While the explanation of all the above non-Mendelian features was based on the partial epigenetic stability in somatic cells, there is also an interesting perspective on the role and outcomes of the partial epigenetic stability during the maturation of the germline (Fig. 3). The meiotic epigenetic metastability allows for re-thinking on the issue of familiality (minor proportion of all cases) and sporadicity (overwhelming majority of the cases) in complex disease. Extrapolating from a single experimental finding of intergenerational dynamics of epigenetic regulation (Allen et al. 1990), it can be hypothesized that disease epimutations may develop in two possible ways: (1) regress towards the norm in the germline of an affected individual, and his/her offspring will not be affected, or (2) may persist across generations and become

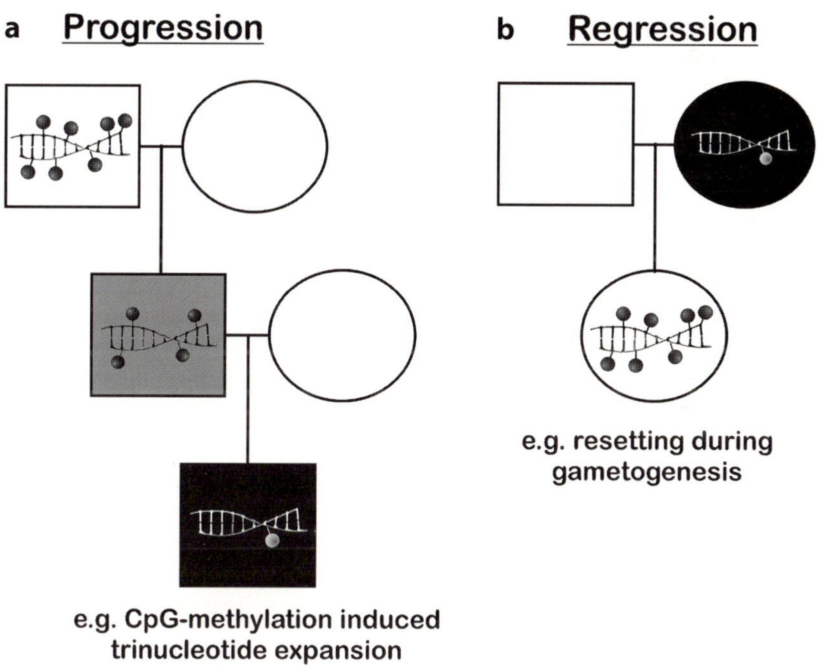

Fig. 3a, b Epigenetic perspective on the familial and sporadic cases of psychiatric disease. **a** Some epimutations may persist across generations and become even more pathogenic, which results in increased clinical severity and earlier age of onset. In some diseases it can occur that the symptoms get progressively worse every generation. **b** Other epimutations may regress towards the normal in the germline of a psychiatric patient, and his/her offspring will not be affected

even more pathogenic (Petronis 2004). Such meiotically persistent and progressing epimutations result in increasing clinical severity and earlier age of onset, which is characteristic of genetic anticipation (Petronis 2004). Genetic anticipation is a pattern of inheritance of genetic diseases where each successive generation seems to contract a more severe form of the genetic disease. Genetic anticipation has been widely investigated in psychiatric diseases (McInnis 1996; Petronis and Kennedy 1995) but is very difficult to prove due to various ascertainment biases (Heiman et al. 1996; Hodge and Wickramaratne 1995). Epigenetic studies of the intergeneration epigenetic dynamics relevant to the disease may also shed a new light on the issue of genetic anticipation.

2.5
The Epigenetic Model of Complex Disease

The epigenetic model of complex disease could be imagined as a result of a chain of aberrant epigenetic events that begins with a pre-epimutation, a primary epigenetic problem that takes place during the maturation of the germline. Yet, pre-epimutation increases the risk for the disease but is not sufficient to cause the disease per se. The epigenetic misregulation can be tolerated to some extent, and the age of disease onset may depend on the multidirectional effects of tissue differentiation, stochastic factors, hormones and probably some external environmental factors (nutrition, infections, medications, addictions, etc.; Jaenisch and Bird 2003; Sutherland and Costa 2003; Petronis 2004). It may take decades until the epigenetic misregulation reaches a critical threshold beyond which the cell (tissue, structure) is no longer able to function normally. The phenotypic outcome depends on the overall effect of the series of pre- and post-natal impacts on the pre-epimutation. Only some predisposed individuals will reach the "threshold" of epigenetic misregulation that causes the phenotypic changes that meet the diagnostic criteria for a clinical disorder (Fig. 4). Severity of epigenetic misregulation may fluctuate over time, which in clinical terms is called remission and relapse. In some cases, "ageing" epimutations may start slowly regressing back to the norm. For example, in major psychosis, this is seen as fading psychopathology or even partial recovery, which is consistent with age-dependent epigenetic changes in the genome (Fuke et al. 2004).

In conclusion, it can be argued that epigenetic mechanisms have the potential to explain a number of non-Mendelian features of complex disease. The advantages of the epigenetic scenario in comparison to the DNA sequence-based model is that the former is consistent with long years of ostensible health, critical susceptibility periods, fluctuating course and even clinical improvement after decades of being affected with a debilitating disease. The

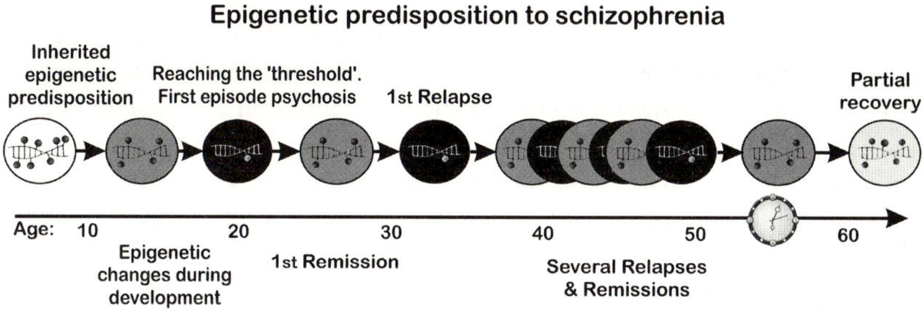

Fig. 4 Disease-development due to pre-epimutations. A pre-epimutation changes during development and may be influenced by various factors such as cell differentiation, intra- and extracellular environment, hormones and stochastic factors. Reaching a certain epigenetic threshold results in psychosis

epigenetic theory does not reject the role of DNA sequence variation but rather suggests that, in complex diseases, contribution of epigenetic factors may be substantial, and DNA sequence variation within genes should be investigated in parallel with the epigenetic regulation of genes.

3
Strategies for Detection of Epigenetic Differences in Complex Genomes

3.1
Methodological Issues in Epigenetic Studies of Complex Disease

The aforementioned theoretical speculations provide the basis for numerous working hypotheses of epigenetic differences between individuals affected with complex disease and controls. Despite significant heuristic value, epigenetic studies of complex disease are confounded by a number of factors which should be taken into account when designing experiments or interpreting data.

First, for an epigenetic study, specific tissues—the primary sites of the disease manifestation—are required, unlike the traditional genetic (linkage and association) studies where the cell/tissue source of the DNA sample is not critical. Disease-associated epigenetic differences are more likely to be detected in the disease-related tissue than the unrelated one, so in the case of major psychiatric disease, post-mortem brain samples are necessary.

Second, when dealing with complex organs such as brain, cellular heterogeneity should be taken into account as well, because there is little doubt that different types of neurons and glial cells may exhibit epigenetic differences—or colon epithelial cells versus the whole tissue. An ideal epigenetic experiment would investigate homogeneous cells collected by laser capture microdissection or fluorescence-activated cell sorting. However, the microdissection approach can only yield several hundred cells (and nanograms of DNA), which may limit the application of further methods (that require large amounts of DNA) for mapping of epigenetic differences.

Third, human tissues usually become available after a relatively long (12 h–30 h) post-mortem interval, which causes degradation of the nuclear epigenomic components, primarily histones. The same applies to the archived pathology samples, such as paraffin-embedded or formalin-fixed tissue sections. In comparison to histone modification, DNA modification is relatively stable (this is one of the main reasons why the below review of specific methods and the majority of epigenetic research to date is focused on DNA methylation analysis).

The fourth complexity is related to our ignorance of what specific genes and genomic loci could be aetiologically important in epigenetic studies of complex disease. It is not clear, for example, if in major psychosis the genes encoding dopamine or serotonin receptors (that are among the most popular molecules in psychiatric research) are really the best targets for epigenomic analyses. The focus on such genes may just be a reflection of our biased and superficial understanding of the etiopathogenic mechanisms of the disturbed brain.

Fifth, it is not clear what size of epimutation—the disease-specific epigenetic change—can be expected in complex disease: Is it a major difference—"black or white" case (such as seen in imprinted genes)—between the affected individuals and controls, or rather the more "shades of grey" scenario, where the epigenetic differences between the affected and unaffected subjects are rather subtle.

Sixth, detected associations of epigenetic changes and disease phenotype do not automatically imply the cause–effect relationship, as disease process can be caused not only by epigenetic changes but could be the cause of some epigenetic changes. Longitudinal studies, especially the ones that include the premorbid conditions (before the presentation of the clinical symptoms) are practically limited to experimental animals. In human studies, analysis of the tissues that are not involved in the disease process may provide some insights on the cause–effect relationship. The expectation is that at least some epigenetic changes can also be detected in other tissues, which may reveal tissue non-specific epigenetic changes reflecting events that took place before the tissues were formed, such as in early embryogenesis. Precedents for this type of research are evidenced by the epimutations at *IGF2* (Cui et al. 2003) in lymphocytes of colon cancer patients, and at *KCNQ1OT1* in the lymphocytes and skin fibroblasts of Beckwith–Wiedemann syndrome individuals (Weksberg et al. 2002).

3.2 Experimental Techniques in DNA Methylation Analysis

A myriad of techniques exist for the identification of methylated cytosines (Table 1). Despite their diversity, most of the analytical approaches can be divided into the bisulphite modification-based techniques and the ones that use methylation-sensitive restriction enzymes (MSREs).

3.2.1 Bisulphite Modification-Based Approaches

The bisulphite modification method has been one of the most significant developments in methylation analysis. The key advantage of this method is sensitivity, as the degree of methylation in each position of cytosines can be identified with great precision. Although there are different permutations of the bisulphite technology, all of them are based on the conversion of cytosine into uracil under conditions where 5-^{m}C remains unaltered (Frommer et al. 1992). A number of published protocols differ in the way the chemical modification is performed; however, the approach remains one of the more demanding techniques of molecular biology. The most commonly encountered artefact arises due to the high salt concentrations in the bisulphite reaction, which favours reannealing of DNA and, in turn, inhibits the sulphonation. The incomplete conversion will then be detected as false methylated cytosines (Rein et al. 1998). In addition, a small portion of 5-methylcytosines (5-^{m}C) may be converted to thymine, which results in false negatives (Thomassin et al. 1999). Furthermore, treatment with bisulphite, especially at high temperatures, leads to DNA degradation due to partial acid-catalysed depurination. Consequently, a high proportion of the template DNA is too fragmented to be analysed. This problem is important when only limited amounts of starting material is available, e.g. when using post-mortem brain samples, small amounts of body fluids or microdissected tissues. This predicament intensifies further if the remaining DNA fragments are lost during the subsequent purification (desalting), which has to be extremely stringent. Since sodium bisulphite is a very effective pH buffer, residual bisulphite will prevent the complete alkalinization of the DNA solution during desulphonation. However, if the reaction intermediate uracil-sulphonates are not converted, any DNA polymerase will be unable to replicate the template. Additionally, due to the 3-dimensional nature of the single-stranded (ss)DNA template, it may occur that some cytosines are not converted because they are included in hairpin structures that contain double-stranded regions (Rother et al. 1995). When analysing hypermethylated sequences, this effect can be even more severe,

Table 1 Methylation profiling technologies

Category	Method	BS	MSR	PCR	Gene	Diff	Global	Resolution[a]	Reference(s)
BS-based	Standard bisulphite genomic sequencing	•	–	•	•	–	–	High	Frommer et al. 1992
	RNase1 T1-MALDI-TOF	•	–	•	•	–	–	Medium	Schatz et al. 2004
	MethylQuant	•	–	•	–	–	–	Very high	Thomassin et al. 2004
	MethyLight/HeavyMethyl-MethyLight/ConLight-MSP	•	–	•	•	–	–	High	Cottrell et al. 2004; Eads et al. 2000; Rand et al. 2002
	Pyrosequencing	•	–	•	•	–	–	Very high	Dupont et al. 2004; Uhlmann et al. 2002
	Chloroacetaldehyde assay	•	–	–	–	–	•	Very low	Oakeley et al. 1999
	MALDI-mass spectrometry	•	–	•	•	–	–	Very high	Tost et al. 2003
	Quantitative analysis of methylated alleles (QAMA)	•	–	•	•	–	–	Medium	Zeschnigk et al. 2004
	Methylation-specific PCR (MSP)	•	–	•	•	–	–	High	Herman et al. 1996
	MSP-DHPLC	•	–	•	•	–	–	High	Baumer et al. 2001
	Methylation-sensitive single-nucleotide primer extension (Ms-SnuPE)	•	–	•	•	–	–	High	Gonzalgo and Jones 2002
	Primer extension and ion pair reverse phase HPLC	•	–	•	•	–	–	High	Matin et al. 2002
	Quantitative real-time PCR	•	–	•	•	–	–	High	Lo et al. 1999
	Methylation-sensitive dot blot assay (MS-DBA)	•	–	•	•	–	–	Medium	Clement and Benhattar 2005
	Microplate-based quantitative methylation assay (MANIC)	•	–	•	•	–	–	Low	Yamamoto et al. 2004

Table 1 (continued)

Category	Method	BS	MSR	PCR	Gene	Diff	Global	Resolution[a]	Reference(s)
	Enzymatic regional methylation assay (ERMA)	•	–	•	•	–	–	Medium	Galm et al. 2002; Zhang et al. 2004
	Denaturing gradient gel electrophoresis (DGGE)	•	–	•	•	–	–	Very low	Guldberg et al. 2002
	In-tube melting curve analysis and melting curve methylation specific PCR (McMSP)	•	–	•	•	–	–	Very low	Akey et al. 2002; Guldberg et al. 2002; Worm et al. 2001
	Methylation-sensitive single-strand conformation analysis (MS-SSCA)	•	–	•	•	–	–	Low	Dobrovic et al. 2002
	Methylation-specific oligonucleotide microarray	•	–	•	•	•	•	High	Adorjan et al. 2002; Balog et al. 2002; Gitan et al. 2002
	Combined bisulphite restriction analysis (COBRA)	•	•	•	•	–	–	High	Xiong and Laird 1997
	Melting curve combined bisulphite restriction analysis (McCOBRA)	•	•	•	•	–	–	Medium	Akey et al. 2002
MSR-based	Arbitrarily primed (AP)-PCR/restriction fingerprinting (MSRF)	–	•	•	–	•	–	Low	Gonzalgo et al. 1997; Huang et al. 1997; Liang et al. 2002
	*Mcr*BC-PCR	–	•	•	•	–	–	Low	Chotai and Payne 1998
	MS-RDA/methylated CpG island amplification (MCA)-RDA	–	•	•	–	•	–	Low	Toyota et al. 1999; Ushijima et al. 1997
	Amplification of inter-methylated sites (AIMS)	–	•	•	–	•	•	Low	Frigola et al. 2002

Table 1 (continued)

Category	Method	BS	MSR	PCR	Gene	Diff	Global	Resolution[a]	Reference(s)
	Methylation-sensitive amplicon subtraction (MS-AS)	–	•	•	–	•	–	Low	Mueller and Doerfler 2000
	MS-AFLP	–	•	•	–	•	–	Low	Xiong et al. 1999
	Electrochemical sensoring LM-PCR	–	•	•	•	–	–	Low	Hou et al. 2003a
	MS-amplicon epigenetic-oligonucleotide microarray	–	•	•	•	•	•	High	Schumacher et al. 2006
	*Not*I-CODE microarrays	–	•	•	–	•	•	Very low	Li et al. 2002
	*Not*I-MS-AFLP microarray	–	•	•	•	•	•	Very low	Yamamoto and Yamamoto 2004
	Hypermethylated fraction amplicon microarray	–	•	•	•	•	•	Medium	Hatada et al. 2002; Yan et al. 2002
	Methylation-sensitive restriction (MSR)-Southern	–	•	–	–	–	•	High	Rein et al. 1998
	*Mcr*BC-Southern	–	•	–	–	–	•	Very low	Sutherland et al. 1992
	Restriction landmark genomic scanning (RLGS)	–	•	–	•	•	–	Very low	Hayashizaki et al. 1993
	Photocrosslinking oligonucleotide hybridization assay	–	•	–	•	–	–	Medium	Peoples et al. 2004
	Size fractionation MSR microarray	–	•	–	•	•	•	High	Tompa et al. 2002
Other	High-performance liquid chromatography (HPLC)	–	–	–	–	–	•	Very low	Kuo et al. 1980
	High-performance or micellar electrokinetic capillary electrophoresis (HPCE)/(MECE)	–	–	–	–	–	•	Very low	Fraga et al. 2002; Stach et al. 2003

Table 1 (continued)

Category	Method	BS	MSR	PCR	Gene	Diff	Global	Resolution[a]	Reference(s)
	Thin-layer chromatography (TLC)	–	–	–	–	–	•	Very low	Schmitt et al. 1997
	*Sss*I acceptance assay	–	–	–	–	–	•	Very low	Wu et al. 1993
	5^{m}C-antibody immunochemical assay	–	–	–	–	–	•	Low	Oakeley et al. 1997
	Nearest-neighbour [^{3}H-CH_3]-analysis	–	–	–	–	–	•	Low	Hubrich-Kuhner et al. 1989
	Hydrazine (N_2H_4)-sequencing	–	–	•	•	–	–	High	Rev. Thomassin et al. 1999
	Permanganate (MnO_4^-)-sequencing	–	–	•	•	–	–	High	Rev. Thomassin et al. 1999
	Ligation-mediated (LM)-PCR	–	–	•	•	–	–	High	Pfeifer et al. 1989
	Methyl-CpG binding domain (MBD) column assays/ICEAMP	–	–	–	–	•	–	Very low	Brock et al. 2001; Cross et al. 1994

BS, bisulphite conversion; MSR, methyl-sensitive restriction enzymes; gene, method is suitable to interrogate larger DNA stretches or whole genes; Diff., enables the direct genome-wide comparison of two test-samples without a priori knowledge of the target; global, enables assessment of the total methylation level of a genome

[a] Very high, quantitative single CpGs; high, nucleotide level; medium, several CpGs; low, many CpGs; very low: global

since methyl groups stabilize the double-helix structure. Finally, the amplification of bisulphite-treated DNA may result in preferential amplification of either the methylated or unmethylated alleles. The "classical" bisulphite modification approach requires cloning and sequencing of the PCR products amplified from the bisulphite-treated DNA. Since the average number of such clones to be sequenced is at least 20, it is evident that this step makes the entire bisulphite modification approach a very labour-intensive procedure. Several approaches that by-pass the cloning and sequencing step by means of direct interrogation of C/T ratio (^{met}C/C before bisulphite modification) in the amplicon have been developed over the last decade: for example, pyrosequencing and methylation-sensitive single-nucleotide primer extension.

Pyrosequencing This method is based on an indirect bioluminometric assay of pyrophosphate (PPi) that is released from each deoxynucleotide (dNTP) upon DNA-chain elongation. In the first step, a primer is hybridized to a bisulphite-treated single-stranded, PCR-amplified template DNA and incubated with dNTP in the presence of exonuclease-deficient Klenow DNA polymerase (see Fig. 5a). Nucleotides are sequentially added to the reaction mix in a predetermined order. If a nucleotide is complementary to the template base and thus incorporated, PPi is released. PPi is then used as a substrate by an ATP sulphurylase, which converts quantitatively adenosine 5′-phosphosulphate (APS) to adenosine triphosphate (ATP). This ATP drives a luciferase-mediated conversion of luciferin to oxyluciferin that generates visible light in amounts that are proportional to the amount of ATP. The emitted light is detected by a charge-coupled device (CCD) and finally results in a peak, indicating the number and type of nucleotide incorporated in the form of a pyrogram.

Pyrosequencing can be applied to quantitate several CpG dinucleotides in one reaction and it works well with very short PCR fragments (Dupont et al. 2004; Uhlmann et al. 2002). This is particularly important in the analysis of samples that may contain moderately degraded DNA, such as paraffin-embedded tissues or post-mortem brain-samples. Pyrosequencing is a relatively rapid analysis once bisulphite-converted DNA is prepared. Additionally, cytosines outside of the analysed CpG positions can be used as internal controls to confirm completion of the bisulphite treatment. Nevertheless, an inherent problem with the described method is sequencing of polymorphic regions in heterogeneous DNA material. Polymorphisms may cause the sequencing reaction to become out of phase, making the interpretation of the succeeding sequence difficult. Another problem is the difficulty in determining the number of incorporated nucleotides in homopolymeric regions, which are present at a high frequency in bisulphite-treated DNA, due to the

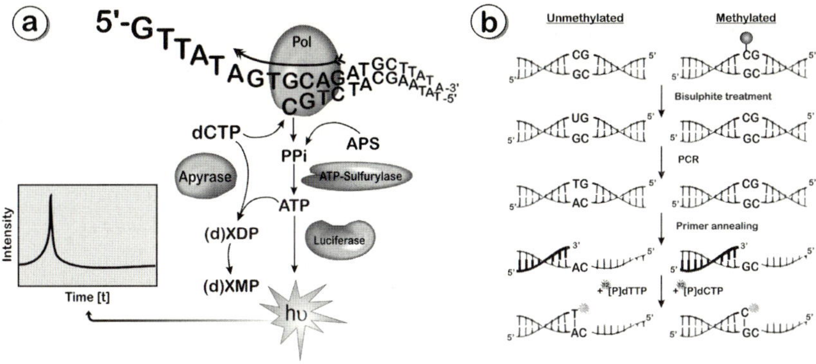

Fig. 5a, b Fine-mapping and confirmation of methylation abnormalities. **a** Pyrosequencing of bisulphite DNA. In this reaction, dCTP (the nucleotide added) is incorporated by a DNA polymerase complementary to the next unpaired nucleotide (*G*) on the bisulphite-treated template DNA. The pyrophosphate (*PPi*) released is converted to ATP and then to a light signal via an enzyme cascade, including ATP-sulphurylase and luciferase. Before the addition of the next nucleotide starts, any excess of nucleotide and ATP is degraded by apyrase, which regenerates the reaction solution. **b** Schematic outline of methylation-sensitive single-nucleotide primer extension (MS-SNuPE). After bisulphite treatment, the DNA is amplified with a strand-specific primer that does not discriminate between methylated and unmethylated alleles. Then, a primer is annealed upstream of the target sequence, immediately terminating 5′ to the original CpG. Finally, a single-nucleotide extension reaction is performed using radioactively or fluorescently labelled triphosphates, and the incorporation of the nucleotides is quantified

nonlinear light response following incorporation of more than 5–6 identical nucleotides.

Methylation-Sensitive Single-Nucleotide Primer Extension This PCR-based technique allows the quantitative analysis of several CpG sites in parallel. After bisulphite conversion, the region of interest is amplified by strand-specific PCR (Fig. 5b). The MS-SNuPE assay utilizes internal primers that anneal to the amplified template and terminate immediately 5′ of the single nucleotide(s) to be assayed. The annealed bisulphite-specific primers, which should not preferentially discriminate between methylated and unmethylated alleles, are extended by a DNA polymerase. Quantitation of the ratio of methylated versus unmethylated cytosine (C versus T) at the original CpG sites can then be determined by incubating the PCR product, primer(s) and DNA polymerase with either [^{32}P]dCTP or [^{32}P]TTP or fluorescent-labelled nucleotides followed by denaturing polyacrylamide gel electrophoresis or quantitation in a sequencer. Opposite-strand Ms-SNuPE primers can also be designed which would incorporate either [^{32}P]dATP or [^{32}P]dGTP. The amount of methylation at multiple CpG sites can be analysed in a single reaction by using a multiplex oligonucleotide strategy without the need for restriction enzymes. Advantages of the Ms-SNuPE technique are that it can be performed in a quantitative manner and with a small amount of starting material. However, it is likely that cytosines which are located close to a CpG dinucleotide of interest do affect the results (Kaminsky et al. 2005). This problem is particularly important when analysing CpG islands. Until recently, many such regions have been investigated using Ms-SNuPe, but CpG dinucleotides in the area of primer design had to be avoided (Dahl and Guldberg 2003), which is a noticeable limitation of the technology. Recent attempts have been made to overcome the problem of Ms-SNuPe primer mismatch effects in order to interrogate CpG sites independent of sequence context, including GC-rich regions, using matrix-assisted laser desorption ionization (MALDI) mass spectrometry (Tost et al. 2003).

3.2.2 Methylation-Sensitive Restriction Enzyme-Based Approaches

Methylation sensitive restriction enzymes were first applied to epigenetic studies at least three decades ago and for a long time were the primary tools for DNA methylation analysis until the fine mapping using the bisulphite modification approach was developed. The interest in MSREs is now resurging as these enzymes are the key tools for large-scale epigenomic profiling using microarrays (see Sect. 3.3). Classical examples of methods using MSREs are MS-Southern and the restriction landmark genomic scanning (RLGS), a method that was used to detect genomic regions with alterations in DNA methylation associated with tumourigenesis (Hayashizaki et al. 1993). RLGS employs direct end labelling of the genomic DNA digested with a methylation-sensitive restriction enzyme (usually *Not*I) and two-dimensional gel-electrophoresis. The status of DNA methylation can then be determined by monitoring the appearance or disappearance of spots in the gel. Other methods use the restriction endonuclease *Mcr*BC to compare DNA sample pairs (Chotai and Payne 1998; Sutherland et al. 1992). *Mcr*BC cleaves DNA containing ^{m}C on one or both strands but will not act upon unmethylated DNA. *Mcr*BC will act upon a pair of Pu^{m}CG sequence elements, thereby detecting a high proportion of methylated CpGs, but will not recognize *Hpa*II/*Msp*I sites (5′-C^{m}CGG-3′) in which the internal cytosine is methylated. *Mcr*BC digestion was used as a diagnostic test for Prader–Willi and Angelman syndromes based on differential digestion of repressed (maternally imprinted) *SNRPN* sequences by *Mcr*BC, followed by PCR amplification of the SNRPN promoter (Chotai and Payne 1998).

Over 250 different methylation-sensitive restriction enzymes (including isoschizomers) are now available (see also Table 2), but which of these enzymes are really useful and informative for methylation profiling? Informative MSREs are defined by the number of cleavage fragments in the range of approximately 75 bp to 2,000 bp that can be ligated to adaptors and efficiently amplified, and are not lost during column-purification steps. Some enzymes, although they cut frequently in the genome, produce fewer informative fragments compared to other enzymes that do not cut as frequently. For example, the non-palindromic *Aci*I (5′-CCGC-3′) recognizes more than twice as many CpG sites in CpG island regions compared to *Hpa*II, but on the other hand produces fewer fragments in the size range that can be detected by PCR or amplified fragment length polymorphism (AFLP) methods (Table 2). Other important enzyme features are their digestion- and ligation-efficiency, non-specific ("star") activity (e.g. *Eco*72I), costs and alternate CpG recognition sequences. An example for the latter is *Tau*I (5′-GC^{C}/$_{G}$GC −3′) that covers approximately 11% of all CpG dinucleotides in CpG islands; however, the results can be ambiguous, since this enzyme recognizes two different CpG-containing sequences. Many other enzymes might be useful for specific purposes, but may be exchanged with enzymes of higher CpG coverage. For example, *Kpn*2I has the recognition sequence 5′-TCCGGA-3′, which is already covered by the 4-base cutter *Hpa*II (5′-CCGG-3′). Other MSREs, such as *Fnu*4HI (5′-GCNGC-3′), will also cut sequences that do not contain a CpG dinucleotide, hence they are relatively inadequate in methylation analysis. All of these requirements for the enzyme reduces the list of potentially useful and informative MSREs to about 17 (Table 2), which would cover up to 85% of all CpG island CpG dinucleotides but less than 50% of all CpG dinucleotides in other genomic regions. The number of CpG sites that could be interrogated would even increase dramatically if a methylation-sensitive enzyme was developed that could cut the palindromic 4-base sequence 5′-TCGA-3′.

To gain the most out of restriction analyses, it is crucial to choose the right enzyme combination for the targets to be interrogated. For example, some MSREs cut relatively frequently in CpG islands but rarely recognize a sequence outside of a CpG island region, as it is the case for *Hin*6I (5′-GCGC-3′) or *Bsp*143II (5′-PuGCGCPy-3′). In contrast, enzymes such as *Hpy*CH4IV (5′-ACGT-3′) cut predominantly outside of CpG island sequences and are less useful in the interrogation of CpG islands, for instance in CpG island microarray-based studies (see Sect. 3.3). Several other methods rely on the specific methylation-sensitive cleavage of the rare cutter *Not*I (5′-GCGGCCGC-3′), for example RLGS and AFLP methods and a couple of microarray approaches (Li et al. 2002; Yamamoto and Yamamoto 2004). However, *Not*I-sites are not well represented in the genome and will only provide a very rough overview

Table 2 Methylation-sensitive restriction enzymes

MSRE	Cut site 5′-3′	Overhang[b]	CpGs in CpG islands[a] (∼%)	CpGs in non-CpG islands[a] (∼%)	Fragments/kb in CpG islands[a]	Fragments/kb in non-CpG islands[a]
*Aci*I (*Ssi*I)	CCGC, GCGG	5′-CG-3′	30.60	17.36	3.23	1.79
*Hin*6I (*Hin*P1I)	GCGC	5′-CG-3′	14.40	5.05	3.98	0.61
*Bsh*1236I (*Acc*II)	CGCG	Blunt end	12.06	2.45	3.57	0.25
*Hpa*II (*Bsi*SI)	CCGG	5′-CG-3′	11.70	9.33	3.98	1.18
*Tau*I	GCSGC	3′-GGC-5′	11.30	2.07	3.25	0.18
*Hin*1I (*Bsa*HI)	GRCGYC	5′-CG-3′	2.6	0.9	1.92	0.11
*Bsh*1285I (*Bsa*O)	CGRYCG	3′-Py-Pu-5′	2.4	<0.05	1.82	<0.02
*Eco*52I (*Eag*I)	CGGCCG	5′-GGCC-3′	1.7	<0.05	1.17	<0.02
*Hpy*CH4IV (*Tai*I)	ACGT	5′-CG-3′	1.66	6.73	1.24	0.97
*Esp*3I (*Bsm*BI)	CGTCTC	Blunt end	1.3	1.93	0.59	0.18
*Cpo*I (*Csp*I)	CGGWCCG	5′-GAC-3′	0.24	<0.05	0.1	<0.02
*Bsp*119I (*Asu*II)	TTCGAA	5′-CG-3′	0.11	0.13	0.11	<0.02
*Pvu*I (*Bsp*CI)	CGATCG	3′-TA-5′	0.11	0.13	0.02	<0.02
*Sal*I	GTCGAC	5′-TCGA-3′	0.09	0.26	<0.02	<0.02
*Bsu*15I (*Cla*II)	ATCGAT	5′-CG-3′	<0.05	0.39	<0.02	0.02
*Not*I (*Cci*NI)	GCGGCCGC	5′-GGCC-3′	<0.05	<0.05	0.1	<0.02
*Pfl*23II (*Sun*I)	CGTACG	5′-GTAC-3′	<0.05	0.13	<0.02	<0.02

R = A/G; Y = C/T; W = A/T; S = C/G

[a] The number of 75-bp- to 2-kb-long ("informative") fragments, derived from several Mbp of randomly selected CpG island and non-CpG island sequences on chromosomes 1, 2, 4, 5, 6, 9, 17, 19 and 20

[b] The isoschizomers may produce different overhangs

of methylation patterns. Hence, it is not advisable to include *Not*I in genome-wide analyses of complex diseases. All the above MSRE aspects are directly relevant to their application in the large-scale high-throughput microarray-based DNA methylation profiling.

3.3 Methods for Large-Scale DNA Modification Analysis: Microarrays

Microarrays constitute a significant advance in methylation analysis of complex disease because they may interrogate a very large number of loci in a highly parallel fashion. The principle of "epigenomic" microarrays is the same as in other kinds of arrays: Fluorescently labelled fragments of the tested nucleic acids hybridize to the complementary DNA sequences on the microarray, and intensity of fluorescent signal at each specific spot represents the amount of a specific fragment in the tested sample. Thus far, enzyme-based "epigenomic" microarray approaches have focussed predominantly on the enrichment and analysis of the hypermethylated DNA fraction of the genome. This technology was used in several studies for the identification of abnormally methylated CpG islands in tumour cells (Hatada et al. 2002; Hou et al. 2004; Huang et al. 1999; Shi et al. 2003b; Yan et al. 2002). Using the hypermethylated DNA fragments for methylation analyses seems to be practical for detection of major epigenetic changes in some regions of the genome. However, the overall proportion of CpG dinucleotides that can be interrogated is substantially lower compared to a potential analysis using the unmethylated DNA fraction. Also, unmethylated cytosines represent a much smaller part of cytosines in comparison to the methylated one (depending on the tissue, over 70% of cytosines in the human genome are methylated). Analysis of this smaller unmethylated fraction is more sensitive to detect subtle methylation abnormalities. For example, if 20% of all CpGs in a given tissue are unmethylated, a de novo methylation of 10% would result in 100% (decrease of from 20% to 10%) difference in the unmethylated fraction. In the same scenario, only a 12% change (from 80% to 90%) would be detected for the hypermethylated fraction of genomic DNA. An approach using the hypomethylated DNA fraction was suggested (Tompa et al. 2002), where a fragmentation by a methyl-sensitive restriction endonuclease is followed by a sucrose gradient size-fractionation. The small fragments (<2.5 kb) will predominantly contain hypomethylated fragments and can then be labelled and hybridized to microarrays (Fig. 6c). In the original protocol, *Msp*I was used for DNA cleavage; however, this enzyme is only blocked by methylation of the outer cytosine in the 5′-CCGG-′3 sequence, a form of methylation that is encountered in plants but usually not in hu-

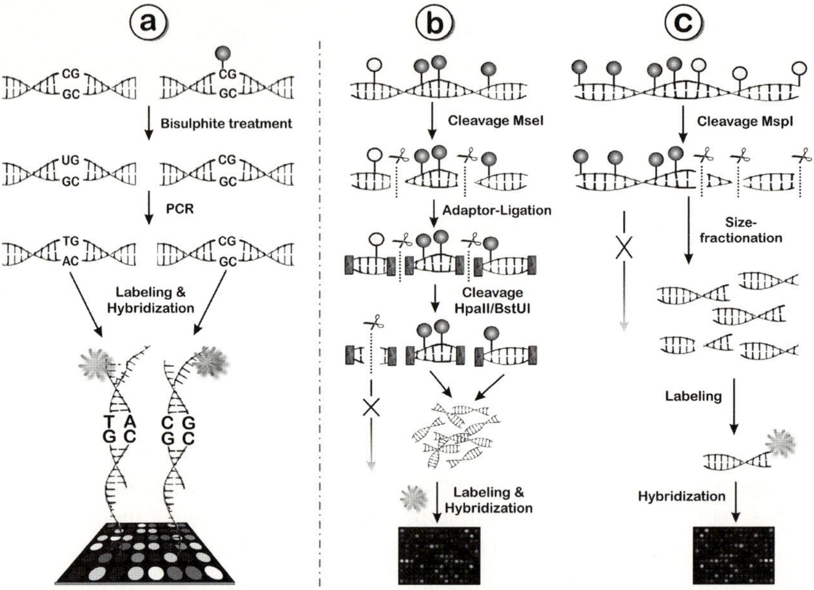

Fig. 6a–c Microarray strategies for methylation profiling. **a** Typical bisulphite approach. An amplified bisulphite-treated sample is hybridized to a set of oligonucleotides (19–25 nucleotides in length) that discriminate methylated and unmethylated cytosine at specific nucleotide positions, and quantitative differences in hybridization are determined by fluorescence analysis. **b** Restriction-based approach that uses the hypermethylated fraction of the genome. Tester and control are cleaved and adaptors specific for the cut-sites are ligated to the fragments. Unmethylated sequences are eliminated by cleavage with MSREs. The remaining hypermethylated fragments are labelled and hybridized to a microarray. **c** After methylation-sensitive cleavage, small fragments (< 2.5 kb; mostly unmethylated) are size-fractionated in a sucrose gradient, labelled and hybridized

mans. Hence, to apply this technique for the study of complex human disease, the endonuclease has to be replaced by another MSRE, such as *Hpa*II or *Aci*I (see Table 2).

For detailed DNA methylation profiling, high-resolution oligonucleotide arrays are recommended, for example microarrays that are based on 25-nucleotide perfect match–mismatch oligomers that have been generated by Affymetrix for transcriptome studies (Kapranov et al. 2002). At this time, microarrays that cover all the non-repetitive regions of human chromosomes 21 and 22 and the regions selected for the ENCODE project (http://www.genome.gov/page.cfm?pageID=10005107) of the human genome are commercially available. There is a very good chance that in the next

5 years high-resolution oligonucleotides-based microarrays for the entire human genome will be manufactured.

Since restriction enzymes are used in many methylation assays, DNA sequence variation (single-nucleotide polymorphisms, SNPs) may simulate epigenetic differences. However, until now most methods used in epigenetic studies have not been differentiating between methylation changes and the presence of SNPs within the restriction sites of the applied restriction enzymes. In order to exclude the impact of DNA sequence variation, it is suggested to check the available SNP databases and identify the DNA sequence variation within the restriction sites of the used enzymes. From CpG island microarray studies, the estimate is that 10% to 30% of methylation variation detected between individuals could be in fact due to DNA sequence variation (Schumacher et al. 2006). For comparison, in a pilot study for the Human Epigenome Project (HEP), interrogation of 3,273 unique CpG sites within the human major histocompatibility complex (MHC) on chromosome 6 revealed that 101 CpGs overlapped with known SNPs, all representing sites at which the CpG was lost (Rakyan et al. 2004).

Microarrays can also be used to interrogate C→T transitions in bisulphite-modified DNA sequences (Adorjan et al. 2002; Balog et al. 2002; Gitan et al. 2002; Hou et al. 2003a, b; Shi et al. 2003a, b). Bisulphite arrays contain oligonucleotides that measure the C(G)/T(A) ratio in the bisulphite-treated DNA, which correspond to ^{met}C/C in the native DNA (Fig. 6a). Bisulphite-based microarray technologies have the advantage that they are not limited to specific recognition sequences, as in cleavage-based approaches. However, although informative and precise, microarrays can contain only a limited number of oligonucleotides because treatment with bisulphite degenerates the four-nucleotide code, which results in the loss of specificity of a large portion of the genome. For example, after bisulphite treatment all of the possible 16 permutations of a four-base sequence containing unmethylated C and T (CCCC, CTCT, CCCT, CCTT, TCTC, TTTC, TTTT, etc.) will become identical TTTT

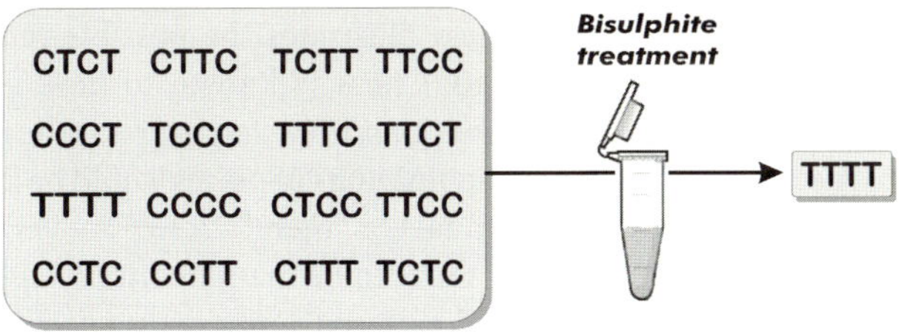

Fig. 7 After bisulphite treatment, all of the possible 16 permutations of a four-base DNA sequence containing unmethylated C and T will become identical

(Fig. 7). This degeneration will predominantly affect unmethylated regions, such as CpG islands.

4 Study of the Epigenetic Norm

In addition to human morbid epigenetics, research of the "normal" epigenome is also of significant interest, as such information may be crucial in understanding of the molecular mechanisms of development, ageing, tissue specificity and sex differences, among other systemic aspects in human biology. Documentation of normal epigenome patterns, however, is not a trivial task.

What would it be to accomplish a comprehensive annotation of the human epigenome? A conservative approach would require epigenomic profiling of DNA methylation and various kinds of histone modifications (at least 10 types) of roughly 16 million nucleosomes (as the basic structural unit of chromosomes) in approximately 260 different cell-types in the human body, at, let us say, approximately 20 different time-points (from the zygote stage through embryogenesis and post-natal development, adolescence, youth, adulthood and ageing) in 100 males and 100 females, each measurement performed in duplicate. Taken all together, one would generate $11 \times 260 \times 20 \times 200 \times 16 \times 10^6 \times 2 = 3.6 \times 10^{14}$ data points (bits). Each data point has to be referenced, which means that the chromosomal location, for example, of the modification, its time point in development, the kind of modification (methylation or acetylation) and so on, have to be stored along with it. For the storage of this raw data alone, one would require at least 0.37 PB (petabytes) of memory capacity, which is equivalent to several hundred average computer systems today. Although it is difficult to imagine how all this epigenetic information can be processed, it is very likely that there are numerous levels of redundancy (such as hypermethylated DNA regions will usually exhibit histone hypoacetylation, cells originating from the same stem should exhibit numerous epigenetic similarities).

An analogy with the game of chess illustrates the possible reduction in information content to be processed. Chess is known to have an infinite number of possible positional combinations; however, in praxis the number is finite, since specific positions and combinations of pieces would be illegal (e.g. the king cannot move into check). In fact, a "mere" 2×10^{46} moves (roughly) are theoretically possible, a number that can be mathematically approached. Out of such subsets of data, "normal" patterns, and eventually the collective behaviours of the system, and the algorithms of the system's interaction with its environment, can be identified.

In the field of complexity theory, statistical approaches that reduce the data without loosing the essential features and characteristics of the system have been developed. For example, an average "behaviour" of a large number of components can be considered rather than the "behaviour" of any individual component (e.g. the co-operation of DNA methyl groups with histone modifications), drawing heavily on the laws of probability and aiming at predictions of larger systems on the basis of the properties of their single constituents. Additionally, not all combinations of epigenetic components are unique; there are patterns present in the arrangements that allow us to classify and filter many combinations in the same way. There are also several practical interrelated approaches based on heuristic functions to study a complex system, which do not rely on static algorithms and pre-defined ideas. Heuristic approaches are self-learning or adaptive processes, based on empirical information intended to increase the probability of solving a problem. For example, "heuristic programming" would approach the problem of finding epigenetic patterns by a method of trial and error in which the success of each attempt at solution is used to improve the subsequent attempts, until a solution acceptable within defined limits is reached.

A good starting point in gathering and interpreting epigenetic data would be to understand ways of describing complex systems (especially the need for a uniform epigenetic nomenclature for multicomponent data). Second, we have to understand the interactions of the components giving rise to the pattern of behaviour and the process of formation of epigenetic information (within the cell and on the evolutionary scale). Ultimately, it is unlikely that the human epigenomic databases will consist of traditional raw data; rather, it will be user-friendly profiles, diagrams, and equations describing developmental, intra- and inter-individual variation, and epigenetic "plasticity" Altogether, this effort will provide a much "loftier view of life" (Beck et al. 1999).

References

Abkevich V, Camp NJ, Hensel CH, et al (2003) Predisposition locus for major depression at chromosome 12q22–12q23.2. Am J Hum Genet 73:1271–1281

Adorjan P, Distler J, Lipscher E, et al (2002) Tumour class prediction and discovery by microarray-based DNA methylation analysis. Nucleic Acids Res 30:e21

Akey DT, Akey JM, Zhang K, Jin L (2002) Assaying DNA methylation based on high-throughput melting curve approaches. Genomics 80:376–384

Allen ND, Norris ML, Surani MA (1990) Epigenetic control of transgene expression and imprinting by genotype-specific modifiers. Cell 61:853–861

Allen ND, Logan K, Lally G, et al (1995) Distribution of parthenogenetic cells in the mouse brain and their influence on brain development and behavior. Proc Natl Acad Sci U S A 92:10782–10786

Amir RE, Van den Veyver IB, Wan M, et al (1999) Rett syndrome is caused by mutations in X-linked MECP2, encoding methyl-CpG-binding protein 2. Nat Genet 23:185–188

Balog RP, de Souza YE, Tang HM, et al (2002) Parallel assessment of CpG methylation by two-color hybridization with oligonucleotide arrays. Anal Biochem 309:301–310

Barlow DP (1995) Gametic imprinting in mammals. Science 270:1610–1613

Baumer A, Wiedemann U, Hergersberg M, Schinzel A (2001) A novel MSP/DHPLC method for the investigation of the methylation status of imprinted genes enables the molecular detection of low cell mosaicisms. Hum Mutat 17:423–430

Baylin SB, Herman JG (2000) DNA hypermethylation in tumorigenesis: epigenetics joins genetics. Trends Genet 16:168–174

Beck S, Olek A, Walter J (1999) From genomics to epigenomics: a loftier view of life. Nat Biotechnol 17:1144

Bertelsen A, Harvald B, Hauge M (1977) A Danish twin study of manic-depressive disorders. Br J Psychiatry 130:330–351

Bestor TH, Chandler VL, Feinberg AP (1994) Epigenetic effects in eukaryotic gene expression. Dev Genet 15:458

Brock GJ, Huang TH, Chen CM, Johnson KJ (2001) A novel technique for the identification of CpG islands exhibiting altered methylation patterns (ICEAMP). Nucleic Acids Res 29:E123

Cardno AG, Gottesman II (2000) Twin studies of schizophrenia: from bow-and-arrow concordances to Star Wars Mx and functional genomics. Am J Med Genet 97:12–17

Chotai KA, Payne SJ (1998) A rapid, PCR based test for differential molecular diagnosis of Prader-Willi and Angelman syndromes. J Med Genet 35:472–475

Clement G, Benhattar J (2005) A methylation sensitive dot blot assay (MS-DBA) for the quantitative analysis of DNA methylation in clinical samples. J Clin Pathol 58:155–158

Constancia M, Pickard B, Kelsey G, Reik W (1998) Imprinting mechanisms. Genome Res 8:881–900

Cooney CA, Dave AA, Wolff GL (2002) Maternal methyl supplements in mice affect epigenetic variation and DNA methylation of offspring. J Nutr 132:2393S–2400S

Cottrell SE, Distler J, Goodman NS, et al (2004) A real-time PCR assay for DNA-methylation using methylation-specific blockers. Nucleic Acids Res 32:e10

Cross SH, Charlton JA, Nan X, Bird AP (1994) Purification of CpG islands using a methylated DNA binding column. Nat Genet 6:236–244

Crow TJ, DeLisi LE, Johnstone EC (1989) Concordance by sex in sibling pairs with schizophrenia is paternally inherited. Evidence for a pseudoautosomal locus. Br J Psychiatry 155:92–97

Csordas A, Puschendorf B, Grunicke H (1986) Increased acetylation of histones at an early stage of oestradiol-mediated gene activation in the liver of immature chicks. J Steroid Biochem 24:437–442

Cui H, Cruz-Correa M, Giardiello FM, et al (2003) Loss of IGF2 imprinting: a potential marker of colorectal cancer risk. Science 299:1753–1755

Cyranowski JM, Frank E, Young E, Shear MK (2000) Adolescent onset of the gender difference in lifetime rates of major depression: a theoretical model. Arch Gen Psychiatry 57:21–27

Dahl C, Guldberg P (2003) DNA methylation analysis techniques. Biogerontology 4:233–250

Deb-Rinker P, Klempan TA, O'Reilly RL, et al (1999) Molecular characterization of a MSRV-like sequence identified by RDA from monozygotic twin pairs discordant for schizophrenia. Genomics 61:133–144

Deb-Rinker P, O'Reilly RL, Torrey EF, Singh SM (2002) Molecular characterization of a 2.7 kb, 12q13-specific, retroviral related sequence isolated by RDA from monozygotic twins discordant for schizophrenia. Genome 45:1–10

Dobrovic A, Bianco T, Tan LW, et al (2002) Screening for and analysis of methylation differences using methylation-sensitive single-strand conformation analysis. Methods 27:134–138

Dupont JM, Tost J, Jammes H, Gut IG (2004) De novo quantitative bisulfite sequencing using the pyrosequencing technology. Anal Biochem 333:119–127

Eads CA, Danenberg KD, Kawakami K, et al (2000) MethyLight: a high-throughput assay to measure DNA methylation. Nucleic Acids Res 28:E32

Ehrlich M, Ehrlich K (1993) Effect of DNA methylation and the binding of vertebrate and plant proteins to DNA. In: Jost J, Saluz P (eds) DNA methylation: molecular biology and biological significance. Birkhauser Verlag, Basel, 145–168

Feinberg AP (1999) Imprinting of a genomic domain of 11p15 and loss of imprinting in cancer: an introduction. Cancer Res 59:1743s–1746s

Fraga MF, Uriol E, Borja Diego L, et al (2002) High-performance capillary electrophoretic method for the quantification of 5-methyl 2′-deoxycytidine in genomic DNA: application to plant, animal and human cancer tissues. Electrophoresis 23:1677–1681

Frigola J, Ribas M, Risques RA, Peinado MA (2002) Methylome profiling of cancer cells by amplification of inter-methylated sites (AIMS). Nucleic Acids Res 30:e28

Frommer M, McDonald LE, Millar DS, et al (1992) A genomic sequencing protocol that yields a positive display of 5-methylcytosine residues in individual DNA strands. Proc Natl Acad Sci U S A 89:1827–1831

Fuke C, Shimabukuro M, Petronis A, et al (2004) Age related changes in 5-methylcytosine content in human peripheral leukocytes and placentas: an HPLC-based study. Ann Hum Genet 68:196–204

Galm O, Rountree MR, Bachman KE, et al (2002) Enzymatic regional methylation assay: a novel method to quantify regional CpG methylation density. Genome Res 12:153–157

Gitan RS, Shi H, Chen CM, et al (2002) Methylation-specific oligonucleotide microarray: a new potential for high-throughput methylation analysis. Genome Res 12:158–164

Gonzalgo ML, Jones PA (2002) Quantitative methylation analysis using methylation-sensitive single-nucleotide primer extension (Ms-SNuPE). Methods 27:128–133

Gonzalgo ML, Liang G, Spruck CH 3rd, et al (1997) Identification and characterization of differentially methylated regions of genomic DNA by methylation-sensitive arbitrarily primed PCR. Cancer Res 57:594–599

Guldberg P, Worm J, Gronbaek K (2002) Profiling DNA methylation by melting analysis. Methods 27:121–127

Hall JG (1990) Genomic imprinting: review and relevance to human diseases. Am J Hum Genet 46:857–873

Hatada I, Kato A, Morita S, et al (2002) A microarray-based method for detecting methylated loci. J Hum Genet 47:448–451
Hayashizaki Y, Hatada I, Hirotsune S, et al (1993) Restriction landmark genomic scanning (RLGS) method and its application (in Japanese). Seikagaku 65:109–115
Heiman GA, Hodge SE, Wickramaratne P, Hsu H (1996) Age-at-interview bias in anticipation studies: computer simulations and an example with panic disorder. Psychiatr Genet 6:61–66
Henikoff S, Matzke MA (1997) Exploring and explaining epigenetic effects. Trends Genet 13:293–295
Herman JG, Graff JR, Myohanen S, et al (1996) Methylation-specific PCR: a novel PCR assay for methylation status of CpG islands. Proc Natl Acad Sci U S A 93:9821–9826
Hodge SE, Wickramaratne P (1995) Statistical pitfalls in detecting age-of-onset anticipation: the role of correlation in studying anticipation and detecting ascertainment bias. Psychiatr Genet 5:43–47
Hou P, Ji M, Ge C, et al (2003a) Detection of methylation of human p16(Ink4a) gene 5′-CpG islands by electrochemical method coupled with linker-PCR. Nucleic Acids Res 31:e92
Hou P, Ji M, Liu Z, et al (2003b) A microarray to analyze methylation patterns of p16(Ink4a) gene 5′-CpG islands. Clin Biochem 36:197–202
Hou P, Ji M, Li S, et al (2004) High-throughput method for detecting DNA methylation. J Biochem Biophys Methods 60:139–150
Howard R, Rabins PV, Seeman MV, Jeste DV (2000) Late-onset schizophrenia and very-late-onset schizophrenia-like psychosis: an international consensus. The International Late-Onset Schizophrenia Group. Am J Psychiatry 157:172–178
Huang TH, Laux DE, Hamlin BC, et al (1997) Identification of DNA methylation markers for human breast carcinomas using the methylation-sensitive restriction fingerprinting technique. Cancer Res 57:1030–1034
Huang TH, Perry MR, Laux DE (1999) Methylation profiling of CpG islands in human breast cancer cells. Hum Mol Genet 8:459–470
Hubrich-Kuhner K, Buhk HJ, Wagner H, et al (1989) Non-C-G recognition sequences of DNA cytosine-5-methyltransferase from rat liver. Biochem Biophys Res Commun 160:1175–1182
Ingrosso D, Cimmino A, Perna AF, et al (2003) Folate treatment and unbalanced methylation and changes of allelic expression induced by hyperhomocysteinaemia in patients with uraemia. Lancet 361:1693–1699
Jablonka E, Lamb M (1995) Epigenetic inheritance and evolution. Oxford University Press, New York, pp 1–360
Jaenisch R, Bird A (2003) Epigenetic regulation of gene expression: how the genome integrates intrinsic and environmental signals. Nat Genet 33 Suppl:245–254
Jantzen K, Fritton HP, Igo-Kemenes T, et al (1987) Partial overlapping of binding sequences for steroid hormone receptors and DNaseI hypersensitive sites in the rabbit uteroglobin gene region. Nucleic Acids Res 15:4535–4552
Jones PA, Laird PW (1999) Cancer epigenetics comes of age. Nat Genet 21:163–167
Jones PL, Veenstra GJ, Wade PA, et al (1998) Methylated DNA and MeCP2 recruit histone deacetylase to repress transcription. Nat Genet 19:187–191

Kaminsky ZA, Assadzadeh A, Flanagan J, Petronis A (2005) Single nucleotide extension technology for quantitative site-specific evaluation of metC/C in GC-rich regions. Nucleic Acids Res 33:e95

Kapranov P, Cawley SE, Drenkow J, et al (2002) Large-scale transcriptional activity in chromosomes 21 and 22. Science 296:916–919

Kendler KS, Prescott CA (1999) A population-based twin study of lifetime major depression in men and women. Arch Gen Psychiatry 56:39–44

Keverne EB (1997) Genomic imprinting in the brain. Curr Opin Neurobiol 7:463–468

Kuo KC, McCune RA, Gehrke CW, et al (1980) Quantitative reversed-phase high performance liquid chromatographic determination of major and modified deoxyribonucleosides in DNA. Nucleic Acids Res 8:4763–4776

Lavrentieva I, Broude NE, Lebedev Y, et al (1999) High polymorphism level of genomic sequences flanking insertion sites of human endogenous retroviral long terminal repeats. FEBS Lett 443:341–347

Leonard CM, Williams CA, Nicholls RD, et al (1993) Angelman and Prader-Willi syndrome: a magnetic resonance imaging study of differences in cerebral structure. Am J Med Genet 46:26–33

Li E (2002) Chromatin modification and epigenetic reprogramming in mammalian development. Nat Rev Genet 3:662–673

Li J, Protopopov A, Wang F, et al (2002) NotI subtraction and NotI-specific microarrays to detect copy number and methylation changes in whole genomes. Proc Natl Acad Sci U S A 99:10724–10729

Liang G, Gonzalgo ML, Salem C, Jones PA (2002) Identification of DNA methylation differences during tumorigenesis by methylation-sensitive arbitrarily primed polymerase chain reaction. Methods 27:150–155

Lo YM, Wong IH, Zhang J, et al (1999) Quantitative analysis of aberrant p16 methylation using real-time quantitative methylation-specific polymerase chain reaction. Cancer Res 59:3899–3903

Matin MM, Baumer A, Hornby DP (2002) An analytical method for the detection of methylation differences at specific chromosomal loci using primer extension and ion pair reverse phase HPLC. Hum Mutat 20:305–311

McInnis MG (1996) Anticipation: an old idea in new genes. Am J Hum Genet 59:973–979

McMahon FJ, Stine OC, Meyers DA, et al (1995) Patterns of maternal transmission in bipolar affective disorder. Am J Hum Genet 56:1277–1286

McMahon FJ, Hopkins PJ, Xu J, et al (1997) Linkage of bipolar affective disorder to chromosome 18 markers in a new pedigree series. Am J Hum Genet 61:1397–1404

Mueller K, Doerfler W (2000) Methylation-sensitive amplicon subtraction: a novel method to isolate differentially methylated DNA sequences in complex genomes. Gene Funct Dis 1:154–160

Nan X, Ng HH, Johnson CA, et al (1998) Transcriptional repression by the methyl-CpG-binding protein MeCP2 involves a histone deacetylase complex. Nature 393:386–389

Nicholls RD (2000) The impact of genomic imprinting for neurobehavioral and developmental disorders. J Clin Invest 105:413–418

Nicholls RD, Knepper JL (2001) Genome organization, function, and imprinting in Prader-Willi and Angelman syndromes. Annu Rev Genomics Hum Genet 2:153–175

Numachi Y, Yoshida S, Yamashita M, et al (2004) Psychostimulant alters expression of DNA methyltransferase mRNA in the rat brain. Ann N Y Acad Sci 1025:102–109

Oakeley EJ, Podesta A, Jost JP (1997) Developmental changes in DNA methylation of the two tobacco pollen nuclei during maturation. Proc Natl Acad Sci U S A 94:11721–11725

Oakeley EJ, Schmitt F, Jost JP (1999) Quantification of 5-methylcytosine in DNA by the chloroacetaldehyde reaction. Biotechniques 27:744–746, 748–750, 752

Ohara K, Xu HD, Mori N, et al (1997) Anticipation and imprinting in schizophrenia. Biol Psychiatry 42:760–766

Pasqualini JR, Mercat P, Giambiagi N (1989) Histone acetylation decreased by estradiol in the MCF-7 human mammary cancer cell line. Breast Cancer Res Treat 14:101–105

Peoples R, Wood M, Van Atta R (2004) Photocrosslinking oligonucleotide hybridization assay for concurrent gene dosage and CpG methylation analysis. Methods Mol Biol 287:233–249

Petronis A (1996) Genomic imprinting in unstable DNA diseases. Bioessays 18:587–590

Petronis A (2004) The origin of schizophrenia: genetic thesis, epigenetic antithesis, and resolving synthesis. Biol Psychiatry 55:965–970

Petronis A, Kennedy JL (1995) Unstable genes—unstable mind? Am J Psychiatry 152:164–172

Petronis A, Popendikyte V, Kan P, Sasaki T (2002) Major psychosis and chromosome 22: genetics meets epigenetics. CNS Spectr 7:209–214

Petronis A, Gottesman II, Kan P, et al (2003) Monozygotic twins exhibit numerous epigenetic differences: clues to twin discordance? Schizophr Bull 29:169–178

Pfeifer GP, Steigerwald SD, Mueller PR, et al (1989) Genomic sequencing and methylation analysis by ligation mediated PCR. Science 246:810–813

Pfeifer K (2000) Mechanisms of genomic imprinting. Am J Hum Genet 67:777–787

Piccinelli M, Wilkinson G (2000) Gender differences in depression. Critical review. Br J Psychiatry 177:486–492

Polymeropoulos MH, Xiao H, Torrey EF, et al (1993) Search for a genetic event in monozygotic twins discordant for schizophrenia. Psychiatry Res 48:27–36

Rakyan V, Whitelaw E (2003) Transgenerational epigenetic inheritance. Curr Biol 13:R6

Rakyan VK, Preis J, Morgan HD, Whitelaw E (2001) The marks, mechanisms and memory of epigenetic states in mammals. Biochem J 356:1–10

Rakyan VK, Blewitt ME, Druker R, et al (2002) Metastable epialleles in mammals. Trends Genet 18:348–351

Rakyan VK, Hildmann T, Novik KL, et al (2004) DNA methylation profiling of the human major histocompatibility complex: a pilot study for the human epigenome project. PLoS Biol 2:e405

Rand K, Qu W, Ho T, et al (2002) Conversion-specific detection of DNA methylation using real-time polymerase chain reaction (ConLight-MSP) to avoid false positives. Methods 27:114–120

Razin A, Shemer R (1999) Epigenetic control of gene expression. Results Probl Cell Differ 25:189–204

Rein T, DePamphilis ML, Zorbas H (1998) Identifying 5-methylcytosine and related modifications in DNA genomes. Nucleic Acids Res 26:2255–2264

Reiss D, Plomin R, Hetherington EM (1991) Genetics and psychiatry: an unheralded window on the environment. Am J Psychiatry 148:283–291

Riggs A, Porter T (1996) Overview of epigenetic mechanisms. In: Russo VEA MR, Riggs AD (eds) Epigenetic mechanisms of gene regulation. Cold Spring Harbor Laboratory Press, Cold Spring Harbor, pp 29–45

Riggs A, Xiong Z, Wang L, JM L (1998) Methylation dynamics, epigenetic fidelity and X chromosome structure. In: Wolffe A (ed) Epigenetics. John Wiley and Sons, Chichester, pp 214–227

Risch N (1990) Genetic linkage and complex diseases, with special reference to psychiatric disorders. Genet Epidemiol 7:3-16; discussion 17–45

Ross SA (2003) Diet and DNA methylation interactions in cancer prevention. Ann N Y Acad Sci 983:197–207

Rother KI, Silke J, Georgiev O, et al (1995) Influence of DNA sequence and methylation status on bisulfite conversion of cytosine residues. Anal Biochem 231:263–265

Saluz HP, Jiricny J, Jost JP (1986) Genomic sequencing reveals a positive correlation between the kinetics of strand-specific DNA demethylation of the overlapping estradiol/glucocorticoid-receptor binding sites and the rate of avian vitellogenin mRNA synthesis. Proc Natl Acad Sci U S A 83:7167–7171

Schatz P, Dietrich D, Schuster M (2004) Rapid analysis of CpG methylation patterns using RNase T1 cleavage and MALDI-TOF. Nucleic Acids Res 32:e167

Schmitt F, Oakeley EJ, Jost JP (1997) Antibiotics induce genome-wide hypermethylation in cultured Nicotiana tabacum plants. J Biol Chem 272:1534–1540

Schulze TG, Chen YS, Badner JA, et al (2003) Additional, physically ordered markers increase linkage signal for bipolar disorder on chromosome 18q22. Biol Psychiatry 53:239–243

Schumacher A (2001) Mechanisms and brain specific consequences of genomic imprinting in Prader-Willi and Angelman syndromes. Gene Funct Dis 1:7–25

Schumacher A, Kapranov P, Kaminsky Z, Flanagan J, Assadzadeh A, Yau P, Virtanen C, Winegarden N, Cheng J, Gingeras T, Petronis A (2006) Microarray-based DNA methylation profiling: technology and applications. Nucleic Acids Res 34:528–542

Shi H, Maier S, Nimmrich I, et al (2003a) Oligonucleotide-based microarray for DNA methylation analysis: principles and applications. J Cell Biochem 88:138–143

Shi H, Wei SH, Leu YW, et al (2003b) Triple analysis of the cancer epigenome: an integrated microarray system for assessing gene expression, DNA methylation, and histone acetylation. Cancer Res 63:2164–2171

Siegfried Z, Eden S, Mendelsohn M, et al (1999) DNA methylation represses transcription in vivo. Nat Genet 22:203–206

Stach D, Schmitz OJ, Stilgenbauer S, et al (2003) Capillary electrophoretic analysis of genomic DNA methylation levels. Nucleic Acids Res 31:E2

Sutherland E, Coe L, Raleigh EA (1992) McrBC: a multisubunit GTP-dependent restriction endonuclease. J Mol Biol 225:327–348

Sutherland JE, Costa M (2003) Epigenetics and the environment. Ann N Y Acad Sci 983:151–160

Taubes G (1995) Epidemiology faces its limits. Science 269:164–169

Thomassin H, Oakeley EJ, Grange T (1999) Identification of 5-methylcytosine in complex genomes. Methods 19:465–475
Thomassin H, Kress C, Grange T (2004) MethylQuant: a sensitive method for quantifying methylation of specific cytosines within the genome. Nucleic Acids Res 32:e168
Tompa R, McCallum CM, Delrow J, et al (2002) Genome-wide profiling of DNA methylation reveals transposon targets of CHROMOMETHYLASE3. Curr Biol 12:65–68
Tost J, Schatz P, Schuster M, et al (2003) Analysis and accurate quantification of CpG methylation by MALDI mass spectrometry. Nucleic Acids Res 31:e50
Toyota M, Ho C, Ahuja N, et al (1999) Identification of differentially methylated sequences in colorectal cancer by methylated CpG island amplification. Cancer Res 59:2307–2312
Truss M, Chalepakis G, Pina B, et al (1992) Transcriptional control by steroid hormones. J Steroid Biochem Mol Biol 41:241–248
Tsujita T, Niikawa N, Yamashita H, et al (1998) Genomic discordance between monozygotic twins discordant for schizophrenia. Am J Psychiatry 155:422–424
Uhlmann K, Brinckmann A, Toliat MR, et al (2002) Evaluation of a potential epigenetic biomarker by quantitative methyl-single nucleotide polymorphism analysis. Electrophoresis 23:4072–4079
Ushijima T, Morimura K, Hosoya Y, et al (1997) Establishment of methylation-sensitive-representational difference analysis and isolation of hypo- and hypermethylated genomic fragments in mouse liver tumors. Proc Natl Acad Sci U S A 94:2284–2289
Veldic M, Caruncho HJ, Liu WS, et al (2004) DNA-methyltransferase 1 mRNA is selectively overexpressed in telencephalic GABAergic interneurons of schizophrenia brains. Proc Natl Acad Sci U S A 101:348–353
Vincent JB, Kalsi G, Klempan T, et al (1998) No evidence of expansion of CAG or GAA repeats in schizophrenia families and monozygotic twins. Hum Genet 103:41–47
Walter J, Paulsen M (2003) Imprinting and disease. Semin Cell Dev Biol 14:101–110
Warren MP, Brooks-Gunn J (1989) Mood and behavior at adolescence: evidence for hormonal factors. J Clin Endocrinol Metab 69:77–83
Waterland RA, Jirtle RL (2003) Transposable elements: targets for early nutritional effects on epigenetic gene regulation. Mol Cell Biol 23:5293–5300
Weaver IC, Cervoni N, Champagne FA, et al (2004) Epigenetic programming by maternal behavior. Nat Neurosci 7:847–854
Weissman MM, Olfson M (1995) Depression in women: implications for health care research. Science 269:799–801
Weksberg R, Shuman C, Caluseriu O, et al (2002) Discordant KCNQ1OT1 imprinting in sets of monozygotic twins discordant for Beckwith-Wiedemann syndrome. Hum Mol Genet 11:1317–1325
Weksberg R, Smith AC, Squire J, Sadowski P (2003) Beckwith-Wiedemann syndrome demonstrates a role for epigenetic control of normal development. Hum Mol Genet 12 Spec No 1:R61 68
Williams CA, Hendrickson JE, Cantu ES, Donlon TA (1989) Angelman syndrome in a daughter with del(15) (q11q13) associated with brachycephaly, hearing loss, enlarged foramen magnum, and ataxia in the mother. Am J Med Genet 32:333–338
Wolff GL, Kodell RL, Moore SR, Cooney CA (1998) Maternal epigenetics and methyl supplements affect agouti gene expression in Avy/a mice. Faseb J 12:949–957

Worm J, Aggerholm A, Guldberg P (2001) In-tube DNA methylation profiling by fluorescence melting curve analysis. Clin Chem 47:1183–1189

Wu J, Issa JP, Herman J, et al (1993) Expression of an exogenous eukaryotic DNA methyltransferase gene induces transformation of NIH 3T3 cells. Proc Natl Acad Sci U S A 90:8891–8895

Xiong LZ, Xu CG, Saghai Maroof MA, Zhang Q (1999) Patterns of cytosine methylation in an elite rice hybrid and its parental lines, detected by a methylation-sensitive amplification polymorphism technique. Mol Gen Genet 261:439–446

Xiong Z, Laird PW (1997) COBRA: a sensitive and quantitative DNA methylation assay. Nucleic Acids Res 25:2532–2534

Yamamoto F, Yamamoto M (2004) A DNA microarray-based methylation-sensitive (MS)-AFLP hybridization method for genetic and epigenetic analyses. Mol Genet Genomics 271:678–686

Yamamoto T, Nagasaka T, Notohara K, et al (2004) Methylation assay by nucleotide incorporation: a quantitative assay for regional CpG methylation density. Biotechniques 36:846–850, 852, 854

Yan PS, Efferth T, Chen HL, et al (2002) Use of CpG island microarrays to identify colorectal tumors with a high degree of concurrent methylation. Methods 27:162–169

Yang AS JP, Shibata A (1996) The mutational burden of 5-methylcytosine. In: Russo V, Riggs A (eds) Epigenetic mechanisms of gene regulation. Cold Spring Harbor Laboratory Press, Cold Spring Harbor, pp 77–94

Yokomori N, Moore R, Negishi M (1995) Sexually dimorphic DNA demethylation in the promoter of the Slp (sex-limited protein) gene in mouse liver. Proc Natl Acad Sci U S A 92:1302–1306

Zeschnigk M, Bohringer S, Price EA, et al (2004) A novel real-time PCR assay for quantitative analysis of methylated alleles (QAMA): analysis of the retinoblastoma locus. Nucleic Acids Res 32:e125

Zhang Z, Chen CQ, Manev H (2004) Enzymatic regional methylation assay for determination of CpG methylation density. Anal Chem 76:6829–6832

Zubenko GS, Maher B, Hughes HB 3rd, et al (2003) Genome-wide linkage survey for genetic loci that influence the development of depressive disorders in families with recurrent, early-onset, major depression. Am J Med Genet 123B:1–18

CTMI (2006) 310:117–140

MSL Proteins and the Regulation of Gene Expression

S. Rea · A. Akhtar (✉)

Gene expression programme, European Molecular Biology Laboratory,
Meyerhofstrasse 1, Heidelberg, Germany
akhtar@embl.de

Abstract Epigenetics describes changes in genome function that occur without a change in the DNA sequence. Dosage compensation is a prime example of the regulation of gene expression by an epigenetic mechanism. Dosage compensation has evolved to balance the expression of sex-linked genes in males and females, which possess different numbers of sex chromosomes. However, the genetic sequence of the chromosomes is the same in both sexes. This mechanism therefore needs (1) to function in a sex-specific manner, (2) to target the sex chromosome from amongst the autosomes and (3) to establish and maintain through development a precise, equalised level of gene expression in one sex compared to the other. The process by which dosage compensation is orchestrated has been well characterised in fruit flies and mammals. Although each has evolved a specific dosage-compensation mechanism, these systems share some underlying themes; the molecular components that mediate dosage compensation in both include non-coding RNA molecules, which act as nucleation points for the compensation process. Both systems utilise chromatin-modifying enzymes to remodel large domains of a chromosome. This review will discuss the mechanism of dosage compensation in *Drosophila* in light of recent developments that have brought into question the previous model of dosage compensation.

1 Introduction

Sexual reproduction occurs in most eukaryotes, and the increased genetic variability thus conferred on the organism gives it an evolutionary advantage to adapt to changing environments. This advantage, however, comes at a cost. In some cases sexual reproduction has led to the generation of sex chromosomes, which evolved from a pair of autosomes following their recruitment into a chromosomal system for sex determination. A barrier to meiotic recombination then developed on the heterogametic (e.g. Y) chromosome to prevent the creation of mixtures of alleles and intersex organisms. This insulation of the sex-determining region eventually spread throughout the chromosome and, in the absence of recombination, the accumulation of mutation events subsequently led to the degeneration of this chromosome (Ohno 1967). Since failure to balance the resulting twofold difference in the homogametic (e.g. X) chromosome's gene expression between the sexes would be lethal, certain mechanisms have evolved to compensate for this difference.

In the three systems—mammals, nemotode worms and fruit flies—in which such processes have been studied at a molecular level, it is clear that this mechanism has evolved independently several times (Pannuti and Lucchesi 2000).

In eutherian mammals, one X chromosome in females (XX) is inactivated to equalise gene expression from the XY males (Lyon 1961). In *Caenorhabditis elegans* hermaphrodites (XX), gene expression from both X chromosomes is reduced by half in comparison to XO males (Meyer and Casson 1986). In *Drosophila melanogaster*, transcription is upregulated twofold in XY males to equalise that from XX females (Mukherjee and Beermann 1965). Since the X chromosomes of males and females are identical at the DNA level, the cell is posed with several gene-regulation problems. First, a chromosome-wide control of genes is required, and this needs to override local gene control. Second, this regulation needs to function in a sex-specific manner. Third, the X chromosome needs to be distinguished from the autosomes, and in the case of mammals one of the X chromosomes needs to be chosen from its identical counterpart. Finally, this regulation needs to be quantitative, establishing and maintaining a specific level of gene expression relative to the other sex throughout the lifetime of the organism. The three above-mentioned systems achieve this regulation of gene expression through modulation of their chromatin organisation in a process termed dosage compensation.

2
The DCC of *Drosophila melanogaster*

The process of dosage compensation is best understood in *Drosophila*, where genetic manipulation has allowed the identification of many of the components involved. Several genetic screens, scoring for lethality occurring specifically in males, have identified a number of genes essential for dosage compensation. The encoded proteins have been termed male specific lethals (MSL) and include MSL-1, MSL-2, MSL-3, maleless (MLE, an RNA helicase) and males absent on the first [MOF, a histone acetyltransferase (HAT); Bashaw and Baker 1995; Hilfiker et al. 1997; Kelley et al. 1995; Kuroda et al. 1991; Zhou et al. 1995; for review see Taipale and Akhtar 2005]. In combination with these proteins, two non-coding RNAs, RNA on the X chromosome 1 (*roX1*) and *roX2*, constitute the dosage-compensation complex (DCC) (Franke and Baker 1999). This complex associates with hundreds of discrete sites on the X chromosome in males but not in females. This localisation can be visualised in immunofluorescence microscopy by the painting of the male X chromosome at specific bands by antibodies against members of the complex in larval polytene chromosome squashes (Bashaw and Baker 1995; Franke and Baker 1999; Kuroda et al. 1991). The dosage-compensated chromosome is also characterised by a more decondensed chromatin structure that correlates with acetylation of lysine 16 of the histone H4 tail (H4K16Ac). This acetylation is mediated by the MOF histone acetyltransferase and is thought to be partly responsible for the twofold increase in gene expression (Akhtar and Becker 2000; Hilfiker et al. 1997; Smith et al. 2000; Turner et al. 1992). Another probable member of the dosage-compensation complex is the kinase JIL-1. Although this protein does not have a male-specific phenotype, weak alleles of *jil-1* cause distortions of the sex ratio, with males more susceptible to partial loss of the protein (Wang et al. 2001). Moreover, it associates with the DCC and leads to enrichment of phosphorylation at serine 10 of histone H3 (H3S10P) (Jin et al. 2000; Wang et al. 2001).

It is worth noting that all the MSL proteins in *Drosophila* have orthologues in mammals (Marin 2003). The human orthologues are hMSL1, hMSL2, hMSL3, RNA helicase A (MLE orthologue) and MYST1 (MOF orthologue). To date only the transcriptional coactivator RNA helicase A has been well characterised (Nakajima et al. 1997). Interestingly, some of these MSL orthologues also interact in human cells (Taipale et al 2005; Bouazoune et al.

2004). They do not, however, appear to be involved in dosage compensation in mammals, since neither MYST1 nor hMSL3 localise specifically to the silenced X chromosome in female cells (M. Taipale and A. Akhtar, unpublished observations). Moreover, it has been shown that the specific enzymatic activity of *Drosophila* MOF for H4K16 is conserved in its human counterpart (M. Taipale and A. Akhtar, unpublished observations), suggesting that the mammalian MSL complex may also have a gene-activating function.

3 Initiation of Dosage Compensation

The first step in dosage compensation is for a cell to determine if it is male or female. In *Drosophila*, sex is determined and the dosage compensation pathway initiated by the master switch gene *sex lethal* (*sxl*) (Bashaw and Baker 1997; Kelley et al. 1997). In females where the ratio of X chromosomes to autosomes is one to one, functional SXL protein is produced. This ratio is detected by a mechanism where activating numerator loci on the X chromosome and negative denominator loci on the autosomes encode transcription factors that compete for control of the *sxl* promoter (Erickson and Cline 1998). The presence of SXL in females inhibits translation of MSL-2 messenger (m)RNA by binding to its 3′- and 5′-untranslated regions and thus prevents its stable association with the ribosome (Gebauer et al. 2003; Grskovic et al. 2003). So although both males and females express the other four MSL proteins, it is only in males, where MSL-2 is translated, that the functional DCC will be formed.

4 Assembly of the DCC

Much of the knowledge we have regarding the assembly of the chromatin modifying DCC has been gained from genetic manipulations combined with immunofluorescence microscopy analysis of the giant polytene chromosomes of larval salivary glands. These studies clearly show that formation of a functional DCC is subject to some degree of both hierarchy and interdependence between the components.

MSL-2 is the primary determinant of DCC assembly. In the absence of MSL-2 the other MSL proteins no longer localise to the X chromosome. Transgenic flies in which MSL-2 is expressed in females are able to assemble the MSL complex on both X chromosomes. This leads to developmental delays and lethality, most likely due to overexpression of X-linked genes (Kelley et al.

1995). MSL-2 interacts with and stabilises MSL-1, and together these proteins form the central core of the DCC, which can localise to the X chromosome in the absence of the other MSLs (Chang and Kuroda 1998; Copps et al. 1998). Similar to MSL-2 fly mutants, depletion of MSL-2 in Schneider cells destabilises the whole complex; however, MSL-1 and MSL-2 can still be detected on the X chromosome in MSL-3 or MOF-depleted cells (Buscaino et al. 2003). The subsequent stages in the formation of the complex are unclear, as they may not occur in obvious steps, but instead appear to be interdependent and probably happen concurrently.

MSL-1 is also able to interact with MSL-2, MSL-3 and MOF, possibly coordinating their assembly into the DCC (Gorman et al. 1995; Scott et al. 2000). MLE is also likely to be involved in DCC assembly, since in *mle* mutants, MOF and MSL-3 do not localise to the X chromosome (Gu et al. 2000). However, it has been shown in Schneider cells that depletion of MSL-3 or MOF by RNA interference causes dissociation of MLE from the complex (Buscaino et al. 2003). This suggests interplay between the different factors to maintain a stable DCC.

RNA is also important for assembly of the complex. Both *roX1* and *roX2* genes encode non-translated RNAs that are expressed specifically in males. They are redundant in function although they are very different in size and sequence (Amrein and Axel 1997; Kageyama et al. 2001; Meller et al. 1997). They are required for association of the MSL proteins with the X chromosome, and loss of both genes results in male-specific lethality, with very few male escapers (Franke and Baker 1999; Meller and Rattner 2002). Further interdependence between the roX RNAs and the MSL proteins has been demonstrated, since both MSL-3 and MOF have also been shown to interact with RNA (Akhtar et al. 2000; Buscaino et al. 2003). Moreover, association of MLE, MSL-3 and MOF to the X chromosome is sensitive to treatment with RNase (Buscaino et al. 2003; Richter et al. 1996).

JIL-1 can interact in vitro with both MSL-1 and MSL-3 via its kinase domains (Jin et al. 2000). Epitope-tagged JIL-1 can also be co-immunoprecipitated with MSL-1, MSL-2 and MSL-3 in *Drosophila* cultured cells (Jin et al. 2000). In contrast to the other MSL proteins, MSL-2 association with the X chromosome is not lost in JIL-1 mutants, suggesting that it is not involved in assembly of the DCC (Wang et al. 2001).

5 Targeting and Distribution of the DCC

Previously it was thought that the characteristic distribution of the DCC along the X chromosome was achieved in a two-step, targeting and spreading fashion. In this model, the core components MSL-1 and MSL-2 are targeted to approximately 35 discrete sites, termed "chromatin entry sites". These sites serve as nucleation centres on which a mature DCC assembles and then spreads *in cis* to associate with hundreds of sites along the rest of the X chromosome, where it mediates a twofold increase in transcription of X-linked genes (Kageyama et al. 2001; Kelley et al. 1999). Although this was a reasonable model based on existing data, it is now clear that both the aspects of "chromatin entry sites" and "spreading" need to be revisited.

The concept of chromatin entry sites came from the observation that MSL-1 and MSL-2 together are able to bind reproducibly to around 35 sites on the male X chromosome in the absence of the other MSL proteins (Kelley et al. 1999; Lyman et al. 1997). The converse is not true, in *msl-1* or *msl-2* mutants the rest of the MSL proteins cannot bind (Lyman et al. 1997). The complete DCC was found to localise to these approximately 35 sites when the concentration of the individual MSL proteins was reduced (Demakova et al. 2003). Taken together, these findings led to the proposal that the core DCC, MSL-1 and MSL-2, would nucleate at these 35 or so sites only. A complete DCC could then assemble at these sites.

However, until recently, only two of the roughly 35 chromatin entry sites have been characterised, and remarkably they contain the genes encoding roX1 and roX2 (Kelley et al. 1999; Meller et al. 2000). In both cases, a 200–300 bp DNA fragment is able to recruit the MSL complex to the autosomes, and these sequences are sensitive to DNase I treatment only in males (Kageyama et al. 2001; Kelley et al. 1999; Meller et al. 2000; Park et al. 2002). Transgenic insertion of this roX DNA fragment on an autosome not only resulted in attraction of the DCC but also resulted in spreading of the complex from the insertion site into flanking chromatin (Kageyama et al. 2001; Kelley et al. 1999). This led to the spreading model in which the assembled DCC would spread *in cis* from the chromatin entry sites to coat the X chromosome.

Work from the laboratories of Kuroda (characterising a third chromatin entry site) and Baker (testing the validity of the spreading model) has generated strong evidence to dispute the targeting and spreading model (Fagegaltier and Baker 2004; Oh et al. 2004). Oh et al. identified a third entry site that maps to the cytological location 18D10 of the X chromosome. This region encompasses a DNase I hypersensitive site that includes a 510-bp sub-region which is sufficient to attract core and assembled MSL complexes in vivo (Oh et al. 2004). However, while mapping the 18D region, they discovered that DNA sequences lacking any chromatin entry sites—when inserted into an autosome—were also able to attract an MSL complex. Both groups verified this surprising result when they looked at the consequences of transposition of regions of the X chromosome onto an autosome. Interestingly, they found that transposed pieces of the X chromosome lacking any entry site attracted the MSL complex when inserted into an autosome (Fagegaltier and Baker 2004; Oh et al. 2004). Fagegaltier and Baker showed that for each of the 11 transpositions studied the binding pattern of the transposed piece of X chromosome was the same as that on the endogenous X (Fagegaltier and Baker 2004). This implies that each of the hundreds of sites to which the MSL complex binds is necessary and sufficient for attracting the DCC, irrespective of the presence of any nearby entry sites.

This group then went on to show that none of the transpositions displayed any spreading of DCC, either for X into autosome or for autosome into X translocations. This was true regardless of the number of high affinity/chromatin entry sites in the transposed section. Moreover, they observed no spreading even in transpositions that contained the *roX1* or the *roX2* entry sites (Fagegaltier and Baker 2004). This is in contrast to earlier work that led to the spreading model, where spreading was observed from *roX* transgenes inserted on an autosome (Kageyama et al. 2001). Oh et al. also found some degree of spreading from a cosmid containing the 18D10 entry site, although this was very rare (<1%; Oh et al. 2004). Fagegaltier and Baker postulate that spreading may be a phenomenon specific to transgenes. Overall, these studies imply that spreading *in cis* along the length of the X chromosome is not part of the dosage compensation mechanism in *Drosophila*.

It is of course still possible for spreading to occur on a smaller scale at each individual site to mediate the upregulation of transcription of individual genes. However, studying spreading at this resolution is not straightforward and will require development of a special experimental system.

Since the hypothesis of chromatin entry sites and the idea of spreading are now put into question, a new model for how the DCC achieves its distribution along the length of the X chromosome at hundreds of specific sites needs to be postulated.

Any part of the X chromosome that can attract the endogenous MSL complex is also able to attract the complex to an autosome (Fagegaltier and Baker 2004). Therefore, the approximately 35 entry sites are now simply thought to be sites of higher affinity for the MSL complex and there is no functional difference between them and the other hundreds of sites. This is in keeping with previous work where the number and position of the DCC was recorded in relation to varying amounts of MSL complex. By titrating the amount of MSL-2 expression in females, the concentration of the rest of the MSL complex could be controlled. In the mutants with low levels of MSL-2 they found binding at only four sites on the X chromosome. Expressing more MSL-2 led to binding at more sites, up to a certain point where 40 binding sites were observed. These binding sites exactly correlate with the roughly 35 so-called entry sites (Demakova et al. 2003).

Important in this model is the role of *roX* RNAs for proper targeting and distributing of the DCC. In the absence of *roX* RNAs the MSL complex is not properly targeted to the X chromosome, and low levels of DCC can be seen to associate with the autosomes, implying that *roX1* and *roX2* play a role in the correct targeting of the MSL complex to its appropriate locations on the X chromosome (Franke and Baker 1999; Meller and Rattner 2002).

Although probably an artefact, at least at the resolution studied so far, the phenomenon of spreading of DCC from transgenes can give some insights into how the MSL complex is targeted to its endogenous loci. Spreading of the DCC from a *roX* transgene into flanking chromatin of an autosome occurs only rarely. However, when the level of endogenous *roX* RNA is reduced, by deleting the *roX* genes from the X chromosome, the amount of spreading dramatically increases (Park et al. 2002). Moreover, MSL proteins are required for *roX* stability, and it has been shown that overexpression of MSL-1 and MSL-2 increases the extent of DCC spreading from a *roX* transgene (Demakova et al. 2003; Meller et al. 2000). These results suggest that the *roX* RNAs compete for a limiting pool of MSL proteins. Furthermore, the MSL complexes that spread from these *roX* transgenes have been shown to contain *roX* RNA and to co-localise with H4K16Ac, implying that they are mature complexes (Kelley et al. 1999). It has also been shown that transcription from a *roX* transgene is required for spreading to occur (Park et al. 2002, 2003).

Recent work from the Becker laboratory has generated important ground rules when considering a possible model. They used fluorescence recovery after photobleaching (FRAP) techniques to determine the dynamics of the association of the DCC with the X chromosome. In cells expressing fluorescently labelled MSL-2 they bleached a region of the X chromosome. If the MSL complex is dynamic then unbleached fluorescent MSL-2 should move to the bleached region of the X chromosome after some time. Surprisingly, it was found that the bleached MSL region remained bleached. The DCC therefore has a very stable binding to the X chromosome; once bound it does not reposition (Straub et al 2005; Bouazoune et al. 2004). This observation also provides evidence to contradict the spreading model, where one would expect a large degree of mobility from the high-affinity sites to facilitate spreading. The stability of binding to the X chromosome could reflect the importance of keeping the DCC on the X chromosome to maintain the twofold increase in transcription, or it could be to prevent the DCC from straying onto the autosomes and perhaps misregulating gene expression there.

With all these observations in mind, one can now revise the dosage-compensation model (Fig. 1). It has been proposed that the *roX* genes are

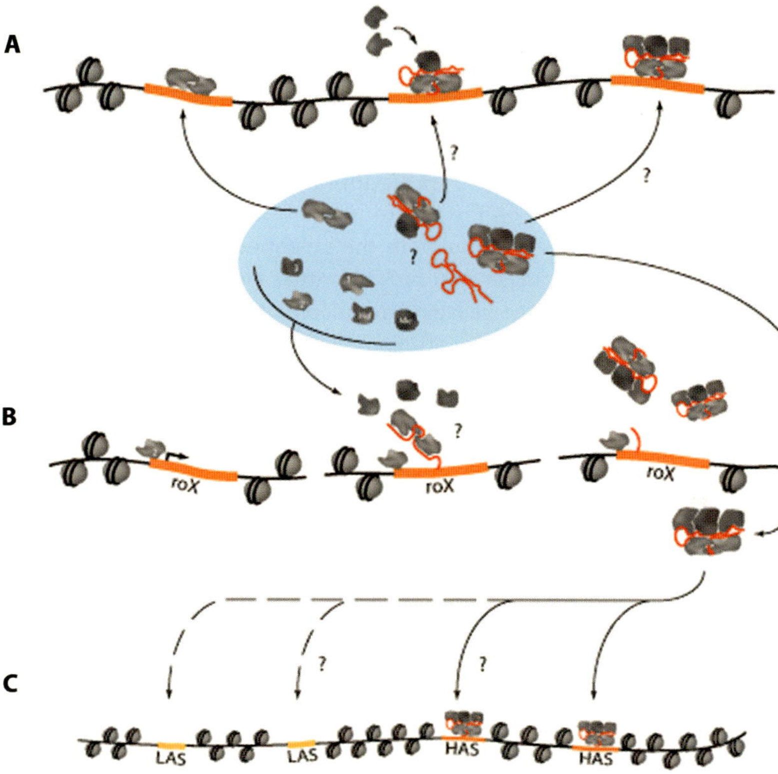

Fig. 1A–C A model of how the dosage-compensation complex (*DCC*) may localise to the hundreds of sites on the X chromosome. Partial complexes formed in the nucleus are able to associate with specific sites (**A**). Mature complexes could then form at these locations. However, partial complex binding appears to be limited to 35–70 high-affinity sites (Gilfillan et al. 2004); fully functional complexes would therefore also need to form in the vicinity of the X chromosome and bind in a hierarchical manner to mediate DCC binding along the entire chromosome (**C**). Alternatively, it has been proposed that the *roX* genes are the sites of MSL complex formation (Fagegaltier and Baker 2004; Park et al. 2002). MSL2 is able to regulate transcription from the *roX* loci (Rattner and Meller 2004); here the MSL proteins could be integrated into mature complexes by binding newly synthesized *roX* RNA (**B**). This would result in a high local concentration of mature MSL complexes at the X chromosome, from where they would then diffuse to high affinity sites (*HAS*). **C** Once the high-affinity sites are all bound, the MSL complex would then begin to associate with lower affinity sites (*LAS*) to eventually bind to the hundreds of sites along the X chromosome and thereby mediate the twofold upregulation of transcription. The nature of the MSL binding sites, and their difference in affinities for MSL complex binding is not well understood. Targeting of the DCC could be to consensus DNA sequences, specific chromatin modifications, sites of active transcription or by an as-yet-undescribed mechanism

the sites of assembly of the DCC (Fagegaltier and Baker 2004; Park et al. 2002), where MSL proteins assemble into mature active complexes by binding nascent *roX* transcripts. When the concentration of MSL proteins is low, *roX* RNAs diffuse away from their site of transcription and are degraded in the nucleoplasm. When there is the correct balance of MSL proteins present they are able to bind nascent *roX* transcripts and thus stabilise them. Since both *roX* genes are located on the X chromosome, this would ensure a high concentration of functional complex in the neighbourhood where it is required to operate. From these *roX* sites a mature DCC could then diffuse to sites of high affinity in the nearby chromosome territory and, once these sites are occupied, to sites of lower affinity. The DCC would eventually tightly associate with the hundreds of discrete sites that coat the X chromosome and mediate the twofold upregulation of transcription required for dosage compensation.

6
MSL Binding Sites

The last issue to address in relation to targeting the DCC is the nature of the MSL binding sites themselves and how they mediate attraction of the DCC. It is unclear what size these binding sites are. By immunofluorescence microscopy these sites can be visualised as bands on stained polytene chromosomes of salivary glands. The size of these bands has been estimated to span up to several hundred kilobases in length (Fagegaltier and Baker 2004). However, these bands may contain several individual binding sites that cannot be separated at this level of resolution, or the size may be exaggerated by the enhancement of the fluorescent signal. A more detailed approach using chromatin immunoprecipitations followed by whole genome array analysis (ChIP on chip) with MSL-1 or MSL-2 antibodies should help in determining the size of these sites and whether there is any consensus sequence within them.

As already mentioned, only three of these sites have been characterised in any detail. These sites include regions spanning the *roX1* and *roX2* genes as well as a region at 18D10. Although a short functional conserved sequence was found in the two *roX* genes, no such sequence was found in the 18D10 site, nor indeed in the rest of the genome (Oh et al. 2004). This conserved sequence may function specifically in the putative role of the *roX* sites in MSL complex assembly. Although a clear-cut consensus MSL binding sequence has not been identified, some have put forward the idea that a form of consensus sequence may have evolved with dosage compensation (Marin and Baker 2000; Marin et al. 1996).

During sex chromosome evolution in *Drosophila*, the Y chromosome gradually degenerated. Depending on whether a specific gene has yet degenerated (through mutation) and whether it is important in males, it can be more important or less important to compensate for its degree of expression from the single X chromosome. Consequently, during evolution, one may expect that different genes (or regions of chromosomes) acquired different affinities for the compensation machinery as determined by their individual importance (Marin and Baker 2000; Marin et al. 1996). These affinities may be reflected in consensus binding sites at the nucleotide level. Interestingly, the X chromosome has been shown to have molecular characteristics distinct from the autosomes. In particular it shows some enrichment in certain mono- and dinucleotide repeats and also certain satellite related repeats (DiBartolomeis et al. 1992; Lowenhaupt et al. 1989; Pardue et al. 1987; Waring and Pollack 1987). It is more likely that these specific repeats have a role outside of targeting the MSL complex. Complete mapping of the DCC binding sites should answer this question.

It is also possible that these affinities are mediated at the chromatin level and that there is a consensus epigenetic mark that can attract the DCC to specific regions. Both DNA methylation and the covalent modification of histone amino termini have been shown to act as specific binding sites for various complexes (de la Cruz et al. 2005; Hung and Shen 2003). Indeed, methylation of histone H3 is required for dosage compensation in mammals (Plath et al. 2003; Mermoud et al. 2002). Such epigenetic marks could be set during embryogenesis, as in mammals, and maintained throughout development.

It is also feasible that there are no X chromosome specific consensus sites, either at the nucleotide or the chromatin level. The DCC could be targeted to sites of ongoing transcription on the X. There is some evidence to support this idea.

Ectopic activation of transcription on the X or on an autosome using an upstream activation sequence (UAS)-Gal4 system is able to attract the MSL complex to the site of transcription (Sass et al. 2003). The reason that the MSL complex then preferentially targets the X and not the autosomes may simply be due to the proximity of its site of assembly (*roX* genes) to the rest of the chromosome. It is possible that more highly expressed genes (e.g. housekeeping genes) would recruit more MSL complex and so may represent the high-affinity sites discussed earlier. Further mapping of these sites together with expression analysis should clarify this possibility.

The aforementioned alternatives for binding site recognition are not mutually exclusive and indeed, may work collaboratively.

7 Modification of Chromatin by the DCC

Eukaryotic DNA is packaged with histones to form chromatin. The resulting compaction is necessary to fit a large genome into a small cell. However, transcription of the genetic material is dependent on its accessibility. Certain molecular mechanisms therefore exist to control the accessibility of this chromatin. First, covalent modification of the histones, by enzymatic addition of acetyl, methyl, phosphate or ubiquitin moieties, can mediate the subsequent binding of effector molecules that can either promote or repress transcription (Jenuwein and Allis 2001). Alternatively, enzymes that remodel chromatin using the energy from ATP hydrolysis can alter the interaction between the DNA and histones (Kingston and Narlikar 1999). Both of these mechanisms are involved in the control of gene expression during dosage compensation.

Covalent histone modifications and their role in chromatin organisation have been the subject of much attention in recent years. This has led to the histone code hypothesis, which explains how chromatin structure can be established and maintained. Basically, this hypothesis predicts that there are (1) enzymes that covalently modify the amino terminal tails of histones and (2) protein modules that specifically recognise these modifications and translate them to functional states (Jenuwein and Allis 2001; Strahl and Allis 2000). At least two of the components of the DCC possess enzymatic activity towards a chromatin substrate. MOF is a histone acetyltransferase (HAT) and JIL-1 is a histone kinase.

MOF is a member of the MYST (MOZ/YBF2/SAS2/TIP60) family, a group of acetyltransferases characterised by a C_2HC-type zinc finger in their catalytic domain. MOF also possesses a chromo domain (Hilfiker et al. 1997). A single amino acid change that renders the enzyme inactive also results in male-specific lethality. Histone acetylation is generally regarded as an activating mark, resulting in a more open chromatin structure that is permissive to transcription (Eberharter and Becker 2002). Indeed, it has been shown that the acetylation of lysine 16 on histone H4 (H4K16Ac) is enriched on the hyperactive male X chromosome (Turner et al. 1992). In contrast to many acetyltransferases, MOF is very specific, acetylating only the lysine at position 16 on the amino terminal tail of histone H4 (H4K16Ac) (Akhtar and Becker 2000; Smith et al. 2000).

JIL-1 contains two tandemly arranged serine/threonine kinase domains and can phosphorylate histone H3 in vitro (Jin et al. 1999). It is required for the maintenance of chromatin structure in flies, and null mutants are not viable. Reduced levels of JIL-1 kinase cause an aberrant condensation of chromatin and lower levels of histone H3 serine 10 phosphorylation (H3S10P) (Wang et al. 2001). JIL-1 is enriched approximately twofold on the X chromosome, and this localisation correlates with S10P and phosphoacetylated S10P/K14Ac histone H3 (Wang et al. 2001). The above observations strongly argue for a role of the chromatin marks set by these enzymes in the mechanism by which the DCC leads to a twofold upregulation in gene expression from the male X chromosome.

Another member of the DCC may have enzymatic capabilities. MSL-2 contains a RING finger, which is essential for its function in dosage compensation. Two of the original *msl-2* mutants contained mutations in their RING finger and constructs with point mutations in their RING finger were unable to rescue *msl-2* mutant flies (Lyman et al. 1997; Zhou et al. 1995). Many RING finger proteins possess E3 ligase activity; this means that they can catalyse the ligation of an activated ubiquitin moiety to a lysine residue of another protein (Joazeiro and Weissman 2000). Interestingly, ubiquitylation of histones is also important in transcriptional regulation (Zhang 2003). Ubiquitylation followed by deubiquitylation of histone H2B is important for gene transcription by the SAGA complex and this works together with histone acetylation and histone methylation with some kind of dynamic interdependence (Henry et al. 2003). Whether MSL-2 possesses ubiquitin ligase activity remains to be seen; if it does, it may provide another chromatin mark to facilitate dosage compensation.

8
Interpreting the Histone Modifications Placed by the DCC

There still exists a chicken-and-egg type question over the presence of H4K16Ac on the compensated X chromosome. Whether it is a consequence of the hypertranscription of the X or a cause for specific upregulation is still unclear. Most experimental evidence supports the latter option. Immunofluorescence studies on polytene chromosomes have shown H4K16Ac to be enriched only on the male X chromosome and absent from transcriptionally active autosomes (Bone et al. 1994; Turner et al. 1992). Furthermore, H4K16Ac mediated by MOF can lead to derepression of transcription both in vitro and in vivo (Akhtar and Becker 2000).

However, things may not be so simple; H4K16Ac may play a role in other aspects of dosage compensation. Overexpression of MOF in Schneider cells leads to ectopic association of the MSL complex with the autosomes (Gu et al. 2000), suggesting that acetylation plays a role in targeting of the complex. These functions are not mutually exclusive, and may even collaborate to mediate dosage compensation.

It is clear that MOF is the enzyme that sets this acetylation mark on the hyperactive X chromosome, but which module is responsible for recognising it? The most obvious candidate is a bromodomain-containing protein. Several examples of bromodomain proteins binding to specifically acetylated histone tails exist (de la Cruz et al. 2005). A number of transcriptional coactivators contain bromodomains; however, so far no binding protein for H4K16Ac has been identified (de la Cruz et al. 2005). It is also possible that H4K16Ac is not a binding site but, on the contrary, may function to inhibit the binding of a protein. One likely victim for this action would be the chromatin remodelling factor imitation switch (ISWI) (Deuring et al. 2000).

ISWI is an ATPase of the SWI2/SNF2 family and is the catalytic subunit of three chromatin remodelling complexes, NURF, CHRAC and ACF (Deuring et al. 2000). On polytene chromosomes it localises to RNA polymerase II-poor regions, implying that its in vivo function is repressive (Deuring et al. 2000). Male flies mutant for *ISWI* show dramatic changes in the organisation of their X chromosome; this abnormal phenotype can be suppressed by disruption of the DCC (Corona et al. 2002). Furthermore, overexpression of MOF enhances the phenotype of *ISWI* mutants, and the presence of H4K12Ac or H4K16Ac inhibits the binding of ISWI and its ATPase activity (Corona et al. 2002). It therefore appears that one role for the acetylation of histone H4 K16 by MOF is to antagonise the repressive effects of the chromatin remodeller ISWI.

The role of the JIL-1 kinase is unclear. Weak alleles of *jil-1* show a distortion in sex ratio, with males being more vulnerable to loss of the protein. It also does not appear to be required for proper assembly or localisation of the complex. However, *jil-1* mutants show defects in chromatin structure, with the X chromosome being more severely affected than the autosomes (Wang et al. 2001). Phosphorylation of histone H3 at serine 10 correlates with activation of immediate-early genes, especially in combination with H3 acetylation (Thomson et al. 2001). It is possible then that JIL-1 co-operates with the DCC to generate a more open chromatin structure to facilitate gene expression.

Another possibility is that these chromatin modifications, H4K16Ac and H3S10, together act as an epigenetic mark to maintain the hyperactive X chromosome's expression pattern during development. Chromatin modifications such as DNA methylation and histone methylation are known to confer heritable repressive states (Jenuwein and Allis 2001). But, acetylation and phosphorylation are generally regarded as more transient modifications and thus would not be suitable for a role in the long-term maintenance of a chromatin state (Jenuwein and Allis 2001). Methylation of histone H3 at lysine 4 has been shown to correlate with transcriptionally active regions (Schneider et al. 2004). However, there is to date no evidence for this more stable mark being involved in dosage compensation in *Drosophila*. Further experiments are required to clarify the chromatin environment on the hyperactive X chromosome.

9
Molecular Mechanism of Dosage Compensation

It is still the subject of much speculation as to how the DCC achieves the twofold transcriptional upregulation of the large number of genes on the X chromosome. Individual genes are expressed at different levels and can be expressed or silenced in different cell types at different stages of development. The dosage compensation machinery, therefore, needs to adapt to each particular gene and also to respond to different developmental stimuli. It is highly unlikely that the DCC can deal with the abundance of dynamic developmental signals on a gene-by-gene basis but instead must co-operate with the transcription machinery.

The DCC could exert its function either by increasing the rate of transcription initiation or through enhancing transcription elongation. It is well known that several genes require acetylation at their promoters to allow transcription to initiate (Kuo et al. 1998, 2000; Reid et al. 2000). However, it is more likely that the DCC operates at the stage of transcription elongation than being required for transcription initiation. Mutant male embryos lacking the MSL complex can develop to larval stages without any significant developmental defects. The eventual lethality in these mutants is not due to an absence, but to an imbalance of X-linked to autosome gene products (Demakova et al. 2003).

The enzymatic activity of MOF is believed to be of central importance in dosage compensation. To address the role of MOF-mediated histone H4 lysine 16 acetylation, Smith et al. performed ChIP experiments with antibodies against H4K16Ac (Smith et al. 2001). They found that H4K16Ac levels were enriched in the coding regions of genes, whereas the promoters had relatively low levels of acetylation. They also found that a gene that is dosage compensated, but not by the MSL complex, is void of H4K16Ac. Although only three genes were used in this study, it suggests that the MSL complex is involved in elongation rather than initiation.

There is further evidence for the involvement of histone acetylation in transcription elongation. In vitro studies on elongation rates using T7 bacteriophage polymerase have shown that the acetylation of the histone tails greatly increases the rate of transcription elongation (Protacio et al. 2000). Winkler et al. have shown in *Saccharomyces cerevisiae* that the function of the Elongator complex, which associates with elongating RNA polymerase II, is dependent on its histone acetyltransferase activity (Winkler et al. 2002).

Furthermore, it has been demonstrated that the elongation ability of RNA polymerase III was increased in the environment of acetylated histones which correlated with an unfolded chromatin state (Tse et al. 1998). But what is the function of this histone acetylation at the molecular level?

Recent unpublished data from Tamkun's lab defines an interesting new role for the ISWI ATPase. Immunostaining of *ISWI* mutant flies revealed that the chromatin of the X chromosome, but not the autosomes, lacks histone H1. Its presence on the autosomes is due to the low levels of maternal ISWI and the lack of H4K16Ac on the autosomes. These remarkable results reveal a new role for ISWI in the stabilisation of higher order chromatin by genome-wide incorporation of histone H1 (Bouazoune et al., 2004; J.W. Tamkun, unpublished observation). One of MOF's more important roles could be to counteract the repressive abilities of ISWI by inhibiting its binding to the X chromosome.

Based on the recent evidence and observations summarised above, we can come up with a speculative model for dosage compensation. First, the MSL complex is assembled and stabilised at the *roX* genes on the male X chromosome. The MSL complex is then targeted to high-affinity sites/transcriptionally active regions in the vicinity. MOF acetylates histone H4 at lysine 16 in the coding region of genes; this modification inhibits the binding of the ATPase ISWI. Lack of ISWI results in a reduction in the amount of histone H1; this in turn interferes with the formation of repressive higher order chromatin. This less-condensed, acetylated chromatin template allows for enhanced transcriptional elongation to take place.

It should be noted that not all genes on the X chromosome are dosage compensated by the MSL complex. Several genes escape compensation (e.g. larval serum protein, *LSPα*) and are thus expressed more in females than in males (Chiang and Kurnit 2003; Ghosh et al. 1989). Yet others are compensated by another mechanism; the X-linked *runt* gene is dosage compensated by some mechanism (possibly directly by the master gene *sxl*) not involving the MSL complex (Smith et al. 2001). Moreover, this gene is located in a large chromosomal region that lacks H4K16Ac (Smith et al. 2001). This suggests that use of H4K16Ac has been monopolised by the MSL complex to mediate its role of dosage compensation.

Since there are already several striking similarities between mammals and flies in their execution of dosage compensation, another interesting possibility is the use of chromosomal territories. The X chromosome in flies appears close to the nuclear periphery in Schneider cells (J. Kind and A. Akhtar, unpublished data), this is reminiscent of the dosage compensated X chromosome in female mammals. Elegant screens conducted in the Laemmli laboratory identified several genes whose encoded proteins could protect a reporter gene from a surrounding heterochromatin environment, allowing its expression. This activation ability was mediated by physical tethering of a chromatin region to the nuclear pore complex (NPC) (Ishii et al. 2002). It would be fascinating if the conspicuous location of the hyperactive X chromosome in flies reflected an interaction with the NPC, which somehow helps mediate regulation of gene expression.

10
The Inverse Effect Hypothesis

It is broadly accepted that the MSL complex functions to specifically upregulate the expression of genes from the single male X chromosome. However, there is another school of thought that believes this is an oversimplified model of dosage compensation. There are several reasons for this. For example, dosage compensation occurs not only in males (1X:2A), but also in metamales (1X:3A), metafemales (3X:2A) and triploid intersexes (2X:3A; Birchler 1992; Birchler et al. 2003). The dosage compensation mechanism described above cannot account for these scenarios. This and other inconsistencies have led the Birchler group to use the inverse effect hypothesis to explain dosage compensation.

It was observed that when large sections of chromosomes were deleted (segmental aneuploidy) the most common effect observed was a genome-wide upregulation of genes (Birchler and Schwartz 1979; Sabl and Birchler 1993). This phenomenon is called inverse dosage effect, where the deleted region often contains negative regulators and loss thereof results in the upregulation of many unlinked genes. The larger the deletion is, the more global the inverse effect. Importantly, this also means that the deleted segment (or parts thereof) may not be downregulated, as the global upregulation feeds back on its non-deleted homologue and returns its expression to normal levels; however, much of the rest of the genome suffers an approximate twofold increase in gene expression (Birchler et al. 2001).

The similarity to *Drosophila*, where the presence of a single X chromosome in males essentially generates an aneuploid state, is obvious. Such an aneuploid state could result in an inverse dosage effect on the autosomes, whereas the single X chromosome is automatically compensated by the upregulation feedback. Such a large-scale twofold upregulation from the autosomes would be lethal for a cell. Birchler and colleagues have proposed that the function of the DCC is to sequester the MOF histone acetyltransferase to the X chromosome (Birchler et al. 2001, 2003; Hiebert and Birchler 1994). This sequestration would remove H4K16Ac and its transcriptional activation effects from the autosomes to the X chromosome. This loss of H4K16Ac would cause a drop in gene expression to near that in females and so rescue the autosomes from an otherwise hyperactive state.

Some experimental support for this model comes from work by Bhadra et al. In some *msl* mutant males, binding of the MSL complex to the X chromosome is disrupted and the complex becomes associated with all the chromosomes. This results in an increase in the level of acetylation on the autosomes and a corresponding increase in the level of gene expression (Bhadra et al. 1999, 2000). Moreover, gene expression experiments in *mle* or *mof* mutants, which lack an MSL complex, revealed that transgenes on the X chromosome remained dosage compensated; however, many autosomal transgenes are upregulated (Birchler 1996). Finally, ectopic expression of MSL-2 in females induces MSL complex formation, which, according to the previous model, should cause an increase in expression of genes on the X chromosome. This, however, is not so. Bhadra and coworkers did not detect any increased expression from X-chromosomal transgenes tested (Bhadra et al. 1999, 2000).

In this model, most of the X chromosome is automatically compensated, but having attracted excessive amounts of H4K16Ac, a mechanism is needed to counteract its activating effects. Similar to results obtained by Corona et al., Birchler's group has shown that X-linked genes show increased expression in *ISWI* mutant individuals (Corona et al. 2002; Pal Bhadra et al. 2005). This has led them to propose that it is the repressive actions of ISWI that limit the hyperactivating effects of hyperacetylation of the X chromosome (Pal Bhadra et al. 2005).

It is not completely clear whether there is one true dosage compensation system. There is experimental evidence to support both possibilities. Further characterisation of the members of the DCC and identification of interacting proteins will help to elucidate this important mechanism of large-scale gene regulation.

Acknowledgements We would like to thank members of the lab for critical reading of the manuscript. Mikko Taipale for use of his template in Fig. 1. S.R. was supported by an EMBO long term fellowship and is currently supported by an HFSP fellowship. Research in our lab is partially funded by DFG; SFB "Transregio 5" and EU NoE "Epigenome".

References

Akhtar A, Becker PB (2000) Activation of transcription through histone H4 acetylation by MOF, an acetyltransferase essential for dosage compensation in Drosophila. Mol Cell 5:367–375

Akhtar A, Zink D, Becker PB (2000) Chromodomains are protein-RNA interaction modules. Nature 407:405–409

Amrein H, Axel R (1997) Genes expressed in neurons of adult male Drosophila. Cell 88:459–469

Bashaw GJ, Baker BS (1995) The msl-2 dosage compensation gene of Drosophila encodes a putative DNA-binding protein whose expression is sex specifically regulated by Sex-lethal. Development 121:3245–3258

Bashaw GJ, Baker BS (1997) The regulation of the Drosophila msl-2 gene reveals a function for Sex-lethal in translational control. Cell 89:789–798

Bhadra U, Pal-Bhadra M, Birchler JA (1999) Role of the male specific lethal (msl) genes in modifying the effects of sex chromosomal dosage in Drosophila. Genetics 152:249–268

Bhadra U, Pal-Bhadra M, Birchler JA (2000) Histone acetylation and gene expression analysis of sex lethal mutants in Drosophila. Genetics 155:753–763

Birchler JA (1992) Expression of cis-regulatory mutations of the white locus in metafemales of Drosophila melanogaster. Genet Res 59:11–18

Birchler JA (1996) X chromosome dosage compensation in Drosophila. Science 272:1190–1191

Birchler JA, Schwartz D (1979) Mutational study of the alcohol dehydrogenase-1 FCm duplication in maize. Biochem Genet 17:1173–1180

Birchler JA, Bhadra U, Bhadra MP, Auger DL (2001) Dosage-dependent gene regulation in multicellular eukaryotes: implications for dosage compensation, aneuploid syndromes, and quantitative traits. Dev Biol 234:275–288

Birchler JA, Pal-Bhadra M, Bhadra U (2003) Dosage dependent gene regulation and the compensation of the X chromosome in Drosophila males. Genetica 117:179–190

Bone JR, Lavender J, Richman R, Palmer MJ, Turner BM, Kuroda MI (1994) Acetylated histone H4 on the male X chromosome is associated with dosage compensation in Drosophila. Genes Dev 8:96–104

Bouazoune K, Korenjak M, Brehm A (2004) The dosage-compensation complex in flies and humans. Genome Biol 5:352

Buscaino A, Kocher T, Kind JH, Holz H, Taipale M, Wagner K, Wilm M, Akhtar A (2003) MOF-regulated acetylation of MSL-3 in the Drosophila dosage compensation complex. Mol Cell 11:1265–1277

Chang KA, Kuroda MI (1998) Modulation of MSL1 abundance in female Drosophila contributes to the sex specificity of dosage compensation. Genetics 150:699–709

Chiang PW, Kurnit DM (2003) Study of dosage compensation in Drosophila. Genetics 165:1167–1181

Copps K, Richman R, Lyman LM, Chang KA, Rampersad-Ammons J, Kuroda MI (1998) Complex formation by the Drosophila MSL proteins: role of the MSL2 RING finger in protein complex assembly. EMBO J 17:5409–5417

Corona DF, Clapier CR, Becker PB, Tamkun JW (2002) Modulation of ISWI function by site-specific histone acetylation. EMBO Rep 3:242–247

de la Cruz X, Lois S, Sanchez-Molina S, Martinez-Balbas MA (2005) Do protein motifs read the histone code? Bioessays 27:164–175

Demakova OV, Kotlikova IV, Gordadze PR, Alekseyenko AA, Kuroda MI, Zhimulev IF (2003) The MSL complex levels are critical for its correct targeting to the chromosomes in Drosophila melanogaster. Chromosoma 112:103–115

Deuring R, Fanti L, Armstrong JA, Sarte M, Papoulas O, Prestel M, Daubresse G, Verardo M, Moseley SL, Berloco M, et al (2000) The ISWI chromatin-remodeling protein is required for gene expression and the maintenance of higher order chromatin structure in vivo. Mol Cell 5:355–365

DiBartolomeis SM, Tartof KD, Jackson FR (1992) A superfamily of Drosophila satellite related (SR) DNA repeats restricted to the X chromosome euchromatin. Nucleic Acids Res 20:1113–1116

Eberharter A, Becker PB (2002) Histone acetylation: a switch between repressive and permissive chromatin. Second in review series on chromatin dynamics. EMBO Rep 3:224–229

Erickson JW, Cline TW (1998) Key aspects of the primary sex determination mechanism are conserved across the genus Drosophila. Development 125:3259–3268

Fagegaltier D, Baker BS (2004) X chromosome sites autonomously recruit the dosage compensation complex in Drosophila males. PLoS Biol 2:e341

Franke A, Baker BS (1999) The rox1 and rox2 RNAs are essential components of the compensasome, which mediates dosage compensation in Drosophila. Mol Cell 4:117–122

Gebauer F, Grskovic M, Hentze MW (2003) Drosophila sex-lethal inhibits the stable association of the 40S ribosomal subunit with msl-2 mRNA. Mol Cell 11:1397–1404

Ghosh S, Chatterjee RN, Bunick D, Manning JE, Lucchesi JC (1989) The LSP1-alpha gene of Drosophila melanogaster exhibits dosage compensation when it is relocated to a different site on the X chromosome. EMBO J 8:1191–1196

Gilfillan GD, Dahlsveen IK, Becker PB (2004) Lifting a chromosome: dosage compensation in Drosophila melanogaster. FEBS Lett 567:8–14

Gorman M, Franke A, Baker BS (1995) Molecular characterization of the male-specific lethal-3 gene and investigations of the regulation of dosage compensation in Drosophila. Development 121:463–475

Grskovic M, Hentze MW, Gebauer F (2003) A co-repressor assembly nucleated by Sex-lethal in the 3′UTR mediates translational control of Drosophila msl-2 mRNA. EMBO J 22:5571–5581

Gu W, Wei X, Pannuti A, Lucchesi JC (2000) Targeting the chromatin-remodeling MSL complex of Drosophila to its sites of action on the X chromosome requires both acetyl transferase and ATPase activities. EMBO J 19:5202–5211

Henry KW, Wyce A, Lo WS, Duggan LJ, Emre NC, Kao CF, Pillus L, Shilatifard A, Osley MA, Berger SL (2003) Transcriptional activation via sequential histone H2B ubiquitylation and deubiquitylation, mediated by SAGA-associated Ubp8. Genes Dev 17:2648–2663

Hiebert JC, Birchler JA (1994) Effects of the maleless mutation on X and autosomal gene expression in Drosophila melanogaster. Genetics 136:913–926

Hilfiker A, Hilfiker-Kleiner D, Pannuti A, Lucchesi JC (1997) mof, a putative acetyl transferase gene related to the Tip60 and MOZ human genes and to the SAS genes of yeast, is required for dosage compensation in Drosophila. EMBO J 16:2054–2060

Hung MS, Shen CK (2003) Eukaryotic methyl-CpG-binding domain proteins and chromatin modification. Eukaryot Cell 2:841–846

Ishii K, Arib G, Lin C, Van Houwe G, Laemmli UK (2002) Chromatin boundaries in budding yeast: the nuclear pore connection. Cell 109:551–562

Jenuwein T, Allis CD (2001) Translating the histone code. Science 293:1074–1080

Jin Y, Wang Y, Walker DL, Dong H, Conley C, Johansen J, Johansen KM (1999) JIL-1: a novel chromosomal tandem kinase implicated in transcriptional regulation in Drosophila. Mol Cell 4:129–135

Jin Y, Wang Y, Johansen J, Johansen KM (2000) JIL-1, a chromosomal kinase implicated in regulation of chromatin structure, associates with the male specific lethal (MSL) dosage compensation complex. J Cell Biol 149:1005–1010

Joazeiro CA, Weissman AM (2000) RING finger proteins: mediators of ubiquitin ligase activity. Cell 102:549–552

Kageyama Y, Mengus G, Gilfillan G, Kennedy HG, Stuckenholz C, Kelley RL, Becker PB, Kuroda MI (2001) Association and spreading of the Drosophila dosage compensation complex from a discrete roX1 chromatin entry site. EMBO J 20:2236–2245

Kelley RL, Solovyeva I, Lyman LM, Richman R, Solovyev V, Kuroda MI (1995) Expression of msl-2 causes assembly of dosage compensation regulators on the X chromosomes and female lethality in Drosophila. Cell 81:867–877

Kelley RL, Wang J, Bell L, Kuroda MI (1997) Sex lethal controls dosage compensation in Drosophila by a non-splicing mechanism. Nature 387:195–199

Kelley RL, Meller VH, Gordadze PR, Roman G, Davis RL, Kuroda MI (1999) Epigenetic spreading of the Drosophila dosage compensation complex from roX RNA genes into flanking chromatin. Cell 98:513–522

Kingston RE, Narlikar GJ (1999) ATP-dependent remodeling and acetylation as regulators of chromatin fluidity. Genes Dev 13:2339–2352

Kuo MH, Zhou J, Jambeck P, Churchill ME, Allis CD (1998) Histone acetyltransferase activity of yeast Gcn5p is required for the activation of target genes in vivo. Genes Dev 12:627–639

Kuo MH, vom Baur E, Struhl K, Allis CD (2000) Gcn4 activator targets Gcn5 histone acetyltransferase to specific promoters independently of transcription. Mol Cell 6:1309–1320

Kuroda MI, Kernan MJ, Kreber R, Ganetzky B, Baker BS (1991) The maleless protein associates with the X chromosome to regulate dosage compensation in Drosophila. Cell 66:935–947

Lowenhaupt K, Rich A, Pardue ML (1989) Nonrandom distribution of long mono- and dinucleotide repeats in Drosophila chromosomes: correlations with dosage compensation, heterochromatin, and recombination. Mol Cell Biol 9:1173–1182

Lyman LM, Copps K, Rastelli L, Kelley RL, Kuroda MI (1997) Drosophila male-specific lethal-2 protein: structure/function analysis and dependence on MSL-1 for chromosome association. Genetics 147:1743–1753

Lyon MF (1961) Gene action in the X-chromosome of the mouse (Mus musculus L.). Nature 190:372–373

Marin I (2003) Evolution of chromatin-remodeling complexes: comparative genomics reveals the ancient origin of "novel" compensasome genes. J Mol Evol 56:527–539

Marin I, Baker BS (2000) Origin and evolution of the regulatory gene male-specific lethal-3. Mol Biol Evol 17:1240–1250

Marin I, Franke A, Bashaw GJ, Baker BS (1996) The dosage compensation system of Drosophila is co-opted by newly evolved X chromosomes. Nature 383:160–163

Meller VH, Rattner BP (2002) The roX genes encode redundant male-specific lethal transcripts required for targeting of the MSL complex. EMBO J 21:1084–1091

Meller VH, Wu KH, Roman G, Kuroda MI, Davis RL (1997) roX1 RNA paints the X chromosome of male Drosophila and is regulated by the dosage compensation system. Cell 88:445–457

Meller VH, Gordadze PR, Park Y, Chu X, Stuckenholz C, Kelley RL, Kuroda MI (2000) Ordered assembly of roX RNAs into MSL complexes on the dosage-compensated X chromosome in Drosophila. Curr Biol 10:136–143

Mermoud JE, Popova B, Peters AH, Jenuwein T, Brockdorff N (2002) Histone H3 lysine 9 methylation occurs rapidly at the onset of random X inactivation. Curr Biol 12:247–251

Meyer BJ, Casson LP (1986) Caenorhabditis elegans compensates for the difference in X chromosome dosage between the sexes by regulating transcript levels. Cell 47:871–881

Mukherjee AS, Beermann W (1965) Synthesis of ribonucleic acid by the X-chromosomes of Drosophila melanogaster and the problem of dosage compensation. Nature 207:785–786

Nakajima T, Uchida C, Anderson SF, Lee CG, Hurwitz J, Parvin JD, Montminy M (1997) RNA helicase A mediates association of CBP with RNA polymerase II. Cell 90:1107–1112

Oh H, Bone JR, Kuroda MI (2004) Multiple classes of MSL binding sites target dosage compensation to the X chromosome of Drosophila. Curr Biol 14:481–487

Ohno S (1967) Sex Chromosomes and sex-linked genes. Springer-Verlag, Heidelberg, Berlin, New York

Pal Bhadra M, Bhadra U, Kundu J, Birchler JA (2005) Gene expression analysis of the function of the MSL complex in Drosophila. Genetics 169:2061–2074

Pannuti A, Lucchesi JC (2000) Recycling to remodel: evolution of dosage-compensation complexes. Curr Opin Genet Dev 10:644–650

Pardue ML, Lowenhaupt K, Rich A, Nordheim A (1987) (dC-dA)n.(dG-dT)n sequences have evolutionarily conserved chromosomal locations in Drosophila with implications for roles in chromosome structure and function. EMBO J 6:1781–1789

Park Y, Kelley RL, Oh H, Kuroda MI, Meller VH (2002) Extent of chromatin spreading determined by roX RNA recruitment of MSL proteins. Science 298:1620–1623

Park Y, Mengus G, Bai X, Kageyama Y, Meller VH, Becker PB, Kuroda MI (2003) Sequence-specific targeting of Drosophila roX genes by the MSL dosage compensation complex. Mol Cell 11:977–986

Plath K, Fang J, Mlynarczyk-Evans SK, Cao R, Worringer KA, Wang H, de la Cruz CC, Otte AP, Panning B, Zhang Y (2003) Role of histone H3 lysine 27 methylation in X inactivation. Science 300:131–135

Protacio RU, Li G, Lowary PT, Widom J (2000) Effects of histone tail domains on the rate of transcriptional elongation through a nucleosome. Mol Cell Biol 20:8866–8878

Rattner BP, Meller VH (2004) Drosophila male-specific lethal 2 protein controls sex-specific expression of the roX genes. Genetics 166:1825–1832

Reid JL, Iyer VR, Brown PO, Struhl K (2000) Coordinate regulation of yeast ribosomal protein genes is associated with targeted recruitment of Esa1 histone acetylase. Mol Cell 6:1297–1307

Richter L, Bone JR, Kuroda MI (1996) RNA-dependent association of the Drosophila maleless protein with the male X chromosome. Genes Cells 1:325–336
Sabl JF, Birchler JA (1993) Dosage dependent modifiers of white alleles in Drosophila melanogaster. Genet Res 62:15–22
Sass GL, Pannuti A, Lucchesi JC (2003) Male-specific lethal complex of Drosophila targets activated regions of the X chromosome for chromatin remodeling. Proc Natl Acad Sci U S A 100:8287–8291
Schneider R, Bannister AJ, Myers FA, Thorne AW, Crane-Robinson C, Kouzarides T (2004) Histone H3 lysine 4 methylation patterns in higher eukaryotic genes. Nat Cell Biol 6:73–77
Scott MJ, Pan LL, Cleland SB, Knox AL, Heinrich J (2000) MSL1 plays a central role in assembly of the MSL complex, essential for dosage compensation in Drosophila. EMBO J 19:144–155
Smith ER, Pannuti A, Gu W, Steurnagel A, Cook RG, Allis CD, Lucchesi JC (2000) The drosophila MSL complex acetylates histone H4 at lysine 16, a chromatin modification linked to dosage compensation. Mol Cell Biol 20:312–318
Smith ER, Allis CD, Lucchesi JC (2001) Linking global histone acetylation to the transcription enhancement of X-chromosomal genes in Drosophila males. J Biol Chem 276:31483–31486
Strahl BD, Allis CD (2000) The language of covalent histone modifications. Nature 403:41–45
Straub T, Neumann MF, Prestel M, Kremmer E, Kaether C, Haass C, Becker PB (2005) Stable chromosomal association of MSL2 defines a dosage-compensated nuclear compartment. Chromosoma 114(5):352–364
Taipale M, Rea S, Richter K, Vilar A, Lichter P, Imhof A, Akhtar A (2005) hMOF histone acetyltransferase is required for histone H4 lysine 16 acetylation in mammalian cells. Mol Cell Biol 25(15):6798–6810
Taipale M, Akhtar A (2005) Chromatin mechanisms in Drosophila dosage compensation. Prog Mol Subcell Biol 38:123–149
Thomson S, Clayton AL, Mahadevan LC (2001) Independent dynamic regulation of histone phosphorylation and acetylation during immediate-early gene induction. Mol Cell 8:1231–1241
Tse C, Sera T, Wolffe AP, Hansen JC (1998) Disruption of higher-order folding by core histone acetylation dramatically enhances transcription of nucleosomal arrays by RNA polymerase III. Mol Cell Biol 18:4629–4638
Turner BM, Birley AJ, Lavender J (1992) Histone H4 isoforms acetylated at specific lysine residues define individual chromosomes and chromatin domains in Drosophila polytene nuclei. Cell 69:375–384
Wang Y, Zhang W, Jin Y, Johansen J, Johansen KM (2001) The JIL-1 tandem kinase mediates histone H3 phosphorylation and is required for maintenance of chromatin structure in Drosophila. Cell 105:433–443
Waring GL, Pollack JC (1987) Cloning and characterization of a dispersed, multicopy, X chromosome sequence in Drosophila melanogaster. Proc Natl Acad Sci U S A 84:2843–2847
Winkler GS, Kristjuhan A, Erdjument-Bromage H, Tempst P, Svejstrup JQ (2002) Elongator is a histone H3 and H4 acetyltransferase important for normal histone acetylation levels in vivo. Proc Natl Acad Sci U S A 99:3517–3522

Zhang Y (2003) Transcriptional regulation by histone ubiquitination and deubiquitination. Genes Dev 17:2733–2740

Zhou S, Yang Y, Scott MJ, Pannuti A, Fehr KC, Eisen A, Koonin EV, Fouts DL, Wrightsman R, Manning JE, et al (1995) Male-specific lethal 2, a dosage compensation gene of Drosophila, undergoes sex-specific regulation and encodes a protein with a RING finger and a metallothionein-like cysteine cluster. EMBO J 14:2884–2895

CTMI (2006) 310:141–178

DNA Methylation Profiles of Female Steroid Hormone-Driven Human Malignancies

M. Campan · D. J. Weisenberger · P. W. Laird (✉)

Department of Surgery and Department of Biochemistry and Molecular Biology, Keck School of Medicine, USC/Norris Comprehensive Cancer Center, University of Southern California, 1441 Eastlake Avenue, Room 6418, Los Angeles, CA 90089-9176, USA
plaird@usc.edu

Abstract Tumor DNA contains valuable clues about the origin and pathogenesis of human cancers. Alterations in DNA methylation can lead to silencing of genes associated with distinct tumorigenic pathways. These pathway-specific DNA methylation changes help define tumor-specific DNA methylation profiles that can be used to further our understanding of tumor development, as well as provide tools for molecular diagnosis and early detection of cancer. Female sex hormones have been implicated in the etiology of several of the women's cancers including breast, endometrial, ovarian, and proximal colon cancers. We have reviewed the DNA methylation profiles of these cancers to determine whether the hormonal regulation of these cancers results in specific DNA methylation alterations. Although subsets of tumors in each of these four types of cancers were found to share some DNA methylation alterations, we did not find evidence for global hormone-specific DNA methylation alterations, suggesting that female sex hormones may participate in different tumorigenic pathways that are associated with distinct DNA methylation-based molecular signatures. One such pathway may include *MLH1* methylation in the context of the CpG island methylator phenotype.

1 Introduction

The past decade of cancer research has been characterized by an increased interest in the field of epigenetics. The genetic information encoded in the nucleotide sequence of the genome is selectively decoded during normal development and differentiation of an organism. This is achieved by modulating gene expression in a coordinated spatial and temporal manner. The blueprint for the gene expression pattern in a specific cell is contained in the epigenetic information, which manifests its influence on gene expression without altering the primary nucleotide sequence of the DNA. Covalent modifications of histone proteins, RNA-associated gene silencing processes, and DNA methylation are integral components of the epigenetic mechanism controlling gene expression patterns (Egger et al. 2004). All these systems have the potential to work in concert to modify the spatial structure of the DNA and the proteins associated with it (chromatin), in order to establish structural states that are either favorable (open chromatin), or not favorable (closed chromatin) for gene expression. As a result, transcription factors that mediate gene expression are allowed or denied access to important regulatory regions of genes called promoters, and thus genes can be activated or silenced.

Increased DNA methylation at promoter regions is associated with transcriptional silencing that translates into closed chromatin states. Lack of promoter DNA methylation is generally necessary, but not sufficient, for transcriptional activity and translates into active or open chromatin states. At this time, it is not clear what the sequence of events is that leads to the establishment of specific chromatin states. One proposed scenario is that histone modifications precede the changes in DNA methylation, and represent the initiating event in gene silencing, whereas DNA methylation acts as a lock to maintain a permanent state of gene silencing (Tamaru and Selker 2001). Alternatively, DNA methylation may precede histone modifications and thus may be the initiating event. Understanding the dynamics of epigenetic regulation will help us gain insight into the molecular mechanisms responsible for inappropriate gene expression or silencing of genes in pathological conditions.

The distribution of DNA methylation across the genome is highly compartmentalized, reflecting the role DNA methylation plays in regulating gene expression. Heavily methylated DNA regions, mainly associated with repetitive elements, alternate with DNA regions that are free of methylation, usually associated with promoter regions of genes. However, not all gene promoters are unmethylated. This distribution reflects the role DNA methylation plays to silence the expression of parasitic DNA sequences such as transposon and endogenous retroviruses, and of genes on the inactive chromosome X in women, imprinted genes, and germ-line specific genes, as well as other genes with tissue-specific expression. Since the spectrum of genes associated with promoter DNA methylation may differ from tissue to tissue, specific profiles of DNA methylation are expected to be characteristic for each cell type. The normal DNA methylation profiles that are specific or variable between cell types are largely unknown. More recently, the Human Epigenome Project has been launched with the purpose of characterizing the global epigenetic profiles of human cells, known also as the "human methylome" or "epigenome" (Fazzari and Greally 2004).

The normal DNA profiles become severely altered during the process of malignant transformation and aging, reflecting the dynamic nature of DNA methylation. In these instances, DNA methylation is redistributed across the genome such that much of the genome loses some DNA methylation, while selected gene promoters become abnormally hypermethylated. The overall result of these changes is a net reduction in the genomic content of cytosine-5 methylation. Chromosomal instability, retroviral and transposon reactivation, and loss of imprinting are probably the most important consequences of loss of DNA methylation in cancer cells. Specific promoter hypermethylation events can lead to transcriptional silencing of tumor-suppressor genes, as well as loss of imprinting. While the effects of DNA hypomethylation for cancer development have been less well studied so far, cancer-associated DNA hypermethylation has been the topic of investigation in numerous studies. DNA methylation aberrations have been documented in virtually all types of human malignancies, and these changes appear to be more frequent than genetic alterations.

Accumulating evidence suggests that the occurrence of DNA hypermethylation in cancer is not entirely random (Costello et al. 2000; Esteller et al. 2001a; Feltus et al. 2003; Huang et al. 1999). The spectrum of DNA regions affected by hypermethylation has been shown to differ between tumor types, such that unique profiles of DNA hypermethylation can be defined for each type of human cancer. The differential susceptibility of various DNA regions to DNA hypermethylation suggests that specific cellular pathways may have been disrupted and thus have the potential to influence the development of different types of tumors. The systematic characterization of cancer-specific DNA methylation profiles may provide clues about the molecular mechanisms involved in the tumorigenesis of specific types of cancers. Similarly, correlations can be made between DNA methylation profiles and risk factors associated with specific cancers in order to elucidate the role of these factors in inducing methylation changes. The information gained from characterizing DNA methylation profiles in various human cancers has proved useful for the development of DNA methylation-based tumor markers for diagnostic, prognostic, and therapeutic purposes.

Despite the fact that as few as 15 DNA methylation markers may be sufficient to correctly classify tumors according to their tissue type (Paz et al. 2003), more specific and sensitive tumor markers are needed for clinical applications. Also, better insight into the mechanistic pathways of cancer and other epigenetic diseases (Egger et al. 2004), as well as a better understanding of the normal patterns of DNA methylation, can be achieved if a larger number of DNA methylation markers are investigated. This can only be accomplished by using high-throughput methods of screening and analysis of DNA methylation. In the past several years, our laboratory has established a collection of more than 200 DNA methylation markers for MethyLight analysis using a candidate gene approach. Most of these reactions have been successfully used to acquire detailed DNA methylation profiles of colorectal, esophageal (Eads et al. 2001), gastric (Eads et al. 2001), breast (Widschwendter et al. 2004), and ovarian cancers, as well as of normal and diseased brain tissues.

The scope of this review is to summarize data available in the literature or obtained in our own laboratory to exemplify how analyses of DNA methylation profiles analysis can be exploited to identify methylation markers directly related to known cancer risk factors. For this we examined and compared the most frequent DNA methylation alterations associated with cancers driven by female sex steroid hormones, and we comment on the potential DNA methylation markers that could be directly related to the effect of sex steroid hormones on DNA methylation and that may be relevant for the pathogenesis of gender-specific malignancies.

2 Normal and Aberrant DNA Methylation Profiles

By DNA methylation, we specifically refer to a methylated cytosine immediately 5′ to a guanine. The covalent binding of a methyl group at the 5′-position of the cytosine occurs in almost 70% of all CpG dinucleotides in the human genome and serves as a mark for both the transcription and replication machinery. During DNA replication, the methylation marks from the parental strand serve as a template that will be reproduced on the newly synthesized daughter strand. A family of DNA methyltransferases (DNMTs) consisting of three major enzymes, DNMT1, 3a, and 3b, and their alternatively spliced isoforms are responsible for the establishment and maintenance of the methylation information. DNA methylation is established early during embryogenesis and is stably maintained throughout the replicative life of a cell.

The DNA methylation profile of a cell is defined by the DNA methylation status across many sites in the genome (Laird 2003). Normal DNA methylation profiles are related to the CpG dinucleotide distribution throughout the genome, and reflect the spectrum of gene expression in a certain tissue and the age of the organism. The presence of methylation at CpG sites has led to the depletion of 80% of these dinucleotides during the course of evolution due to the spontaneous deamination of the methylated cytosine to thymine. The remaining CpGs are dispersed throughout the genome, except at the 5′ regions of genes where discrete clusters of CpGs can be found, termed CpG islands. A CpG island usually encompasses the promoter and the first exon of a gene transcript. Different transcripts of the same gene can have separate promoters and alternatively spliced first exons, and thus distinct CpG islands can be associated with these promoters. Some tissue-specific genes lack CpG islands at their 5′-end region. With few exceptions, CpG islands are unmethylated in normal cells, whereas the dispersed CpGs are usually methylated. The tissue-specific genes with CpG-poor promoters are methylated in the tissues where these genes are silenced, and unmethylated in the tissues where they are expressed (Eden and Cedar 1994). Most of the tissue-specific genes with CpG islands remain methylation-free even when their associated genes are repressed (Bird 2002). Several notable exceptions have been described, however, including the methylation of the *adenomatosis polyposis coli* (*APC*) gene CpG island in normal gastric mucosa (Eads et al. 2001), and the methylation of the *DNAJC15(MCJ)* gene CpG island in normal ovarian tissue and other normal cells of epithelial origin (Strathdee et al. 2004). DNA methylation profiles of older individuals are different from those of younger ones due to progressive methylation of selected CpG islands accompanied by global loss of methylation in the dispersed CpGs. Genes that undergo age-related methylation have been designated as "type A" genes (Issa 2000).

The methylation changes detected in cancer cells mimic those in aging cells but the former occur to a greater extent. Genes that become methylated specifically in cancers have been designated as "type C." Tumor DNA methylation profiles reflect, in part, the methylation profile of the cell of origin for a specific tumor and the age-specific methylation changes of type A genes, as well as the abnormal methylation of type C genes. As many as 3,500 CpG islands (Markl et al. 2001) or an average 600 CpG islands per genome (Costello et al. 2000) have been estimated to become abnormally methylated in bladder cancer or various other types of cancers, respectively. It is highly unlikely that all cancer-associated methylation changes play a causative role in tumor development. However, DNA methylation abnormalities associated with some genes can lead to tumorigenesis if the proteins encoded by these genes are involved in important regulatory pathways such as cellular proliferation or apoptosis. The existence of distinct tumor-specific DNA methylation profiles has been reported (Costello et al. 2000; Esteller et al. 2001a; Feltus et al. 2003; Huang et al. 1999), suggesting that at least some of these methylation changes may be pathway specific rather than random events.

DNA methylation-based molecular signatures of individual tumor types can thus be defined by the combination of three or four DNA methylation events in tissue-specific, type A, and pathway-specific type C genes. Such DNA methylation markers have been used successfully to detect tumor DNA in a variety of biological fluids including serum and plasma (Laird 2003) or biopsy specimens (Pellise et al. 2004). Subtypes of the same tumor type can also have distinct DNA methylation profiles. The most notable example is the molecular signature of proximal colon cancer, a subset of colorectal cancers with distinct clinicopathological and molecular features (Iacopetta 2003). Methylation of a unique subset of genes that includes *MLH1* [*mutL homolog 1, colon cancer, nonpolyposis type 2 (E. coli)*] and $p16^{INK4a}$ (*CDKN2A*) occurs frequently in these cancers in the context of the CpG island mutator phenotype (CIMP). CIMP represents a generalized methylation defect defined by the presence of concordant methylation of multiple type C genes (Kondo and Issa 2004). CIMP has also been described in a subset of gastric tumors, suggesting that this phenomenon may occur in other tumors as well.

Distinct DNA methylation profiles in a given tumor have already been shown to predict tumor behavior. Unique DNA methylation profiles may also be characteristic for tumors that are under the influence of common risk factors such as exposure to particular carcinogens or regulation by specific hormonal pathways. The challenge of finding such profiles still lies ahead. The latest advances in new methylation marker development will speed up this endeavor.

3
Gender-Specific Gene Methylation Profiles in Four Hormone-Driven Cancers

Sex steroid hormones play a pivotal role in the development of several female and male malignancies. Women experiencing prolonged exposure to or having high circulating levels of estrogen are at increased risk for breast, endometrial, and ovarian cancers. Various aspects of the ovarian reproductive function represent well-established risk factors for these types of cancers. Aside from these cancers that develop from cells that are established physiologic targets of female hormones, epidemiological studies have also described several other human cancers that have higher frequencies in women, including the cancers that develop in the proximal colon. The hormonal link of this malignancy is just starting to be elucidated.

CpG island hypermethylation is one of the earliest somatic alterations to occur during cancer development. Many genes are also abnormally methylated in female cancers, and specific DNA methylation profiles of these cancers have already been described. In order to compare and contrast possible common or divergent tumorigenic pathways, and to assess the association between the female sexual hormones and DNA methylation changes in these four hormone-driven cancers, we performed a review of the literature with respect to the methylation patterns of 12 genes. We chose these genes based on their potential functional association with female hormonal regulatory pathways or the presence of methylation in these genes in female-associated cancers. These genes include the steroid receptor genes for estrogen (*ESR1*, *ESR2*) and progesterone (*PGR*), genes involved in DNA damage prevention and repair (*BRCA1*, *GSTP1*, *MGMT*, *MLH1*), genes involved in the cell cycle control (*ARF*, *CDKN2A*, *RASSF1*), and tumor invasion and metastasis gene *E-cadherin* (*CDH1*), as well as the tumor suppressor gene *APC* and the pro-inflammatory gene *COX-2* (*PTGS2*).

3.1
Breast Cancer

The incidence rates for the three most common gynecological malignancies—breast cancer, endometrial adenocarcinomas, and ovarian adenocarcinomas—increase sharply at menarche and decline abruptly at menopause (Pike et al. 2004). This was a clear indication that risk factors associated with the ovarian function are involved in the etiology of these tumors. Menopause is the most effective protective factor against each of these cancers (Pike et al. 2004). Early age at menarche, late age at menopause, and late age at first full-term pregnancy increase the risk for breast cancer, while removal of the ovaries at a younger age has a protective effect against breast cancer (Kelsey and Bernstein 1996). Ultimately, these risk factors modulate the status of the two ovarian hormones, estrogen and progesterone, that are known to control the normal development of the mammary gland and also induce breast tumorigenesis. Estrogen is thought to serve either as a preinitiator or initiator of breast tumorigenesis, or as a growth promoter of existing breast malignancies (Hilakivi-Clarke 2000). The estrogen antagonist synthetic compounds such as tamoxifen, which block the action of estrogen, are very efficient in preventing the breast tumorigenesis and the recurrence of the disease (Gelber et al. 1996).

All sexual hormones exert their effects on their target tissues through steroid receptors. The mechanism by which sexual hormones can influence the pathogenesis of various hormonal-driven cancers may be related to the function of these receptors. The female sex steroid receptors, estrogen receptor α (ERα), estrogen receptor β (ERβ), and progesterone receptor (PR), are DNA binding molecules that act as transcription factors. Following their activation by binding to their specific ligands, estrogen and progesterone steroid receptors recognize and bind specific hormone-responsive DNA elements situated in the promoter regions of the hormonally regulated genes. The steroid receptors are among the few receptors that interact directly with components of the transcriptional machinery and chromatin structure to regulate gene expression (Kinyamu and Archer 2004). The steroid receptors are expressed in a tissue- and cell-specific manner, and their expression can be affected by methylation during aging and tumorigenesis.

3.1.1
ESR1 and *ESR2*

Most of the biological effects of estrogen and its therapeutic synthetic antagonists are mediated via two distinct estrogen receptors called ERα and ERβ, which are encoded by the *ESR1* and *ESR2* genes, respectively. Although they recognize the same estrogen-responsive elements, they have been shown to have opposing activities at activating protein-1 (AP1) sites (Paech et al. 1997) and to differ in the use of their ligand-independent activation function (AF1) domains for transactivation (Cowley and Parker 1999). In vitro studies have also shown that these two receptors have different responses to tamoxifen and other synthetic antagonists of estrogen (Nilsson et al. 2001). While estrogen-activated ERα stimulates cell proliferation (Nilsson et al. 2001), ERβ has been shown to inhibit the proliferation and invasion of breast cancer cells (Lazennec et al. 2001). Consequently, the differential expression of these receptors in a tissue-specific manner may also explain some of the tissue-specific effects of estrogen. While both ERs are expressed in ductal and lobular breast epithelium (Flototto et al. 2001), ERβ is more abundant than ERα in normal breast epithelium (Widschwendter and Jones 2002).

Both of these receptors have promoter-associated CpG islands that have the potential to become abnormally methylated in cancer. The subsequent loss of expression of these receptors can disrupt the normal estrogen-signaling pathway and result in inactivation of downstream targets of this pathway. Multiple promoters have been described for both *ESR*s corresponding to various isoforms of these receptors. However, most of the methylation analyses have been performed on the A promoter for the *ESR1* and on the promoter associated with exon 1 for the *ESR2*.

Aberrant methylation of the *ESR1* gene promoter A has been documented in various normal epithelial tissues as an age-dependent modification, as well as in many types of cancers, including colon and breast cancers (Kondo and Issa 2004; Table 1). Due to the increase in methylation of the *ESR1* gene with age, it has been hypothesized that hypermethylation of *ESR1* in cancers may simply reflect the stochastic predisposition of the *ESR1* gene to become methylated with progressive rounds of DNA replication (Velicescu et al. 2002), and thus may not be of consequence for the tumorigenic process (Kondo and Issa 2004). Indeed, breast tumors from older individuals are more likely to have *ESR1* gene promoter methylation, whereas *ESR1* methylation is less frequently methylated in women that develop breast cancer at younger age (M. Campan, D.J. Weisenberger, Q. Feng, S.E. Hawes, N.B. Kiviat, P.W. Laird, manuscript in preparation). Interestingly, the accumulation of *ESR1* methylation does not appear to continue after cells become malignantly transformed. The majority of breast cancers do not have high levels of *ESR1* methylation, despite the high rate of proliferation, characteristic for tumor cells. Loss of *ESR1* expression has been recently shown to induce changes in the chromatin structure of the *PGR* gene and of many other downstream targets of the estrogen-signaling pathway, with accompanying promoter hypermethylation and transcriptional silencing (Leu et al. 2004). These results suggest that epigenetic inactivation of *ESR1*, even due to age-related stochastic events, can have important biological consequences that can result in tumorigenesis, by disrupting important growth regulatory pathways. These results also suggest that DNA methylation changes can be pathway specific, rather than as a consequence of stochastic processes, and this may help to explain the existence of tumor type-specific DNA methylation profiles. Based on these findings, it would be interesting to determine if the observed age-related methylation of the *ESR1* gene is caused by similar mechanisms, as a consequence of reduction in the estrogen levels, as occurs during menopause in women. Almost all breast cancers show some degree of DNA methylation at the *ESR1* gene promoter (Wid-

Table 1 DNA methylation frequencies of select genes in breast, endometrial, ovarian, colon, and proximal colon cancers. The methylation frequency for each gene represents a weighted average of methylation frequencies when multiple reports were available

HUGO gene symbol	Chromosomal location	Frequency (%) Breast cancer	Endometrial cancer	Ovarian cancer	Colon cancer	Proximal colon cancer	References[a]
APC	5q21	28	37	14	21	52	[9, 33, 11, 16, 25, 53, 22, 8, 21, 35, 54–56]
ARF	9p21	20	16	13	30	29	[9, 11, 19, 7, 41, 42, 14, 22, 4, 21, 40, 54–56]
BRCA1	17q21	16	ND	17	0	ND	[9, 11, 49, 33, 2, 5, 43, 36, 14, 3, 47, 54, 56]
CDH1	16q22	41	26	26	49	64	[49, 9, 11, 2, 33, 35, 26, 30, 36, 22, 21, 13, 54–56]
CDKN2A	9p21	17	16	8	30	27	[9, 11, 19, 7, 41, 42, 14, 22, 4, 21, 40, 9, 11, 33, 19, 42, 48, 51, 43, 14, 23, 29, 45, 22, 8, 39, 31, 4, 21, 50, 56]
ESR1 promoter A	6q25.1	49	1	29	81	ND	[49, 20, 33, 2, 28, 38, 32, 8, 15, 54–56, 54–56[b]]
ESR1 promoter C	6q.25.1	ND	94				
ND	ND	ND	[38]				
ESR2	14q21	52	0	22	22	ND	[38, 49, 54–56[b]]
GSTP1	11q13	29	0	3	4	9	[2, 9, 11, 33, 19, 10, 36, 21, 54–56]
MGMT	10q26	8	0	4	38	29	[9, 11, 49, 21, 22, 54–56]
MLH1	3p21	29	41	10	20	40	[9, 11, 49, 34, 27, 12, 48, 44, 43, 17, 45, 22, 8, 24, 21, 55, 56]
PGR promoter A	11q22	ND	0				
ND	37	ND	[37, 56]				
PGR promoter B	11q22	66	75	0	80	ND	[37, 49, 54, 55]

Table 1 (continued)

HUGO gene symbol	Chromosomal location	Frequency (%) Breast cancer	Endometrial cancer	Ovarian cancer	Colon cancer	Proximal colon cancer	References[a]
PTGS2	1p25	99	ND	0	13	38	[49, 46, 21, 54–56]
RASSF1	3p21	77	ND	31	19	ND	[2, 18, 34, 49, 6, 1, 19, 36, 14, 52, 21, 54–56]

HUGO, The Human Genome Organisation; ND, not determined

[a] References: [1]Agathanggelou et al. 2001; [2]Bae et al. 2004; [3]Baldwin et al. 2000; [4]Burri et al. 2001; [5]Catteau et al. 1999; [6]Dammann et al. 2001; [7]Dominguez et al. 2003; [8]Eads et al. 1999; [9]Esteller et al. 2001a; [10]Esteller et al. 1998a; [11]Esteller et al. 2001b; [12]Esteller et al. 1998b; [13]Garinis et al. 2002; [14]Ibanez de Caceres et al. 2004; [15]Issa et al. 1994; [16]Jin et al. 2001; [17]Kane et al. 1997; [18]Kondo and Issa 2004; [19]Krassenstein et al. 2004; [20]Lapidus et al. 1996; [21]Lee et al. 2004; [22]Lind et al. 2004; [23]McCluskey et al. 1999; [24]Miyakura et al. 2001; [25]Moreno-Bueno et al. 2002; [26]Moreno-Bueno et al. 2003; [27]Murata et al. 2002; [28]Navari et al. 2000; [29]Niederacher et al. 1999; [30]Nishimura et al. 2003; [31]Norrie et al. 2003; [32]O'Doherty et al. 2002; [33]Parrella et al. 2004; [34]Paz et al. 2003; [35]Pijnenborg et al. 2004; [36]Rathi et al. 2002; [37]Sasaki et al. 2001a; [38]Sasaki et al. 2001; [39]Schneider-Stock et al. 2003; [40]Shen et al. 2003; [41]Silva et al. 2001; [42]Silva et al. 2003; [43]Strathdee et al. 2001; [44]Strathdee et al. 1999; [45]Toyota et al. 1999; [46]Toyota et al. 2000b; [47]Wang et al. 2004; [48]Whitcomb et al. 2003; [49]Widschwendter et al. 2004; [50]Wiencke et al. 1999; [51]Wong et al. 1999; [52]Yoon et al. 2001; [53]Zysman et al. 2002; [54]M. Campan, D.J. Weisenberger, Q. Feng, S.E. Hawes, N.B. Kiviat, P.W. Laird, manuscript in preparation; [55]Ehrlich et al. 2006; [56]D.J. Weisenberger, K. Siegmund, M. Campan, J. Young, T.I. Long, M.A. Faasse, G.H. Kang, M. Widschwendter, D. Weener, D. Buchanan, H. Koh, L. Simms, M. Barker, B. Leggett, J. Levine, A.J. French, S.N. Thibodeau, J. Jass, R. Haile, P.W. Laird, submitted

[b] Methylation frequencies from cell lines have been included in the analysis

schwendter et al. 2004). However, only 30% of these tumors may have high enough levels of DNA methylation at this locus (Bae et al. 2004; Lapidus et al. 1996; Parrella et al. 2004) that can result in loss of *ESR1* expression. This is in agreement with the finding that two-thirds of breast cancers express ERα.

The hormonal receptor (HR) status of breast tumors, defined by the presence or absence of ER and PR, constitutes an important indicator of response to therapy and survival. For instance, patients with ER^+ breast tumors have better survival rates, respond better to anti-estrogenic therapy, and are less likely to have tumor recurrence than those with ER^- or ER^-/PR^- breast tumors (Li et al. 2003). We have recently shown an association between DNA methylation changes of *ESR1* and *PGR* in breast tumors and the HR status and response to anti-estrogenic therapy (Widschwendter et al. 2004). A molecular profiling of breast tumors, using DNA methylation profiles, identified two distinct groups of tumors that differed with respect to their HR status. DNA methylation of neither of the two HR genes was the best predictor of the overall HR status, suggesting either a complex interplay between hormone receptor gene methylation and hormone receptor status, or that DNA methylation markers may not always correlate with the gene expression status, due to threshold effects. Nevertheless, we found that *ESR1* methylation was the best predictor of PR status and of response to tamoxifen treatment, whereas the methylation of *PGR* was the best predictor of ER status. Interestingly, the association between *ESR1* methylation and PR status was not inversely correlated, even though estrogen signaling is known to activate PGR expression (Widschwendter et al. 2004). It is clear that a full understanding of the relationship between hormone receptor gene expression and DNA methylation, and its role in breast carcinogenesis will require further investigation. *ESR1* methylation was also shown to be associated with the methylation of *CDH1, GSTP1, CCND2*, and *TRβ1*, suggesting that this association may represent a molecular signature of a specific subset of breast cancers with yet-unidentified common phenotypic characteristics (Nass et al. 2000; Parrella et al. 2004).

The methylation status of the *ESR2* gene promoter in normal or various cancerous tissues has not yet been investigated to a large extent. As in the case of *ESR1*, low levels of *ESR2* gene promoter methylation are detected in the majority (79%) of breast cancers (Widschwendter et al. 2004), although only about 10% may have high levels of methylation (M. Campan, D.J. Weisenberger, Q. Feng, S.E. Hawes, N.B. Kiviat, P.W. Laird, manuscript in preparation).

3.1.2
PGR

The progesterone receptor (PR) is a member of the steroid receptor superfamily responsible for mediating the physiological effects of progesterone. The *PGR* gene encodes two protein isoforms, PR-A and PR-B, which are regulated independently (Kastner et al. 1990). PR-B can also be induced by the estrogen-signaling pathway, following the activation of the ER by estrogen, whereas the shorter isoform PR-A is not induced by ER (Widschwendter et al. 2002). PR-B acts as a transcription factor, whereas PR-A acts as a trans-dominant repressor of other nuclear receptor activities, including PR-B and ER (Vegeto et al. 1993). Both PRs are expressed in hormonally receptive tissues of the breast, endometrium, and ovary. In mice, PR-A was shown to be important for progesterone-dependent reproductive responses required for female fertility associated with the function of the ovary and endometrium, while the PR-B isoform was required for the normal proliferation and differentiation of the mammary gland (Conneely et al. 2003). Progesterone, through its nuclear receptors, is believed to play a role in the development of breast, endometrial, and ovarian cancers. PRs may also be involved in tumorigenesis as a consequence of its activation by the estrogen-signaling pathway. The presence of PR in breast tumors is used to predict functional ER status and therefore also to predict the response to endocrine therapies and disease prognosis (Clarke and Sutherland 1990; Widschwendter et al. 2002). PR status is concordant with that of the ER status in breast tumors.

The presence of CpG islands in the promoters of both PR isoforms suggests that their expression can be modulated by DNA methylation changes. Despite evidence for loss of PR expression, or changes in the ratio between the two isoforms in various cancers, very few studies have tried to assess the role of DNA methylation in these processes. Low levels of *PGR-B* methylation are frequently detected in almost all breast tumors (Widschwendter et al. 2004). It is not clear whether or not this is age related. However, high levels of methylation that may result in gene silencing are detected in only 16% of the breast tumors (M. Campan, D.J. Weisenberger, Q. Feng, S.E. Hawes, N.B. Kiviat, P.W. Laird, manuscript in preparation). Leu et al. have recently shown that methylation of the *PGR* promoter can result from the disruption of the estrogen-signaling pathway (Leu et al. 2004).

3.1.3
BRCA1

BRCA1 is thought to act as a tumor suppressor gene for both breast and ovarian cancers. The BRCA protein is involved in important cellular processes such as transcriptional regulation, control of homologous recombination, and repair of DNA damage (Welcsh and King 2001). The expression of BRCA1 is suggested to be under the control of sexual hormones. During puberty and pregnancy, when the levels of estrogen increase considerably, BRCA1 protein expression is also sharply upregulated. The induction of BRCA1 by estrogen during these periods of intense breast epithelial proliferation may reduce the risk of breast cancer by preventing the accumulation of genetic and chromosomal abnormalities. However, no estrogen-responsive element has been identified in the promoter of the *BRCA1* gene, suggesting that BRCA1 regulation by sexual hormones may occur at the RNA or protein levels (Marks et al. 1997). Alternatively, BRCA1 may be required to modulate the effects of estrogens on breast epithelium. BRCA1 was shown to inhibit estrogen-dependent transcriptional pathways related to mammary epithelial proliferation (Fan et al. 1999). *BRCA1* mutations are very common in the inherited breast and ovarian cancers, but they are rare in the sporadic form of these diseases. Inactivation of *BRCA1* in these circumstances can occur via chromosomal deletions and epigenetic silencing (Esteller et al. 2000a).

The frequency of *BRCA1* methylation in sporadic breast cancers is around 17% (Catteau et al. 1999; Esteller et al. 2000a, 2001b; Parrella et al. 2004; Widschwendter et al. 2004); however, the vast majority of the tumors with *BRCA1* promoter methylation belong to specific subtypes of breast cancers—medullary and mucinous—which are also the most common types of hereditary breast cancers (Esteller et al. 2000a). *BRCA1* gene promoter methylation is more frequent in breast tumors that lack ER and PR expression (Catteau et al. 1999). Hormone receptor negative tumors occur more frequently in younger women (Althuis et al. 2004), suggesting that *BRCA1* inactivation by DNA methylation may be characteristic of a distinct tumorigenic pathway that is not yet fully understood. However, we did not find *BRCA1* methylation to be significantly correlated with either the ER status or the PR status in breast cancers (Widschwendter et al. 2004). At this point, it is unclear how DNA methylation of the *BRCA1* gene promoter can be influenced by estrogen- or progesterone-mediated processes in the breast and ovary. The *BRCA1* state in hereditary and sporadic breast and ovarian cancers parallels the circumstances of another DNA repair gene, *MLH1*, in hereditary nonpolyposis colon cancer (HNPCC) and sporadic colon and endometrial cancer, which are discussed in more detail below (see Sects. 3.1.9 and 3.2, and 3.4)

3.1.4
RASSF1

The protein encoded by this gene is a member of the Ras family of GTPases involved in the cell cycle regulation of progression through mitosis, DNA repair, and Ras-induced apoptosis (Vos et al. 2000). Several aspects of the tumorigenic process can be associated with the Ras signaling pathways, including regulation of anchorage-dependent and anchorage-independent growth, tumor initiation, and invasion (Malaney and Daly 2001). Loss of *RASSF1* gene expression occurs in many types of cancer, suggesting a tumor-suppressor function for this gene (Pfeifer et al. 2002). The mechanisms responsible for the loss of *RASSF1* expression in human cancers involve chromosomal deletions but also epigenetic inactivation by hypermethylation of its gene promoter (Pfeifer et al. 2002). Hypermethylation of the *RASSF1* gene promoter occurs in a large percentage of breast cancers (77%) as an early event during breast tumorigenic process (Agathanggelou et al. 2001; Bae et al. 2004; Dammann et al. 2001; Krassenstein et al. 2004; Lehmann et al. 2002; Widschwendter et al. 2004). More recently, we have also shown that *RASSF1* methylation is a significant predictor of the hormonal status of advanced breast tumors (Widschwendter et al. 2004). The relevance of this epigenetic process to the hormonal regulation of tumor development in the mammary gland requires further investigation. *RASSF1* methylation can also be detected in less than 10% of normal tissues, possibly because of age-related methylation changes (Widschwendter and Jones 2002).

3.1.5
PTGS2

Prostaglandin-endoperoxide synthase 2 (PTGS2), also known as cyclooxygenase-2 or COX-2, is the key enzyme in prostaglandin biosynthesis involved in inflammation and mitogenesis. Prostaglandin levels are increased at sites of inflammation and in cancers. These molecules stimulate cell proliferation (Nolan et al. 1988), inhibit the proliferation of the immune cells (Huang et al. 1996), and alter the antigen processing functions of the dendritic cells (Sharma et al. 2003). Prostaglandin can also promote mitogenesis of mammary epithelial cells (Bandyopadhyay et al. 1987) and can increase the local estrogen synthesis in the breast by increasing the aromatase expression, which is responsible for estrogen synthesis in the adipose tissue adjacent to the breast tumor (Zhao et al. 1997).

Growth factors, proinflammatory cytokines, and tumor promoters (Hla and Neilson 1992) regulate the expression of *PTGS2*, and *PTGS2* expression is altered in many human malignancies. Many tumors including breast cancers overexpress PTGS2 (Soslow et al. 2000). However, low PTGS2 expression has been reported in colon and gastric cancers. The loss of PTGS2 expression has been shown to be associated with aberrant methylation of its gene promoter in CIMP$^+$ proximal *PTGS2* methylation!colon cancerscolon cancers (Toyota et al. 2000b) that occur more frequently in women. Despite the fact that PTGS2 is overexpressed in breast tumors, we found that the frequency of methylation of the *PTGS2* gene promoter is almost 100% in breast tumors (Widschwendter et al. 2004). *PTGS2* gene promoter methylation appears to be tumor-specific, since it was not detected in DNA extracted from non-neoplastic breasts. Some possible explanations for this discrepancy are that the region of the *PTGS2* gene investigated by our MethyLight assay may not be important for the transcriptional regulation of this gene or that the methylation may be restricted to one allele of the gene while the gene can still be expressed from the remaining unmethylated allele. Alternatively, methylation may occur in other cell types than the mammary epithelium, such as inflammatory cells that can be common in some tumors. The relationship between *PTGS2* expression and the methylation of its gene promoter in breast cancer remains to be investigated in future studies. Nevertheless, *PTGS2* methylationbreast cancers can be used as a powerful marker for diagnosis and prognosis of breast cancers.

3.1.6 *CDH1*

The *CDH1* gene encodes a protein known as E-cadherin, which belongs to the cadherin superfamily comprising calcium-dependent glycoproteins involved in cell–cell adhesion. Loss of CDH1 function occurs in many human cancers, and is believed to contribute to tumor progression by facilitating tumor invasion and metastasis. Mutation-related loss of function is common in breast, ovarian, and colorectal cancers. In many cancers, however, *CDH1* gene expression has been shown to be suppressed by DNA methylation (Yoshiura et al. 1995).

In breast cancers, loss of CDH1 expression is more common in poorly differentiated and advanced-staged tumors (Widschwendter and Jones 2002). Approximately 41% of the breast cancers have tumor-specific methylation of the *CDH1* gene promoter (Bae et al. 2004; Esteller et al. 2001a, b; Parrella et al. 2004; Widschwendter et al. 2004). No detectable methylation of this gene was found in benign breast tumors, attesting to the role of this protein in tumor progression (Parrella et al. 2004). An association between the methylation of *CDH1* and *ESR1* was also established, suggesting that it may represent a molecular feature of a subset of breast tumors (Parrella et al. 2004).

3.1.7
CDKN2A and *ARF*

The short arm of human chromosome 9, which is lost in many human malignancies, harbors the *CDKN2A* tumor-suppressor gene encoding two structurally and functionally unrelated tumor suppressors, the $p16^{INK4A}$ and the ARF protein, also referred to as $p14^{ARF}$. The two alternative transcripts of the *CDKN2A* locus are initiated from separate promoters, and have distinct first exons, which splice into common downstream exons that are translated in alternative reading frames. Both of these proteins are involved in cell cycle regulation, and exert their effects on two of the most important tumor suppressors, RB1 and TP53. $p16^{INK4A}$ prevents the inactivation of RB1 by blocking the ability of cyclin D-dependent kinases to phosphorylate RB1. The ARF protein activates TP53 by binding to the MDM 2 protein, thus preventing the MDM 2-mediated degradation of TP53. Despite their shared genetic material and close proximity in the genome, these two isoforms are independently regulated, and may be epigenetically silenced in various types of cancers. In a recent study, the methylation of the *ARF* gene promoter was found to be higher than that of the $p16^{INK4A}$ promoter [the gene is designated in this review by its official Human Genome Organisation (HUGO) symbol *CDKN2A*], although both methylation events were associated with poor prognosis (Dominguez et al. 2003; Silva et al. 2003). In breast cancer, the overall frequency of methylation of the *CDKN2A* and *ARF* promoters is relatively low, around 17% (Esteller et al. 2001a, b; Krassenstein et al. 2004; Parrella et al. 2004; Silva et al. 2003) and 20% (Esteller et al. 2001a; Krassenstein et al. 2004; Silva et al. 2001, 2003), respectively. However, the methylation changes in these genes may still be important for the tumorigenic process or can serve as molecular markers for a subset of breast cancers.

3.1.8
APC

Adenomatosis polyposis coli (*APC*) is a tumor suppressor gene whose inactivation by mutations, deletions, or promoter methylation causes a hereditary form of colon cancer called familial adenomatous polyposis (FAP). The APC protein is involved in Wnt-signaling, regulation of chromosomal and cytoskeletal integrity, cell–cell adhesion, and cell migration (Narayan and Roy 2003). Its inactivation results in β-catenin stabilization, *c-myc* gene activation, impairment of the RB1 pathway and consequent cell cycle deregulation and chromosomal instability (Narayan and Roy 2003). APC is believed to also play a role in breast carcinogenesis. Mutations of the *APC* gene and loss of heterozygosity at this gene locus are frequent events in sporadic breast cancers, occurring in 18% and 25% of tumors respectively (Furuuchi et al. 2000; Medeiros et al. 1994). Methylation of the *APC* gene promoter is detected in approximately 28% of all sporadic breast cancers (Esteller et al. 2001a; Jin et al. 2001; Parrella et al. 2004); however, no association with any known breast cancer subtypes has been described.

3.1.9
MLH1

This gene encodes the MLH1 protein, also known as the mismatch repair mutL homolog 1, involved in DNA mismatch repair. Defects or loss of expression of this protein causes a form of genetic instability known as microsatellite instability (MSI), which is characterized by genome-wide short nucleotide insertions and deletions. Failure to correct errors occurring during DNA replication by MLH1 or other members of the DNA mismatch repair system (MMR) may result in accumulation of mutations in cancer genes that control cell proliferation and survival (Marra and Boland 1996). The *MLH1* gene is frequently inactivated by mutations in the hereditary colon cancer syndrome HNPCC, whereas DNA methylation-mediated gene silencing is more common in sporadic forms of various types of cancers. DNA mismatch repair defects involving MLH1 also occur in sporadic breast cancers (Moreno-Bueno et al. 2003), although one report suggested that these might be rare events (Adem et al. 2003). MSI has been reported in sporadic breast cancers (Chagpar et al. 2004). Breast cancers have also been documented among family members of HNPCC patients, suggesting that this cancer might also be part of the HNPCC syndrome (de Leeuw et al. 2003; Oliveira Ferreira et al. 2004). *MLH1* is inactivated by DNA methylation in about 29% of breast tumors (Esteller et al. 2001a; Murata et al. 2002; Widschwendter et al. 2004). The phenotypic characteristics and other molecular features of the breast tumors subtype with MLH1 methylation still need to be investigated.

3.1.10
GSTP1

Another gene, whose expression is reduced in many types of cancer due to promoter methylation, is the *GSTP1* gene, encoding the glutathione *S*-transferase π protein. This enzyme catalyzes the conjugation of glutathione with electrophilic and hydrophobic compounds, including carcinogens, in order to facilitate normal cellular detoxification reactions (Coles and Ketterer 1990). By preventing the accumulation of genotoxic compounds, GSTP1 acts to protect cells from DNA damage and tumor initiation. Aberrant methylation leading to reduced GSTP1 expression levels has been documented only in a limited number of cancers, including breast cancer. Approximately 29% of all breast cancers have associated *GSTP1* promoter hypermethylation (Bae et al. 2004; Esteller et al. 1998a, 2001a, b; Krassenstein et al. 2004; Parrella et al. 2004). It has been proposed that methylation-induced loss of GSTP1 function can lead to the formation of estrogen-related DNA adducts that, if not repaired, can initiate tumorigenesis in the breast (Esteller 2000). A recent study has also shown that estrogen and its inhibitors can regulate *GSTP1* expression through ESR2 activation, which in turn increases the level of cellular protection against oxidative stress (Montano et al. 2004). The association between the methylation of *GSTP1* and the hormonal status of the breast tumors is even more intriguing. *GSTP1* is expressed only in ER^- breast cancer lines, whereas in the ER^+ cell lines *GSTP1* expression is lost due to promoter hypermethylation (Jhaveri et al. 1998; Moscow et al. 1988). These data suggest the existence of a connection between estrogen and the activity of this enzyme. The relationship between *GSTP1* gene regulation by the estrogen signaling pathway and the methylation of the *GSTP1* gene promoter requires further clarification.

3.1.11
MGMT

The protein encoded by this gene is a DNA repair protein capable of removing mutagenic methyl or alkyl adducts from guanines and thymines. This was demonstrated in vitro as well as in mouse models (Esteller 2000). However, since the MGMT protein was shown not to be essential for DNA replication or cell survival, its role in normal cells is not yet fully understood. Like GSTP1, MGMT provides protection from genotoxic compounds. This protection may be lost when MGMT expression is silenced by DNA hypermethylation, and this event may play a decisive role in the process of malignant transformation. Abnormally generated O^6-methylguanine is recognized as an adenine by the DNA polymerase machinery, thus deficiency in the MGMT enzyme may lead to guanine (G)-to-adenine (A) transitions. Accumulation of such mutations in key regulatory proteins could impact the process of tumor initiation and progression. Interestingly, this gene is rarely mutated in cancers, and only transcriptional silencing by DNA methylation has been shown to cause functional loss of *MGMT* expression. Like *MLH1*, methylation of *MGMT* is another example that epigenetic changes can lead to genetic alterations. *MGMT* gene promoter hypermethylation has been reported in several human cancers (Esteller et al. 1999). However, fewer than 10% of the breast cancers (Esteller et al. 2001a; Widschwendter et al. 2004) have methylation of the *MGMT* gene promoter, suggesting that breast carcinogenesis generally does not involve inactivation of MGMT.

3.2 Endometrial Cancer

Menopause is the most effective protective factor against endometrial cancer, suggesting that female sexual hormones also play an important role in the etiology of this type of cancer (Pike et al. 2004). The risk for endometrial cancer is increased primarily by exposure to estrogen unopposed by progesterone, suggesting that the two female hormones may have opposing effects, such that estrogen increases the risk, while progesterone has a protective effect for endometrial cancer (Cohen and Rahaman 1995). ERs and PRs are highly expressed in endometrial epithelium. The hormone receptor status of endometrial tumors constitutes an important indicator of response to therapy, and survival. Loss of ER expression is common and is associated with poor prognosis in endometrial cancer (Navari et al. 2000), while loss of PR is associated with more advanced disease. Since *PGR* is a downstream target of ER-mediated estrogen activity, these findings support the hypothesis that excessive estrogen stimulation unopposed by PR-mediated progesterone activity is critical for the development and progression of endometrial cancer.

Silencing of these receptors by aberrant DNA methylation has been also documented in endometrial cancer. Interestingly, hypermethylation of an alternative promoter of the *ESR1* gene, the C promoter, has been documented in over 90% of endometrial cancer (Sasaki et al. 2001b), whereas almost no methylation of *ESR1* promoter A (Navari et al. 2000; Sasaki et al. 2001b) or *ESR2* (Sasaki et al. 2001b) genes have been reported. The absence of age-related methylation of the *ESR1* gene promoter A in either normal endometrial epithelium or endometrial cancers, despite the high rates of DNA proliferation that is characteristic for both of these tissues, is also intriguing. The differential epigenetic inactivation of specific *ESR1* gene promoters in different types of cancers suggests that the *ESR1* silencing through these mechanisms may be related to the promoter usage of these receptor isoforms in these tissues, and that the silencing by methylation of these receptors may be important for tumor development. Both PR isoforms are expressed in normal endometrial epithelium. However, only *PGR* promoter B is frequently methylated (75%) and silenced in endometrial cancers but not in normal endometrial epithelium (Sasaki et al. 2001a). The significance of the selective methylation of *PGR-B* for endometrial cancer development might not be fully understood until the specific function of the PRs in the normal endometrium is elucidated.

Endometrial cancers are also the most common type of extracolonic tumors occurring in women with hereditary colon cancer syndrome HNPCC (Watson and Lynch 1993). Defects in mismatch repair genes, including *MLH1*, are found in the majority of tumors with MSI (Liu et al. 1994). Approximately 20% of all endometrial cancers exhibit MSI (Risinger et al. 1993). However, germline and somatic mutations in DNA mismatch repair genes are relatively rare in endometrial cancer (Katabuchi et al. 1995). *MLH1* methylation is a more common finding in endometrial cancers (41%), and the majority of these methylation events have associated MSI (Esteller et al. 1998b, 2001a; Whitcomb et al. 2003). Unlike colorectal cancer, where sporadic MSI^+ cancers have *MLH1* methylation in association with the methylation of a number of other genes, in the context of CIMP, no correlation between the methylation of *MLH1* and the methylation of other CIMP genes, including *CDKN2A* (Esteller et al. 1998b) has been reported in endometrial cancers. Relatively low frequencies of methylation (16%) have been reported for both CDKN2A and ARF gene promoters in endometrial cancers (Esteller et al. 2001a; Whitcomb et al. 2003; Wong et al. 1999).

These results prompted the suggestion that *MLH1* methylation might not occur in the setting of widespread abnormal methylation in these tumors (Matias-Guiu et al. 2001), as is the case in colorectal cancer. However, if CIMP is caused by defects in transacting factors that might be required for the protection of CpG islands from abnormal methylation, different defects might lead to the methylation of distinct groups of structurally similar CpG islands (Laird 2003). Such transacting factors could be tissue-specific. Also, their activity could depend on the tissue-specific usage of the promoters they are protecting. For example, DNA methylation of *MLH1* in endometrial cancer is associated with methylation of a different set of genes from those characterized for colorectal CIMP (Whitcomb et al. 2003), including *thrombospondin 2* (*THBS2*), which is methylated in 62% of all endometrial cancers. In $CIMP^+$ colon cancers, *MLH1* methylation is associated with the methylation of *THBS1*, a gene encoding for an angiogenesis modulator. It is likely that THBS2 rather than THBS1 may be the main modulator of angiogenesis in endometrium, and its inactivation is required for tumor development in endometrium. If methylation of either of these two promoters occurs, only the methylation events at *THBS2* should be selected for in endometrial tumors. DNA methylation of the *APC* gene promoter, another gene not associated with CIMP in colorectal cancer, was reported to occur very frequently in endometrial cancer (37%) in the context of MSI (Moreno-Bueno et al. 2002; Pijnenborg et al. 2004; Zysman et al. 2002). Methylation of *CDH1* occurs in approximately 26% of all endometrial cancers (Moreno-Bueno et al. 2002; Nishimura et al. 2003; Pijnenborg et al. 2004). No abnormal methylation of the *MGMT* and *GSTP1* gene promoters has been detected in endometrial cancers (Esteller et al. 2001a). The methylation status of *BRCA1* has not been reported in endometrial cancers, perhaps in part because female carriers of *BRCA1* mutations do not develop this type of cancer.

3.3 Ovarian Cancer

As in the case of breast and endometrium, menopause is also the most effective protective factor against ovarian cancer (Pike et al. 2004). A high number of pregnancies can also reduce the risk of ovarian cancer. Studies performed on benign ovarian tumors, which have the same cell of origin as ovarian carcinomas, have shown that estrogen may be an etiological factor for ovarian cancer through its growth stimulatory and anti-apoptotic effects, whereas progesterone appears to offer protection against ovarian cancer (Ho 2003; Zhou et al. 2002). All ERs and PRs are expressed at various levels in these tumors. The presence of both ERs and PRs in ovarian cancers is associated with a good prognosis (Leake and Owens 1990). About 80% of all ovarian tumors express *ESR1*, and about 50% express *ESR2*, and *PGR*. Loss of expression of *ESR2* (Bardin et al. 2004) and *PGR* promoter A (Akahira et al. 2002) has been proposed to be an important event leading to the development of ovarian cancer. The methylation status of the HR genes has not been examined in great detail. Promoter hypermethylation of *ESR1* and *ESR2* genes occurred in approximately 29% (O'Doherty et al. 2002) and 22% (Ehrlich et al. 2006) of ovarian cancers, respectively, while no hypermethylation of *PGR-B* was found in these cancers (Ehrlich et al. 2006).

Ovarian cancers can also occur in the context of hereditary cancer syndromes such as HNPCC (Watson and Lynch 1993) and in hereditary breast-ovarian cancer in women with *BRCA1* and *BRCA2* germ-line mutations (Claus et al. 1996). Silencing of *MLH1* and *BRCA1*, respectively, by aberrant DNA methylation occurs in these syndromes as well as in a small number of sporadic ovarian cancers. Hypermethylation and silencing of the *MLH1* gene occurs in 10% of sporadic ovarian cancers (Strathdee et al. 1999, 2001), while approximately 17% of all ovarian cancers have methylation of the *BRCA1* gene (Baldwin et al. 2000; Catteau et al. 1999; Esteller et al. 2000a, 2001a; Rathi et al. 2002; Ibanez de Caceres et al. 2004; Strathdee et al. 2001; Wang et al. 2004). *MLH1* hypermethylation is associated with MSI and resistance to chemotherapy (Strathdee et al. 1999). Concordant methylation of multiple genes has been also reported in ovarian cancers. However, it has been proposed that more than one CIMP phenomena may be characteristic for subtypes of this disease (Strathdee et al. 2001). No association with the methylation of *MLH* or other CIMP genes could be demonstrated for *BRCA1* (Strathdee et al. 2001), suggesting that methylation of this gene is representative for a distinct subtype of breast cancers that may not be part of CIMP. Due to the known regulation by estrogen of the *BRCA1* gene, future studies will be needed to clarify the relationship between silencing by methylation of this gene and the hormonal receptor status of the ovarian tumors.

The frequency of *CDKN2A* methylation reported in the literature varies substantially, in part due to the analysis of different regions of the promoter in various studies. Higher frequencies of methylation are common for the CpG island located in the second exon of the *CDKN2A* gene; however, this methylation event is not associated with gene silencing. The CpG island associated with the promoter and the first exon of *CDKN2A* was more often reported to be unmethylated in most ovarian cancers. The overall frequency of methylation for this gene is less than 10% (Esteller et al. 2001a; Ibanez de Caceres et al. 2004; McCluskey et al. 1999; Niederacher et al. 1999; Strathdee et al. 2001; Wong et al. 1999). Low frequencies of methylation were also reported for the *ARF* and APC genes (Esteller et al. 2001a; Ibanez de Caceres et al. 2004; Rathi et al. 2002). No hypermethylation was found at *MGMT*, *GSTP1*, and *PTGS2* gene promoters in ovarian cancers (Esteller et al. 2001a; Rathi et al. 2002). High frequencies of methylation have been shown, however, for *RASSF1* (31%) and for *CDH1* (26%), suggesting that these two genes may play a role in ovarian cancer development (Agathanggelou et al. 2001; Rathi et al. 2002; Ibanez de Caceres et al. 2004; Yoon et al. 2001).

3.4
Proximal Colon Cancer

A large number of epidemiological studies suggest a hormonal basis for the pathogenesis of colorectal cancers. Evidence for the potential role of sex hormones in the etiology of colorectal cancer comes from several observations. Males and females differ in their incidence and mortality rates of colorectal cancer when localization of the tumor and the age of the patient at the time of diagnosis are taken in consideration. Age-adjusted colon cancer incidence rates are slightly higher for men than for women (Wingo et al. 1998). Also, the risk of colon cancer in family members afflicted by the hereditary colon cancer syndrome HNPCC is lower in women than in men (Froggatt et al. 1999). However, colon tumors that develop in the proximal part of the large intestine occur more frequently in older women (Butcher et al. 1985), whereas tumors of the distal colon are more common in men (Alley and McNee 1986). Parity and hormone replacement therapy (HRT) have also been shown to be inversely associated with colon cancer (McMichael and Potter 1980). HRT reduces the risk of colorectal cancer (Davis et al. 1989; Furner et al. 1989) and improves the survival from this disease (Mandelson et al. 2003; Slattery et al. 1999). These data suggest that estrogen may have a protective effect for colorectal cancer development in women, and that when the protective effect is lost during menopause, the proximal colonic epithelium is at greater risk than the distal one for developing colon cancer.

The mechanism by which estrogens may affect the tumorigenic process is not clear. One proposed explanation that has not been fully explored is that estrogens can interfere with bile acid metabolism. Secondary bile acids are carcinogenic in rats (Reddy et al. 1976), and their concentration in feces is increased in patients with colorectal cancer as compared with the non-cancer patients (Reddy et al. 1977). Since the secondary bile acids are absorbed primarily in the proximal colon, it has been postulated that they may play a role in the initiation and promotion of tumorigenesis in the proximal colonic epithelium (McMichael and Potter 1980). Removal of the gallbladder, resulting in an increase in the intestinal load of secondary bile acids (Malagelada et al. 1973; Pomare and Heaton 1973), has also been shown to be associated with a slight increase in the risk of proximal colon cancer (Todoroki et al. 1999; Turunen and Kivilaakso 1981; Vernick et al. 1980). Bile acid synthesis is also reduced following HRT (Everson et al. 1991). However, HRT has been shown to improve the survival only from distal colon cancer and not from proximal colon cancer, which is more common in women (Mandelson et al. 2003).

An alternative mechanism proposed to account for gender-specific differences in colon cancer, including the reduced incidence and improved prognosis from colon cancer following HRT, is that estrogen might manifest its influences on colonic epithelium through changes in DNA methylation. In recent years, the development of colon cancer has been linked to several genetic and epigenetic changes. Proximal and distal colon cancers have been shown to evolve by different genetic pathways, which could be influenced by sex-related factors (Breivik et al. 1997). Approximately 15% of sporadic colon cancers exhibit MSI. Colon tumors with MSI are more common in women than in men and they are more likely to occur in the proximal colon (Breivik et al. 1997). The majority of the sporadic colon cancers with MSI are associated with hypermethylation of the *MLH1* gene (Kane et al. 1997). Moreover, the high incidence of MSI tumors in older women and in the proximal colon has been linked to the methylation of the *MLH1* gene (Malkhosyan et al. 2000), suggesting a direct relationship between female gender and a specific DNA methylation alteration. Interestingly, various levels of *MLH1* methylation were detected in all of the other women's cancers. In addition, hypermethylation of the *MLH1* gene in MSI colon tumors has also been shown to co-exist with the hypermethylation of several other CpG islands in the context of CIMP (Toyota et al.

1999). It has been postulated that CIMP reflects the existence of an underlying molecular mechanism responsible for the protection of CpG islands against abnormal methylation (Laird 2003). Alternatively, it has been proposed that CIMP may be caused by continuous exposure to epimutagens such as viral infections, diet, carcinogens, or chronic inflammation (Kondo and Issa 2004).

The association between *MLH1* methylation in the context of CIMP and the female gender in proximal colon cancers is not yet fully understood. It has been proposed that genetic defects in chromosome X-linked genes may be responsible for this gender-specific susceptibility to DNA methylation (Malkhosyan et al. 2000). Another explanation for this phenomenon is related to the fact that changes in DNA methylation could result following gene inactivation. It is possible that during menopause the absence of female hormones can lead to silencing of many genes that are downstream targets in the estrogen hormone-signaling pathway. DNA methylation could subsequently accumulate in the promoter of some of these genes, especially those that have weak protection against spreading of methylation from sites in the genome designated as "methylation centers" (Kondo and Issa 2004). These centers are normally heavily methylated based on sequence characteristics. This hypothesis is supported by the findings of a recent paper that showed that disruption of the estrogen-signaling pathway results in the silencing of multiple downstream target genes, a process accompanied by extensive chromatin remodeling, including promoter hypermethylation (Leu et al. 2004). Reactivation of these genes required demethylation and could not be achieved by simply reestablishing the estrogen-signaling pathway. This may explain why HRT does not improve the prognosis of proximal cancer (Mandelson et al. 2003). In order to reactivate genes in the estrogen-signaling pathway that have been already silenced by DNA methylation, a demethylation step followed by HRT may be required. In view of that, a combination treatment of DNA methylation inhibitors and HRT may prove to be of clinical benefit for proximal colon cancer in women. A CIMP-like phenomenon has been described in ovarian cancers but not in breast or endometrial cancers.

Based on epigenetic (CIMP) and genetic (MSI) criteria, colorectal cancers can be divided into at least four different molecular entities: MSI^+CIMP^+, MSI^-CIMP^+, MSI^+CIMP^-, and MSI^-CIMP^- tumors (Kondo and Issa 2004). These four types of colorectal cancers appear to evolve through distinct genetic pathways in which defects in two proto-oncogenes, *KRAS* and *BRAF*, and a tumor suppressor gene, *TP53*, play an important role (Rajagopalan et al. 2002; Toyota et al. 2000a). Recently, *BRAF* and *KRAS* mutations have been shown to be mutually exclusive in colon cancers (Nagasaka et al. 2004). *BRAF* mutations appear to associate preferentially with $CIMP^+$ tumors, while *KRAS* mutations are more common in tumors with less extensive promoter hypermethylation

(Nagasaka et al. 2004). Interestingly, *BRAF* mutations are also very frequent in two other female-specific cancers, ovarian and thyroid cancers.

The methylation profile of $CIMP^+MSI^+$ colon cancers has been defined using a limited panel of DNA methylation markers. The concordant methylation of five or six CpG islands associated with three known genes, *MLH1*, *CDKN2A*, and *THBS1*, and three uncharacterized DNA regions, *MINT1*, *MINT2*, and *MINT31*, defines CIMP in colon cancers. For instance, hypermethylation of *MLH1* and *CDKN2A* occurs in approximately 20% (Eads et al. 1999; Esteller et al. 2001b; Kane et al. 1997; Lind et al. 2004; Miyakura et al. 2001; Paz et al. 2003; Toyota et al. 1999) and 30% (Burri et al. 2001; Eads et al. 1999; Esteller et al. 2001a, b; Lind et al. 2004; Norrie et al. 2003; Schneider-Stock et al. 2003; Toyota et al. 1999) of all colon cancers, respectively, usually in association with the methylation of the other CIMP genes. However, in MSI^+ colon cancers, the frequency of these methylation events increases to 69% (Esteller et al. 2001a; Laghi et al. 2003; Lee et al. 2004; Miyakura et al. 2001; Yamamoto et al. 2002) and 60% (Laghi et al. 2003; Yamamoto et al. 2002), respectively. Statistically significant associations between the methylation of these loci and the female gender and the proximal location of the tumor in the colon have been also described (Lee et al. 2004; Norrie et al. 2003; Schneider-Stock et al. 2003; Toyota et al. 2000a, b; Welcsh et al. 2001; Wiencke et al. 1999). Methylation of both *ARF* (30%) and *PTGS2* (13%) gene promoters have been reported in colon cancers, also in the context of CIMP and MSI (Burri et al. 2001; Dominguez et al. 2003; Esteller et al. 2001a, b; Lind et al. 2004; Shen et al. 2003; Toyota et al. 2000a, b). In a recent study, the rate of methylation of the *ARF* promoter was higher than that of the *CDKN2A* promoter, but both methylation events were associated with poor prognosis in colon cancers (Dominguez et al. 2003). The relationship between the specific methylation of these loci and the estrogen-signaling pathway, however, still needs to be elucidated.

The hormone receptor status of the proximal colon tumors may be relevant to our understanding of the tumorigenic process in the colon, as is the case for other hormone-driven cancers. To date, very few studies have examined the patterns of expression of the male or female steroid receptors and their isoforms along the gastrointestinal tract. Interestingly, ERβ, which is also the main estrogen receptor expressed in the male urogenital tract, is the predominant estrogen receptor expressed in the colonic epithelium, and its levels of expression have been shown to be reduced in female colon cancers (Widschwendter et al. 2004). It has been proposed that activation by estrogens of this receptor may mediate antimitogenic signals as was observed in breast cancer cells (Cross et al. 2004).

About 80% of colon cancers show methylation of the *ESR1* gene promoter (Eads et al. 1999; Issa et al. 1994). This is thought to be age-related since methylation of this gene also occurs in normal colonic mucosa of older cancer-free people. *ESR2* methylation can be detected in approximately 20% of colon cancers, but not in the normal colonic mucosa, suggesting that methylation of this locus is a cancer-specific event (D.J. Weisenberger, K. Siegmund, M. Campan, J. Young, T.I. Long, M.A. Faasse, G.H. Kang, M. Widschwendter, D. Weener, D. Buchanan, H. Koh, L. Simms, M. Barker, B. Leggett, J. Levine, A.J. French, S.N. Thibodeau, J. Jass, R. Haile, P.W. Laird, submitted). The pattern of expression of individual PR transcripts in the gastrointestinal (GI) tract or in cancers of the GI tract has not yet been clearly defined. Preliminary data from our laboratory suggest that both *PGR* promoters are frequently methylated in colon cancers. The significance of these methylation changes is not known. In light of a potential role of sex hormones in the pathogenesis of colon cancer, future studies may be needed to clarify whether there are gender, age, or location-specific differences in the expression of any of these sex steroid receptors and their isoforms in the colonic mucosa, and how the patterns of expression change during tumorigenesis.

Methylation of the *BRCA1* or *GSTP1* genes has not been shown in colon cancers. In contrast, methylation of *MGMT*, another gene with DNA repair functions, has been found in 38% of all colon cancers, and this methylation event was associated with mutations in the *KRAS* gene (Esteller et al. 2000b). Unlike *BRAF* mutations, the *KRAS* mutations are very common in the distal colon cancers that occur more often in male patients. The possible association between *MGMT* methylation and male gender is interesting in light of the fact that *MGMT* methylation is also a common finding in gliomas, esophageal adenocarcinomas, and head and neck tumors, which all have a high preponderance in males.

The overall frequency of methylation for *RASSF1*, *APC*, and *CDH1* genes colon cancers is 19%, 21%, and 49% respectively (Eads et al. 1999; Esteller et al. 2001a, b; Garinis et al. 2002; Kondo and Issa 2004; Lind et al. 2004; Paz et al. 2003; Yoon et al. 2001). No association between these methylation events and CIMP, gender, tumor location, or MSI status has been established.

4 Conclusions

A single common methylation signature for all of these four women's cancers could not be demonstrated using this panel of DNA methylation markers, suggesting that multiple tumorigenic pathways controlled by female hormones may be involved in the development of these cancers. This is not totally unexpected, knowing that estrogen protects against proximal colon cancers but is thought to be tumor-promoting in breast, ovarian, and endometrial cancers. However, methylation of five of these genes, *MLH1*, *CDKN2A*, *ARF*, *CDH1*, and *APC*, was detected in a subset of all of these cancers, suggesting that subtypes of these cancers may share common tumorigenic pathways. Since *MLH1* and *CDKN2A* methylation is associated with MSI in the context of CIMP in proximal colon cancers, it would be interesting to determine if these phenomena are also characteristic for the subtypes of breast, endometrial, and ovarian cancers that have methylation of these genes. This possibility is not totally unfounded since all of these types of cancers have also been described in HNPCC patients. MSI is present in a subset of sporadic breast, endometrial, and ovarian cancers, and *MLH1* methylation with associated MSI has been demonstrated in sporadic endometrial and ovarian cancers. A recent study showed high frequencies of DNA methylation for the *ARF*, *CDH1*, and *APC* genes in proximal colon cancers (Lee et al. 2004). However, it is not clear if methylation of these genes is associated with CIMP. More studies will be required to clarify the association between the methylation of these genes, CIMP, and the hormone regulation in these female-specific cancers.

Distinct tissue-specific methylation signatures for subtypes of these cancers also emerged from this review of women's cancers. *GSTP1* methylation was found only in breast tumors, while *BRCA1* methylation was detected only in subsets of breast and ovarian cancers. Both of these genes are regulated by the estrogen pathway and GSTP1 can play a role in estrogen metabolism.

Hormone-driven cancers are also characterized by distinct patterns of methylation of the HR genes, which may reflect the tissue specificity of these receptors. Loss of *ESR1* expression due to epigenetic silencing was detected in all of these cancers. However, this occurred via selective methylation of different *ESR1* promoters in different tumors. Differential methylation of the *ESR2* gene promoter and both *PGR* promoters was also observed in these cancers. The methylation events affecting the expression of these hormone receptors could be of critical importance for tumor development and progression of hormone-driven cancers, and also might serve as potential tumor markers for diagnosis and prognosis. Unfortunately, insufficient data were available in the literature about the methylation status of these receptors to allow for a more detailed comparison between the four types of cancers. Ultimately, complete DNA methylation-based molecular signatures of hormone-driven cancers, reflecting alterations in hormone regulated pathways, might require inclusion of information about the DNA methylation status of all HRs, along with the information about DNA methylation of other relevant genes. Defining tumor-specific DNA methylation profiles of women's cancers—and understanding the role that female hormones play in shaping these profiles—is still a work in progress.

References

Adem C, Soderberg CL, Cunningham JM, Reynolds C, Sebo TJ, Thibodeau SN, Hartmann LC, Jenkins RB (2003) Microsatellite instability in hereditary and sporadic breast cancers. Int J Cancer 107:580–582

Agathanggelou A, Honorio S, Macartney DP, Martinez A, Dallol A, Rader J, Fullwood P, Chauhan A, Walker R, Shaw JA, Hosoe S, Lerman MI, Minna JD, Maher ER, Latif F (2001) Methylation associated inactivation of RASSF1A from region 3p21.3 in lung, breast and ovarian tumours. Oncogene 20:1509–1518

Akahira J, Suzuki T, Ito K, Kaneko C, Darnel AD, Moriya T, Okamura K, Yaegashi N, Sasano H (2002) Differential expression of progesterone receptor isoforms A and B in the normal ovary, and in benign, borderline, and malignant ovarian tumors. Jpn J Cancer Res 93:807–815

Alley PG, McNee RK (1986) Age and sex differences in right colon cancer. Dis Colon Rectum 29:227–229

Althuis MD, Fergenbaum JH, Garcia-Closas M, Brinton LA, Madigan MP, Sherman ME (2004) Etiology of hormone receptor-defined breast cancer: a systematic review of the literature. Cancer Epidemiol Biomarkers Prev 13:1558–1568

Bae YK, Brown A, Garrett E, Bornman D, Fackler MJ, Sukumar S, Herman JG, Gabrielson E (2004) Hypermethylation in histologically distinct classes of breast cancer. Clin Cancer Res 10:5998–6005

Baldwin RL, Nemeth E, Tran H, Shvartsman H, Cass I, Narod S, Karlan BY (2000) BRCA1 promoter region hypermethylation in ovarian carcinoma: a population-based study. Cancer Res 60:5329–5333

Bandyopadhyay GK, Imagawa W, Wallace D, Nandi S (1987) Linoleate metabolites enhance the in vitro proliferative response of mouse mammary epithelial cells to epidermal growth factor. J Biol Chem 262:2750–2756

Bardin A, Hoffmann P, Boulle N, Katsaros D, Vignon F, Pujol P, Lazennec G (2004) Involvement of estrogen receptor beta in ovarian carcinogenesis. Cancer Res 64:5861–5869

Bird A (2002) DNA methylation patterns and epigenetic memory. Genes Dev 16:6–21

Breivik J, Lothe RA, Meling GI, Rognum TO, Borresen-Dale AL, Gaudernack G (1997) Different genetic pathways to proximal and distal colorectal cancer influenced by sex-related factors. Int J Cancer 74:664–669

Burri N, Shaw P, Bouzourene H, Sordat I, Sordat B, Gillet M, Schorderet D, Bosman FT, Chaubert P (2001) Methylation silencing and mutations of the p14ARF and p16INK4a genes in colon cancer. Lab Invest 81:217–229

Butcher D, Hassanein K, Dudgeon M, Rhodes J, Holmes FF (1985) Female gender is a major determinant of changing subsite distribution of colorectal cancer with age. Cancer 56:714–716

Catteau A, Harris WH, Xu CF, Solomon E (1999) Methylation of the BRCA1 promoter region in sporadic breast and ovarian cancer: correlation with disease characteristics. Oncogene 18:1957–1965

Chagpar A, Magliocco A, Kerviche A, Tan L, Walley B, DeCoteau JF (2004) The replication error phenotype is associated with the development of distant metastases in hormonally treated patients with breast carcinoma. Cancer 100:913–919

Clarke CL, Sutherland RL (1990) Progestin regulation of cellular proliferation. Endocr Rev 11:266–301

Claus EB, Schildkraut JM, Thompson WD, Risch NJ (1996) The genetic attributable risk of breast and ovarian cancer. Cancer 77:2318–2324

Cohen CJ, Rahaman J (1995) Endometrial cancer. Management of high risk and recurrence including the tamoxifen controversy. Cancer 76:2044–2052

Coles B, Ketterer B (1990) The role of glutathione and glutathione transferases in chemical carcinogenesis. Crit Rev Biochem Mol Biol 25:47–70

Conneely OM, Mulac-Jericevic B, Lydon JP (2003) Progesterone-dependent regulation of female reproductive activity by two distinct progesterone receptor isoforms. Steroids 68:771–778

Costello JF, Fruhwald MC, Smiraglia DJ, Rush LJ, Robertson GP, Gao X, Wright FA, Feramisco JD, Peltomaki P, Lang JC, Schuller DE, Yu L, Bloomfield CD, Caligiuri MA, Yates A, Nishikawa R, Su Huang H, Petrelli NJ, Zhang X, O'Dorisio MS, Held WA, Cavenee WK, Plass C (2000) Aberrant CpG-island methylation has non-random and tumour-type-specific patterns. Nat Genet 24:132–138

Cowley SM, Parker MG (1999) A comparison of transcriptional activation by ER alpha and ER beta. J Steroid Biochem Mol Biol 69:165–175

Cross HS, Kallay E, Lechner D, Gerdenitsch W, Adlercreutz H, Armbrecht HJ (2004) Phytoestrogens and vitamin D metabolism: a new concept for the prevention and therapy of colorectal, prostate, and mammary carcinomas. J Nutr 134:1207S–1212S

Dammann R, Yang G, Pfeifer GP (2001) Hypermethylation of the cpG island of Ras association domain family 1A (RASSF1A), a putative tumor suppressor gene from the 3p21.3 locus, occurs in a large percentage of human breast cancers. Cancer Res 61:3105–3109

Davis FG, Furner SE, Persky V, Koch M (1989) The influence of parity and exogenous female hormones on the risk of colorectal cancer. Int J Cancer 43:587–590

de Leeuw WJ, van Puijenbroek M, Tollenaar RA, Cornelisse CJ, Vasen HF, Morreau H (2003) Correspondence re: A. Muller et al., Exclusion of breast cancer as an integral tumor of hereditary nonpolyposis colorectal cancer. Cancer Res 62:1014–1019, 2002. Cancer Res 63:1148–1149

Dominguez G, Silva J, Garcia JM, Silva JM, Rodriguez R, Munoz C, Chacon I, Sanchez R, Carballido J, Colas A, Espana P, Bonilla F (2003) Prevalence of aberrant methylation of p14ARF over p16INK4a in some human primary tumors. Mutat Res 530:9–17

Eads CA, Danenberg KD, Kawakami K, Saltz LB, Danenberg PV, Laird PW (1999) CpG island hypermethylation in human colorectal tumors is not associated with DNA methyltransferase overexpression. Cancer Res 59:2302–2306

Eads CA, Lord RV, Wickramasinghe K, Long TI, Kurumboor SK, Bernstein L, Peters JH, DeMeester SR, DeMeester TR, Skinner KA, Laird PW (2001) Epigenetic patterns in the progression of esophageal adenocarcinoma. Cancer Res 61:3410–3418

Eden S, Cedar H (1994) Role of DNA methylation in the regulation of transcription. Curr Opin Genet Dev 4:255–259

Egger G, Liang G, Aparicio A, Jones PA (2004) Epigenetics in human disease and prospects for epigenetic therapy. Nature 429:457–463

Ehrlich M, Woods CB, Yu MC, Dubeau L, Yang F, Campan M, Weisenberger DJ, Long TI, Youn B, Fiala ES, Laird PW (2006) Quantitative analysis of associations between DNA hypermethylation, hypomethylation, and DNMT RNA levels in ovarian tumors. Oncogene (in press)

Esteller M (2000) Epigenetic lesions causing genetic lesions in human cancer: promoter hypermethylation of DNA repair genes. Eur J Cancer 36:2294–2300

Esteller M, Corn PG, Urena JM, Gabrielson E, Baylin SB, Herman JG (1998a) Inactivation of glutathione S-transferase P1 gene by promoter hypermethylation in human neoplasia. Cancer Res 58:4515–4518

Esteller M, Levine R, Baylin SB, Ellenson LH, Herman JG (1998b) MLH1 promoter hypermethylation is associated with the microsatellite instability phenotype in sporadic endometrial carcinomas. Oncogene 17:2413–2417

Esteller M, Hamilton SR, Burger PC, Baylin SB, Herman JG (1999) Inactivation of the DNA repair gene O6-methylguanine-DNA methyltransferase by promoter hypermethylation is a common event in primary human neoplasia. Cancer Res 59:793–797

Esteller M, Silva JM, Dominguez G, Bonilla F, Matias-Guiu X, Lerma E, Bussaglia E, Prat J, Harkes IC, Repasky EA, Gabrielson E, Schutte M, Baylin SB, Herman JG (2000a) Promoter hypermethylation and BRCA1 inactivation in sporadic breast and ovarian tumors. J Natl Cancer Inst 92:564–569

Esteller M, Toyota M, Sanchez-Cespedes M, Capella G, Peinado MA, Watkins DN, Issa JP, Sidransky D, Baylin SB, Herman JG (2000b) Inactivation of the DNA repair gene O6-methylguanine-DNA methyltransferase by promoter hypermethylation is associated with G to A mutations in K-ras in colorectal tumorigenesis. Cancer Res 60:2368–2371

Esteller M, Corn PG, Baylin SB, Herman JG (2001a) A gene hypermethylation profile of human cancer. Cancer Res 61:3225–3229

Esteller M, Fraga MF, Guo M, Garcia-Foncillas J, Hedenfalk I, Godwin AK, Trojan J, Vaurs-Barriere C, Bignon YJ, Ramus S, Benitez J, Caldes T, Akiyama Y, Yuasa Y, Launonen V, Canal MJ, Rodriguez R, Capella G, Peinado MA, Borg A, Aaltonen LA, Ponder BA, Baylin SB, Herman JG (2001b) DNA methylation patterns in hereditary human cancers mimic sporadic tumorigenesis. Hum Mol Genet 10:3001–3007

Everson GT, McKinley C, Kern F Jr (1991) Mechanisms of gallstone formation in women. Effects of exogenous estrogen (Premarin) and dietary cholesterol on hepatic lipid metabolism. J Clin Invest 87:237–246

Fan S, Wang J, Yuan R, Ma Y, Meng Q, Erdos MR, Pestell RG, Yuan F, Auborn KJ, Goldberg ID, Rosen EM (1999) BRCA1 inhibition of estrogen receptor signaling in transfected cells. Science 284:1354–1356

Fazzari MJ, Greally JM (2004) Epigenomics: beyond CpG islands. Nat Rev Genet 5:446–455

Feltus FA, Lee EK, Costello JF, Plass C, Vertino PM (2003) Predicting aberrant CpG island methylation. Proc Natl Acad Sci U S A 100:12253–12258

Flototto T, Djahansouzi S, Glaser M, Hanstein B, Niederacher D, Brumm C, Beckmann MW (2001) Hormones and hormone antagonists: mechanisms of action in carcinogenesis of endometrial and breast cancer. Horm Metab Res 33:451–457

Froggatt NJ, Green J, Brassett C, Evans DG, Bishop DT, Kolodner R, Maher ER (1999) A common MSH2 mutation in English and North American HNPCC families: origin, phenotypic expression, and sex specific differences in colorectal cancer. J Med Genet 36:97–102

Furner SE, Davis FG, Nelson RL, Haenszel W (1989) A case-control study of large bowel cancer and hormone exposure in women. Cancer Res 49:4936–4940

Furuuchi K, Tada M, Yamada H, Kataoka A, Furuuchi N, Hamada J, Takahashi M, Todo S, Moriuchi T (2000) Somatic mutations of the APC gene in primary breast cancers. Am J Pathol 156:1997–2005

Garinis GA, Menounos PG, Spanakis NE, Papadopoulos K, Karavitis G, Parassi I, Christeli E, Patrinos GP, Manolis EN, Peros G (2002) Hypermethylation-associated transcriptional silencing of E-cadherin in primary sporadic colorectal carcinomas. J Pathol 198:442–449

Gelber RD, Cole BF, Goldhirsch A, Rose C, Fisher B, Osborne CK, Boccardo F, Gray R, Gordon NH, Bengtsson NO, Sevelda P (1996) Adjuvant chemotherapy plus tamoxifen compared with tamoxifen alone for postmenopausal breast cancer: meta-analysis of quality-adjusted survival. Lancet 347:1066–1071

Hilakivi-Clarke L (2000) Estrogens, BRCA1, and breast cancer. Cancer Res 60:4993–5001

Hla T, Neilson K (1992) Human cyclooxygenase-2 cDNA. Proc Natl Acad Sci U S A 89:7384–7388

Ho SM (2003) Estrogen, progesterone and epithelial ovarian cancer. Reprod Biol Endocrinol 1:73

Huang M, Sharma S, Mao JT, Dubinett SM (1996) Non-small cell lung cancer-derived soluble mediators and prostaglandin E2 enhance peripheral blood lymphocyte IL-10 transcription and protein production. J Immunol 157:5512–5520

Huang TH, Perry MR, Laux DE (1999) Methylation profiling of CpG islands in human breast cancer cells. Hum Mol Genet 8:459–470

Iacopetta B (2003) Aberrant DNA methylation: have we entered the era of more than one type of colorectal cancer? Am J Pathol 162:1043–1045

Ibanez de Caceres I, Battagli C, Esteller M, Herman JG, Dulaimi E, Edelson MI, Bergman C, Ehya H, Eisenberg BL, Cairns P (2004) Tumor cell-specific BRCA1 and RASSF1A hypermethylation in serum, plasma, and peritoneal fluid from ovarian cancer patients. Cancer Res 64:6476–6481

Issa JP (2000) CpG-island methylation in aging and cancer. Curr Top Microbiol Immunol 249:101–118

Issa JP, Ottaviano YL, Celano P, Hamilton SR, Davidson NE, Baylin SB (1994) Methylation of the oestrogen receptor CpG island links ageing and neoplasia in human colon. Nat Genet 7:536–540

Jhaveri MS, Morrow CS (1998) Methylation-mediated regulation of the glutathione S-transferase P1 gene in human breast cancer cells. Gene 210:1–7

Jin Z, Tamura G, Tsuchiya T, Sakata K, Kashiwaba M, Osakabe M, Motoyama T (2001) Adenomatous polyposis coli (APC) gene promoter hypermethylation in primary breast cancers. Br J Cancer 85:69–73

Kane MF, Loda M, Gaida GM, Lipman J, Mishra R, Goldman H, Jessup JM, Kolodner R (1997) Methylation of the hMLH1 promoter correlates with lack of expression of hMLH1 in sporadic colon tumors and mismatch repair-defective human tumor cell lines. Cancer Res 57:808–811

Kastner P, Krust A, Turcotte B, Stropp U, Tora L, Gronemeyer H, Chambon P (1990) Two distinct estrogen-regulated promoters generate transcripts encoding the two functionally different human progesterone receptor forms A and B. EMBO J 9:1603–1614

Katabuchi H, van Rees B, Lambers AR, Ronnett BM, Blazes MS, Leach FS, Cho KR, Hedrick L (1995) Mutations in DNA mismatch repair genes are not responsible for microsatellite instability in most sporadic endometrial carcinomas. Cancer Res 55:5556–5560

Kelsey JL, Bernstein L (1996) Epidemiology and prevention of breast cancer. Annu Rev Public Health 17:47–67

Kinyamu HK, Archer TK (2004) Modifying chromatin to permit steroid hormone receptor-dependent transcription. Biochim Biophys Acta 1677:30–45

Kondo Y, Issa JP (2004) Epigenetic changes in colorectal cancer. Cancer Metastasis Rev 23:29–39

Krassenstein R, Sauter E, Dulaimi E, Battagli C, Ehya H, Klein-Szanto A, Cairns P (2004) Detection of breast cancer in nipple aspirate fluid by CpG island hypermethylation. Clin Cancer Res 10:28–32

Laghi L, Bianchi P, Malesci A (2003) Gender difference for promoter methylation pattern of hMLH1 and p16 in sporadic MSI colorectal cancer. Gastroenterology 124:1165–1166

Laird PW (2003) The power and the promise of DNA methylation markers. Nat Rev Cancer 3:253–266

Lapidus RG, Ferguson AT, Ottaviano YL, Parl FF, Smith HS, Weitzman SA, Baylin SB, Issa JP, Davidson NE (1996) Methylation of estrogen and progesterone receptor gene 5′ CpG islands correlates with lack of estrogen and progesterone receptor gene expression in breast tumors. Clin Cancer Res 2:805–810

Lazennec G, Bresson D, Lucas A, Chauveau C, Vignon F (2001) ER beta inhibits proliferation and invasion of breast cancer cells. Endocrinology 142:4120–4130

Leake R, Owens O (1990) The prognosis values of steroid receptors, growth factors, and growth factor receptors in ovarian cancer. In: Sharp F, Mason W, Leake R (eds) Ovarian cancer, biological and therapeutic challenges. Chapman and Hall Medical, London, pp 69–75

Lee S, Hwang KS, Lee HJ, Kim JS, Kang GH (2004) Aberrant CpG island hypermethylation of multiple genes in colorectal neoplasia. Lab Invest 84:884–893

Lehmann U, Langer F, Feist H, Glockner S, Hasemeier B, Kreipe H (2002) Quantitative assessment of promoter hypermethylation during breast cancer development. Am J Pathol 160:605–612

Leu YW, Yan PS, Fan M, Jin VX, Liu JC, Curran EM, Welshons WV, Wei SH, Davuluri RV, Plass C, Nephew KP, Huang TH (2004) Loss of estrogen receptor signaling triggers epigenetic silencing of downstream targets in breast cancer. Cancer Res 64:8184–8192

Li CI, Daling JR, Malone KE (2003) Incidence of invasive breast cancer by hormone receptor status from 1992 to 1998. J Clin Oncol 21:28–34

Lind GE, Thorstensen L, Lovig T, Meling GI, Hamelin R, Rognum TO, Esteller M, Lothe RA (2004) A CpG island hypermethylation profile of primary colorectal carcinomas and colon cancer cell lines. Mol Cancer 3:28

Liu B, Parsons RE, Hamilton SR, Petersen GM, Lynch HT, Watson P, Markowitz S, Willson JK, Green J, de la Chapelle A, et al (1994) hMSH2 mutations in hereditary nonpolyposis colorectal cancer kindreds. Cancer Res 54:4590–4594

Malagelada JR, Go VL, Summerskill WH, Gamble WS (1973) Bile acid secretion and biliary bile acid composition altered by cholecystectomy. Am J Dig Dis 18:455–459

Malaney S, Daly RJ (2001) The ras signaling pathway in mammary tumorigenesis and metastasis. J Mammary Gland Biol Neoplasia 6:101–113

Malkhosyan SR, Yamamoto H, Piao Z, Perucho M (2000) Late onset and high incidence of colon cancer of the mutator phenotype with hypermethylated hMLH1 gene in women. Gastroenterology 119:598

Mandelson MT, Miglioretti D, Newcomb PA, Harrison R, Potter JD (2003) Hormone replacement therapy in relation to survival in women diagnosed with colon cancer. Cancer Causes Control 14:979–984

Markl ID, Cheng J, Liang G, Shibata D, Laird PW, Jones PA (2001) Global and gene-specific epigenetic patterns in human bladder cancer genomes are relatively stable in vivo and in vitro over time. Cancer Res 61:5875–5884

Marks JR, Huper G, Vaughn JP, Davis PL, Norris J, McDonnell DP, Wiseman RW, Futreal PA, Iglehart JD (1997) BRCA1 expression is not directly responsive to estrogen. Oncogene 14:115–121

Marra G, Boland CR (1996) DNA repair and colorectal cancer. Gastroenterol Clin North Am 25:755–772

Matias-Guiu X, Catasus L, Bussaglia E, Lagarda H, Garcia A, Pons C, Munoz J, Arguelles R, Machin P, Prat J (2001) Molecular pathology of endometrial hyperplasia and carcinoma. Hum Pathol 32:569–577

McCluskey LL, Chen C, Delgadillo E, Felix JC, Muderspach LI, Dubeau L (1999) Differences in p16 gene methylation and expression in benign and malignant ovarian tumors. Gynecol Oncol 72:87–92

McMichael AJ, Potter JD (1980) Reproduction, endogenous and exogenous sex hormones, and colon cancer: a review and hypothesis. J Natl Cancer Inst 65:1201–1207

Medeiros AC, Nagai MA, Neto MM, Brentani RR (1994) Loss of heterozygosity affecting the APC and MCC genetic loci in patients with primary breast carcinomas. Cancer Epidemiol Biomarkers Prev 3:331–333

Miyakura Y, Sugano K, Konishi F, Ichikawa A, Maekawa M, Shitoh K, Igarashi S, Kotake K, Koyama Y, Nagai H (2001) Extensive methylation of hMLH1 promoter region predominates in proximal colon cancer with microsatellite instability. Gastroenterology 121:1300–1309

Montano MM, Deng H, Liu M, Sun X, Singal R (2004) Transcriptional regulation by the estrogen receptor of antioxidative stress enzymes and its functional implications. Oncogene 23:2442–2453

Moreno-Bueno G, Hardisson D, Sanchez C, Sarrio D, Cassia R, Garcia-Rostan G, Prat J, Guo M, Herman JG, Matias-Guiu X, Esteller M, Palacios J (2002) Abnormalities of the APC/beta-catenin pathway in endometrial cancer. Oncogene 21:7981–7990

Moreno-Bueno G, Hardisson D, Sarrio D, Sanchez C, Cassia R, Prat J, Herman JG, Esteller M, Matias-Guiu X, Palacios J (2003) Abnormalities of E- and P-cadherin and catenin (beta-, gamma-catenin, and p120ctn) expression in endometrial cancer and endometrial atypical hyperplasia. J Pathol 199:471–478

Moscow JA, Townsend AJ, Goldsmith ME, Whang-Peng J, Vickers PJ, Poisson R, Legault-Poisson S, Myers CE, Cowan KH (1988) Isolation of the human anionic glutathione S-transferase cDNA and the relation of its gene expression to estrogen-receptor content in primary breast cancer. Proc Natl Acad Sci U S A 85:6518–6522

Murata H, Khattar NH, Kang Y, Gu L, Li GM (2002) Genetic and epigenetic modification of mismatch repair genes hMSH2 and hMLH1 in sporadic breast cancer with microsatellite instability. Oncogene 21:5696–5703

Nagasaka T, Sasamoto H, Notohara K, Cullings HM, Takeda M, Kimura K, Kambara T, MacPhee DG, Young J, Leggett BA, Jass JR, Tanaka N, Matsubara N (2004) Colorectal cancer with mutation in BRAF, KRAS, and wild-type with respect to both oncogenes showing different patterns of DNA methylation. J Clin Oncol 22:4584–4594

Narayan S, Roy D (2003) Role of APC and DNA mismatch repair genes in the development of colorectal cancers. Mol Cancer 2:41

Nass SJ, Herman JG, Gabrielson E, Iversen PW, Parl FF, Davidson NE, Graff JR (2000) Aberrant methylation of the estrogen receptor and E-cadherin 5′ CpG islands increases with malignant progression in human breast cancer. Cancer Res 60:4346–4348

Navari JR, Roland PY, Keh P, Salvesen HB, Akslen LA, Iversen OE, Das S, Kothari R, Howey S, Phillips B (2000) Loss of estrogen receptor (ER) expression in endometrial tumors is not associated with de novo methylation of the 5′ end of the ER gene. Clin Cancer Res 6:4026–4032

Niederacher D, Yan HY, An HX, Bender HG, Beckmann MW (1999) CDKN2A gene inactivation in epithelial sporadic ovarian cancer. Br J Cancer 80:1920–1926

Nilsson S, Makela S, Treuter E, Tujague M, Thomsen J, Andersson G, Enmark E, Pettersson K, Warner M, Gustafsson JA (2001) Mechanisms of estrogen action. Physiol Rev 81:1535–1565

Nishimura M, Saito T, Yamasaki H, Kudo R (2003) Suppression of gap junctional intercellular communication via 5′ CpG island methylation in promoter region of E-cadherin gene in endometrial cancer cells. Carcinogenesis 24:1615–1623

Nolan RD, Danilowicz RM, Eling TE (1988) Role of arachidonic acid metabolism in the mitogenic response of BALB/c 3T3 fibroblasts to epidermal growth factor. Mol Pharmacol 33:650–656

Norrie MW, Hawkins NJ, Todd AV, Meagher AP, O'Connor TW, Ward RL (2003) Inactivation of p16INK4a by CpG hypermethylation is not a frequent event in colorectal cancer. J Surg Oncol 84:143–150

O'Doherty AM, Church SW, Russell SE, Nelson J, Hickey I (2002) Methylation status of oestrogen receptor-alpha gene promoter sequences in human ovarian epithelial cell lines. Br J Cancer 86:282–284

Oliveira Ferreira F, Napoli Ferreira CC, Rossi BM, Toshihiko Nakagawa W, Aguilar S Jr, Monteiro Santos EM, Vierira Costa ML, Lopes A (2004) Frequency of extra-colonic tumors in hereditary nonpolyposis colorectal cancer (HNPCC) and familial colorectal cancer (FCC) Brazilian families: an analysis by a Brazilian Hereditary Colorectal Cancer Institutional Registry. Fam Cancer 3:41–47

Paech K, Webb P, Kuiper GG, Nilsson S, Gustafsson J, Kushner PJ, Scanlan TS (1997) Differential ligand activation of estrogen receptors ERalpha and ERbeta at AP1 sites. Science 277:1508–1510

Parrella P, Poeta ML, Gallo AP, Prencipe M, Scintu M, Apicella A, Rossiello R, Liguoro G, Seripa D, Gravina C, Rabitti C, Rinaldi M, Nicol T, Tommasi S, Paradiso A, Schittulli F, Altomare V, Fazio VM (2004) Nonrandom distribution of aberrant promoter methylation of cancer-related genes in sporadic breast tumors. Clin Cancer Res 10:5349–5354

Paz MF, Fraga MF, Avila S, Guo M, Pollan M, Herman JG, Esteller M (2003) A systematic profile of DNA methylation in human cancer cell lines. Cancer Res 63:1114–1121

Pellise M, Castells A, Gines A, Agrelo R, Sole M, Castellvi-Bel S, Fernandez-Esparrach G, Llach J, Esteller M, Bordas JM, Pique JM (2004) Detection of lymph node micrometastases by gene promoter hypermethylation in samples obtained by endosonography-guided fine-needle aspiration biopsy. Clin Cancer Res 10:4444–4449

Pfeifer GP, Yoon JH, Liu L, Tommasi S, Wilczynski SP, Dammann R (2002) Methylation of the RASSF1A gene in human cancers. Biol Chem 383:907–914

Pijnenborg JM, Kisters N, van Engeland M, Dunselman GA, de Haan J, de Goeij AF, Groothuis PG (2004) APC, beta-catenin, and E-cadherin and the development of recurrent endometrial carcinoma. Int J Gynecol Cancer 14:947–956

Pike MC, Pearce CL, Wu AH (2004) Prevention of cancers of the breast, endometrium and ovary. Oncogene 23:6379–6391

Pomare EW, Heaton KW (1973) The effect of cholecystectomy on bile salt metabolism. Gut 14:753–762

Rajagopalan H, Bardelli A, Lengauer C, Kinzler KW, Vogelstein B, Velculescu VE (2002) Tumorigenesis: RAF/RAS oncogenes and mismatch-repair status. Nature 418:934

Rathi A, Virmani AK, Schorge JO, Elias KJ, Maruyama R, Minna JD, Mok SC, Girard L, Fishman DA, Gazdar AF (2002) Methylation profiles of sporadic ovarian tumors and nonmalignant ovaries from high-risk women. Clin Cancer Res 8:3324–3331

Reddy BS, Wynder EL (1977) Metabolic epidemiology of colon cancer. Fecal bile acids and neutral sterols in colon cancer patients and patients with adenomatous polyps. Cancer 39:2533–2539

Reddy BS, Narasawa T, Weisburger JH, Wynder EL (1976) Promoting effect of sodium deoxycholate on colon adenocarcinomas in germfree rats. J Natl Cancer Inst 56:441–442

Risinger JI, Berchuck A, Kohler MF, Watson P, Lynch HT, Boyd J (1993) Genetic instability of microsatellites in endometrial carcinoma. Cancer Res 53:5100–5103

Sasaki M, Dharia A, Oh BR, Tanaka Y, Fujimoto S, Dahiya R (2001a) Progesterone receptor B gene inactivation and CpG hypermethylation in human uterine endometrial cancer. Cancer Res 61:97–102

Sasaki M, Kotcherguina L, Dharia A, Fujimoto S, Dahiya R (2001b) Cytosine-phosphoguanine methylation of estrogen receptors in endometrial cancer. Cancer Res 61:3262–3266

Schneider-Stock R, Boltze C, Peters B, Hopfner T, Meyer F, Lippert H, Roessner A (2003) Differences in loss of p16INK4 protein expression by promoter methylation between left- and right-sided primary colorectal carcinomas. Int J Oncol 23:1009–1013

Sharma S, Stolina M, Yang SC, Baratelli F, Lin JF, Atianzar K, Luo J, Zhu L, Lin Y, Huang M, Dohadwala M, Batra RK, Dubinett SM (2003) Tumor cyclooxygenase 2-dependent suppression of dendritic cell function. Clin Cancer Res 9:961–968

Shen L, Kondo Y, Hamilton SR, Rashid A, Issa JP (2003) P14 methylation in human colon cancer is associated with microsatellite instability and wild-type p53. Gastroenterology 124:626–633

Silva J, Dominguez G, Silva JM, Garcia JM, Gallego I, Corbacho C, Provencio M, Espana P, Bonilla F (2001) Analysis of genetic and epigenetic processes that influence p14ARF expression in breast cancer. Oncogene 20:4586–4590

Silva J, Silva JM, Dominguez G, Garcia JM, Cantos B, Rodriguez R, Larrondo FJ, Provencio M, Espana P, Bonilla F (2003) Concomitant expression of p16INK4a and p14ARF in primary breast cancer and analysis of inactivation mechanisms. J Pathol 199:289–297

Slattery ML, Anderson K, Samowitz W, Edwards SL, Curtin K, Caan B, Potter JD (1999) Hormone replacement therapy and improved survival among postmenopausal women diagnosed with colon cancer (USA). Cancer Causes Control 10:467–473

Soslow RA, Dannenberg AJ, Rush D, Woerner BM, Khan KN, Masferrer J, Koki AT (2000) COX-2 is expressed in human pulmonary, colonic, and mammary tumors. Cancer 89:2637–2645

Strathdee G, MacKean MJ, Illand M, Brown R (1999) A role for methylation of the hMLH1 promoter in loss of hMLH1 expression and drug resistance in ovarian cancer. Oncogene 18:2335–2341

Strathdee G, Appleton K, Illand M, Millan DW, Sargent J, Paul J, Brown R (2001) Primary ovarian carcinomas display multiple methylator phenotypes involving known tumor suppressor genes. Am J Pathol 158:1121–1127

Strathdee G, Davies BR, Vass JK, Siddiqui N, Brown R (2004) Cell type-specific methylation of an intronic CpG island controls expression of the MCJ gene. Carcinogenesis 25:693–701

Tamaru H, Selker EU (2001) A histone H3 methyltransferase controls DNA methylation in Neurospora crassa. Nature 414:277–283

Todoroki I, Friedman GD, Slattery ML, Potter JD, Samowitz W (1999) Cholecystectomy and the risk of colon cancer. Am J Gastroenterol 94:41–46

Toyota M, Ahuja N, Ohe-Toyota M, Herman JG, Baylin SB, Issa JP (1999) CpG island methylator phenotype in colorectal cancer. Proc Natl Acad Sci U S A 96:8681–8686

Toyota M, Ohe-Toyota M, Ahuja N, Issa JP (2000a) Distinct genetic profiles in colorectal tumors with or without the CpG island methylator phenotype. Proc Natl Acad Sci U S A 97:710–715

Toyota M, Shen L, Ohe-Toyota M, Hamilton SR, Sinicrope FA, Issa JP (2000b) Aberrant methylation of the Cyclooxygenase 2 CpG island in colorectal tumors. Cancer Res 60:4044–4048

Turunen MJ, Kivilaakso EO (1981) Increased risk of colorectal cancer after cholecystectomy. Ann Surg 194:639–641

Vegeto E, Shahbaz MM, Wen DX, Goldman ME, O'Malley BW, McDonnell DP (1993) Human progesterone receptor A form is a cell- and promoter-specific repressor of human progesterone receptor B function. Mol Endocrinol 7:1244–1255

Velicescu M, Weisenberger DJ, Gonzales FA, Tsai YC, Nguyen CT, Jones PA (2002) Cell division is required for de novo methylation of CpG islands in bladder cancer cells. Cancer Res 62:2378–2384

Vernick LJ, Kuller LH, Lohsoonthorn P, Rycheck RR, Redmond CK (1980) Relationship between cholecystectomy and ascending colon cancer. Cancer 45:392–395

Vos MD, Ellis CA, Bell A, Birrer MJ, Clark GJ (2000) Ras uses the novel tumor suppressor RASSF1 as an effector to mediate apoptosis. J Biol Chem 275:35669–35672

Wang C, Horiuchi A, Imai T, Ohira S, Itoh K, Nikaido T, Katsuyama Y, Konishi I (2004) Expression of BRCA1 protein in benign, borderline, and malignant epithelial ovarian neoplasms and its relationship to methylation and allelic loss of the BRCA1 gene. J Pathol 202:215–223

Watson P, Lynch HT (1993) Extracolonic cancer in hereditary nonpolyposis colorectal cancer. Cancer 71:677–685

Welcsh PL, King MC (2001) BRCA1 and BRCA2 and the genetics of breast and ovarian cancer. Hum Mol Genet 10:705–713

Whitcomb BP, Mutch DG, Herzog TJ, Rader JS, Gibb RK, Goodfellow PJ (2003) Frequent HOXA11 and THBS2 promoter methylation, and a methylator phenotype in endometrial adenocarcinoma. Clin Cancer Res 9:2277–2287

Widschwendter M, Jones PA (2002) DNA methylation and breast carcinogenesis. Oncogene 21:5462–5482

Widschwendter M, Siegmund KD, Muller HM, Fiegl H, Marth C, Muller-Holzner E, Jones PA, Laird PW (2004) Association of breast cancer DNA methylation profiles with hormone receptor status and response to tamoxifen. Cancer Res 64:3807–3813

Wiencke JK, Zheng S, Lafuente A, Lafuente MJ, Grudzen C, Wrensch MR, Miike R, Ballesta A, Trias M (1999) Aberrant methylation of p16INK4a in anatomic and gender-specific subtypes of sporadic colorectal cancer. Cancer Epidemiol Biomarkers Prev 8:501–506

Wingo PA, Ries LA, Rosenberg HM, Miller DS, Edwards BK (1998) Cancer incidence and mortality, 1973–1995: a report card for the U.S. Cancer 82:1197–1207

Wong YF, Chung TK, Cheung TH, Nobori T, Yu AL, Yu J, Batova A, Lai KW, Chang AM (1999) Methylation of p16INK4A in primary gynecologic malignancy. Cancer Lett 136:231–235

Yamamoto H, Min Y, Itoh F, Imsumran A, Horiuchi S, Yoshida M, Iku S, Fukushima H, Imai K (2002) Differential involvement of the hypermethylator phenotype in hereditary and sporadic colorectal cancers with high-frequency microsatellite instability. Genes Chromosomes Cancer 33:322–325

Yoon JH, Dammann R, Pfeifer GP (2001) Hypermethylation of the CpG island of the RASSF1A gene in ovarian and renal cell carcinomas. Int J Cancer 94:212–217

Yoshiura K, Kanai Y, Ochiai A, Shimoyama Y, Sugimura T, Hirohashi S (1995) Silencing of the E-cadherin invasion-suppressor gene by CpG methylation in human carcinomas. Proc Natl Acad Sci U S A 92:7416–7419

Zhao Y, Agarwal VR, Mendelson CR, Simpson ER (1997) Transcriptional regulation of CYP19 gene (aromatase) expression in adipose stromal cells in primary culture. J Steroid Biochem Mol Biol 61:203–210

Zhou H, Luo MP, Schonthal AH, Pike MC, Stallcup MR, Blumenthal M, Zheng W, Dubeau L (2002) Effect of reproductive hormones on ovarian epithelial tumors. I. Effect on cell cycle activity. Cancer Biol Ther 1:300–306

Zysman M, Saka A, Millar A, Knight J, Chapman W, Bapat B (2002) Methylation of adenomatous polyposis coli in endometrial cancer occurs more frequently in tumors with microsatellite instability phenotype. Cancer Res 62:3663–3666

CTMI (2006) 310:179–198

Genome-wide Analysis of DNA Methylation Changes in Human Malignancies

C. Plass[1] (✉) · D. J. Smiraglia[2]

[1]Division of Human Cancer Genetics, The Ohio State University, Tzagournis Medical Research Facility 464A, 420 West 12th Ave., Columbus, OH 43210, USA
Plass-1@medctr.osu.edu

[2]Department of Cancer Genetics, Roswell Park Cancer Institute, Buffalo, NY 14263, USA

Abstract DNA methylation is an epigenetic modification of the DNA sequence and thus does not change the genetic code but affects chromosomal stability and gene expression. DNA methylation patterns are heritable and can be passed on to the daughter cell. In this review, we briefly summarize our current knowledge on normal DNA methylation patterns and move on to discuss the current state of the field with respect to altered DNA methylation in cancer. We make a special attempt to address current questions relating to genome-wide DNA methylation patterns. Since DNA methylation is used as a therapeutic target in clinical studies, it is of utmost importance to define potential target sequences that could be used as diagnostic or prognostic markers. We conclude the review by outlining possible scenarios that may explain tumor type-specific DNA methylation patterns described by assays evaluating genome-wide levels of DNA methylation.

1
Introduction

The human genome project deciphered the whole human genomic sequence and determined that there are roughly 20,000 to 25,000 genes in a human cell responsible for all developmental processes and tissue-specific activities (International Human Genome Sequencing Consortium 2004). The total gene number seems small in the context of a complex human body and all the different functions certain specialized cells have to acquire, even considering the presence of alternatively spliced products and alternative promoter sequences for a number of these genes (Ast 2004). Additional regulatory mechanisms have been identified, including epigenetic modifications of the DNA itself as well as the histones surrounding the DNA. Epigenetic modifications do not change the genetic code of genes; however, they do participate as regulatory switches in gene regulation. Most importantly, epigenetic modifications are heritable. The importance of such epigenetic events for the normal cell is becoming increasingly evident, and with better understanding of normal processes we are learning how epigenetic alterations impact human diseases such as cancer. In this review, we will provide an overview of our current concept of genome-wide epigenetic modifications and how these alterations influence tumorigenesis.

2
DNA Methylation and the Genome

DNA methylation is an epigenetic modification of the DNA and describes the addition of a methyl group to the carbon-5 position of cytosine creating 5-methylcytosine, also known as the fifth base (Doerfler 1983). Targets for methylation are usually cytosines located next to a guanine in CpG dinucleotides, although exceptions have been described in CpNpG or CpCpWpGpG motifs (Clark et al. 1995; Agirre et al. 2003). The majority of CpG dinucleotide DNA methylation is found in inter- and intragenic regions, including repetitive sequences such as satellite sequences and centromeric repeats. It was estimated that over 70% of all CpG dinucleotides are methylated (Cooper and Krawczak 1989). Exceptions are small GC-rich sequences, or CpG islands, which are found preferentially in the promoter regions of genes

(Antequera and Bird 1993). CpG islands are usually unmethylated and were termed HpaII tiny fragments (HTF) islands due to the fact that these sequences contain a high abundance HpaII restriction sites and are digested by *Hpa*II, a methylation-sensitive restriction enzyme (Bird 1986). While most of the estimated 29,000 CpG islands in the human genome are unmethylated, the majority of the CpG islands found on the inactive X chromosome in a female cell are methylated (Goto and Monk 1998). In addition, imprinted genes, expressed either from the paternal or the maternal chromosome, are associated with CpG island regions methylated only on one allele (Li et al. 1993; Razin and Cedar 1994), with these regions referred to as differentially methylated regions (DMRs).

3
The DNA Methylation Machinery

It has been demonstrated that DNA methylation is essential for mammalian development; precise methylation patterns are established during embryonic development (Oswald et al. 2000; Santos et al. 2002). Initially, rapid demethylation is seen in the male pronucleus, followed by progressive demethylation of the female half of the genome, which is completed in the blastocyst stage and by this time has erased most of the methylation marks, with the exception of methylation in imprinted regions (Oswald et al. 2000; Santos et al. 2002). Following this wave of demethylation, DNA methylation patterns are reestablished in a cell lineage-specific fashion. DNA methylation levels increase in ectoderm, whereas DNA methylation seems to be inhibited in primitive endoderm and in the trophoblast (see Li 2002 for a review). DNA methylation patterns are established and maintained by an interplay of several DNA methyltransferase (DNMT) enzymes that transfer a methyl group from the methyl donor, *S*-adenosyl-L-methionine (SAM), to cytosine (Hermann et al. 2004). The importance of these enzymes in mammalian development is demonstrated in mouse mutants lacking their activity (Li et al. 1992; Okano et al. 1999). De novo methyltransferases, DNMT3a and DNMT3b, are enzymes that transfer methyl groups to cytosines of unmethylated DNA with DNMT3a favoring short regions of low CpG density and providing low levels of methylation, but with DNMT3b favoring larger regions of higher CpG density such as pericentromeric regions (Hermann et al. 2004). Mice lacking these enzymes are embryonic or postnatal lethal due to a lack of de novo methylation in early development stages. DNMT1, the maintenance methyltransferase, preferentially detects hemimethylated sequences shortly after the passage of the replication fork where the enzyme copies the methylation pattern of the template strand to the nascent strand, thus maintaining the methylation pattern (Okano et al. 1999). *Dnmt1* mutant mice arrest in late gastrulation and die at embryonic day 9.5 where they exhibit lack of methylation in minor satellite repeats, endogenous C-type retroviruses intracisternal A particles and changes of DNA methylation in imprinted genes (Li et al. 1992).

4
DNA Methylation and Gene Silencing

Work in the past decade has established the involvement of DNA methylation on gene silencing in the context of chromatin changes (Jones and Baylin 2002). Initial clues for the interaction between DNA methylation and histone modifications came from studies in *Neurospora crassa*, a filamentous fungus, in which a mutation in a histone methyltransferase gene, defective in DNA methylation (dim-5), was described (Tamara and Selker 2001). Dim-5 mutants showed a complete loss of DNA methylation in an organism that does not require specific methylation patterns to live. These findings linked histone methylation and the associated chromatin changes with DNA methylation.

The genome is divided into either euchromatin or heterochromatin. Euchromatic regions generally contain actively transcribed regions of the genome and have an open chromatin structure, while heterochromatic regions are transcriptionally silent and are associated with a closed chromatin structure. These chromatin structures are distinguished by a number of characteristic modifications of the histone tails. It was shown that the histone acetylation or deacetylation in lysine residues of histones H2A, H2B, H3, and H4 correlates with the active or inactive state of gene transcription. In addition, histone tail methylation events also have an impact on gene expression. H3 Lys4 methylation is associated with active genes, whereas H3 Lys9 methylation has been found in chromatin of silenced genes. We are only at the beginning of our understanding of such histone tail modifications as highlighted in recent work by Zhang et al. demonstrating the complexity of such modifications using a mass spectrometry assay to measure chromatin modifications on a genome-wide level (Zhang et al. 2003).

Methylated DNA is detected by methyl-CpG-binding domain (MBD) proteins (MECP2, MBD1, MBD2, MBD3, MBD4, and KAISO) (Feng and Zhang 2000; Ng et al. 1999; Prokhortchouk et al. 2001; Rountree et al. 2000). Interestingly, it was shown that several of these MBD proteins are found in protein complexes containing chromatin-modifying enzymes such as histone deacetylase (HDAC). MECP2 complexes with Sin3a and HDACs and has the ability to silence gene activity. MBD2, on the other hand, is found in a complex with NuRD (nucleosome remodeling and histone deacetylase) in addition to HDAC, Mi-2, and ATP-dependent chromatin remodeling factors (Ng et al. 1999). These methyl binding domain proteins, their interactions with protein complexes containing HDACs, and the requirement for the activity of histone methyltransferase for normal DNA methylation in *Neurospora crassa*, all indicate an intimate relationship between DNA methylation and histone tail modifications. The precise nature of this relationship is currently the topic of intensive study.

5 Aberrant DNA Methylation in Cancer

Our current understanding of tumorigenesis describes a cumulative series of genetic and epigenetic events leading to increased cellular proliferation and survival, as well as aberrations in differentiation programs. Genetic and epigenetic events may lead to changes in gene expression resulting in the activation of oncogenes, the silencing of growth-suppressing genes, or to an increased resistance to therapy, providing a selective growth advantage of cells harboring the defect. This concept is well accepted with respect to genetic mutations; however, our understanding of how epigenetic events contribute to tumorigenesis is lagging. For several years it has been known that reduced global levels of 5-methylcytosine characterize human tumors (Feinberg and Vogelstein 1983; Gama-Sosa et al. 1983; Goelz et al. 1985). This net reduction in 5-methylcytosine in cancer is the end result of the combination of two commonly observed major changes in the distribution of 5-methylcytosine: (1) hypomethylation of CpG dinucleotides in large repetitive sequences and (2) regional hypermethylation in gene-regulating CpG islands, with both events occurring in the same cell. Since CpG islands provide only a small fraction of the total CpG dinucleotide targets of methylation as compared to the large fraction of CpG dinucleotides located in repetitive sequences, the contribution of CpG island hypermethylation to the net change in 5-methylcytosine content is negligible. However, due to its association with silencing of known tumor-suppressor genes, research has mainly focused on CpG island or promoter hypermethylation (Baylin et al. 1998).

6 Loss of DNA Methylation in Cancer

The amount of 5-methylcytosine within genomic DNA can be measured directly by HPLC and by liquid chromatography followed by mass spectrometry, or indirectly by using the ability of DNA to accept tritiated methyl groups from a universal methyl donor SAM, or by specific probes detecting methylation changes in restriction sites located within the repeat units (Ehrlich et al. 1982; Ji et al. 1997; Song et al. 2005). These assays demonstrated the general trend of hypomethylation in tumor genomes, data that closely correlate with the degree of malignancy as shown for breast, cervical, and ovarian cancer. These findings led to the conclusion that hypomethylation events may be used as a biological marker with prognostic value (Bernardino et al. 1997; Kim et al. 1994; Narayan et al. 1998; Widschwendter et al. 2004). Several hypotheses have been put forth to explain how hypomethylation may contribute to tumorigenesis and accelerated growth. Early data on *CMYC* and *HRAS* supported the idea that hypomethylation may be involved in the activation of oncogenes (Del Senno et al. 1989; Fang et al. 1996; Vachtenheim et al. 1994). However, although significant correlations have been seen between hypomethylation of these promoters and high-level expression, no direct mechanistic link between these two events has been demonstrated.

Another hypothesis was proposed based on the finding of hypomethylation in pericentromeric heterochromatic repeat sequences described in breast adenocarcinomas, ovarian epithelial tumors, and Wilms tumors (Gama-Sosa et al. 1983; Qu et al. 1999). These hypomethylation events are associated with chromosomal rearrangements such as isochromosomes, unbalanced juxtacentromeric translocations, and whole-arm deletions, plus chromosomal rearrangements found in normal cells treated with demethylating agents 5-azacytidine or 5-aza-2-deoxycytidine (Ji et al. 1997). The idea that hypomethylation might lead to chromosomal instability and subsequent tumor growth supporting gene rearrangements, activations, or both (rearrangements and activations) found additional support with the description of *DNMT3B* mutations in a rare genetic disorder, immunodeficiency–centromeric instability–facial anomalies (ICF) syndrome. ICF syndrome is characterized on the genomic level by abnormal methylation of pericentromeric heterochromatin of chromosomes 1 and 16, similar to the abnormalities seen in sporadic cancers (Hansen et al. 1999; Okano et al. 1999; Xu et al. 1999). Interestingly, however, increased cancer incidence is not associated with ICF syndrome.

Finally, it has been proposed that hypomethylation may lead to the activation of normally silenced retrotransposons such as LINE1 elements. This idea stems from the theory that the normal function of DNA methylation is to defend the genome from the activity of endogenous viruses and invading parasitic sequences. Support for this mechanism comes from the description of loss of LINE1 promoter hypomethylation and transcriptional activation of a subset of these elements in several human malignancies. Activated LINE1 elements are transcribed, reverse transcribed, and subsequently integrate back into the genome at seemingly random sites. Such novel integration sites have been found in the APC gene in colon cancer or CMYC in breast cancer (Miki et al. 1992; Morse et al. 1988).

7 Gain of DNA Methylation in Cancer

Although aberrant DNA methylation of CpG islands represents only a small portion of the sequences that harbor changes in DNA methylation status during tumorigenesis, these changes have received major attention due to the fact that they are involved in gene silencing and thus have similar consequences as gene deletion. Fundamentally different from genetic abnormalities, however, is the fact that these promoter methylation events are reversible by demethylating agents such as 5-azacytidine or 5-aza-2′-deoxycytidine through inhibition of DNA methyltransferases. This has received special attention in hematopoietic malignancies, where these demethylating agents have been tested in clinical trials for their activities in patients with myelodysplastic syndromes (Lubbert et al. 2000; Silverman 2004). In addition, these agents are being tested in lung cancer (Momparler and Ayoub 2001).

Initially, promoter methylation was found in tumor-suppressor genes for which gene mutations had been described (for a review see Baylin et al. 1998). Many of these genes were identified as tumor-suppressor genes mutated in familial cancer syndromes resulting in predisposition to a certain cancer or set of cancers. With the invention of methylation assays based on bisulfite conversion (Clark et al. 1994) and subsequent PCR-based assays such as methylation-sensitive PCR (MSP) (Herman et al. 1996) or combined bisulfite conversion and restriction analysis (COBRA) (Xiong and Laird 1997), the list of genes for which transcriptional silencing was associated with CpG island promoter methylation expanded quickly. It included genes involved in cell cycle regulation (*Rb*, *p14ARF*, *CDKN2A* and *CDKN2B*, *p73* and *p21KIP1*), apoptosis (*DAPK1*, *Caspase 8*, and *TMS1*), differentiation (*WT1*, *PAX6*, and *RAR*) DNA repair (*MGMT* and *hMLH1*), metastasis/invasion (*E-cadherin*), drug resistance (*GSTP1*), and signal transduction (*PTEN*), as well as imprinted genes (*IGF2* and *H19*) (for a review see Costello and Plass 2001).

While most of these genes have initially been identified based on their genetic defects in either familial or sporadic tumors, there is now a fast-growing list of genes identified that are mainly silenced by DNA promoter methylation and for which no, or only a few, gene mutations have been identified. This list is rapidly growing with the inclusion of genome-wide searches for aberrant DNA methylation using various techniques such as methylated CpG island amplification followed by representational difference analysis (MCA-RDA; Toyota et al. 1999), methylation-sensitive arbitrarily primed PCR (MS-AP-PCR; Gonzalgo et al. 1997), differential methylation hybridization (DMH; Yan et al. 2001), a microarray based gene re-expression approach (MGR; Suzuki et al. 2002), and restriction landmark genome scanning (RLGS; Smiraglia and Plass 2002). Examples include *RASSF1A*, a gene identified in a common region of loss of heterozygosity in lung cancer on chromosome 3p21.3 (Agathanggelou et al. 2001; Dammann et al. 2000; Hogg et al. 2002), *hypermethylated in cancer 1* (*HIC1*) on chromosome 17p13.3, a gene identified based on promoter methylation and silencing in leukemia and for which tumor-suppressor function was demonstrated (Chen et al. 2004; Issa et al. 1997), epigenetic silencing of secreted frizzled-related proteins (SFRPs) in colorectal cancer (Suzuki et al. 2004), and four genes, *COE3*, *BMP3B*, *WIT1*, and *SOCS1*, identified in RLGS scans for aberrant DNA methylation (Dai et al. 2004; Plass et al. 1999; Yoshikawa et al. 2001; Zardo et al. 2002). All of these assays identified novel target sequences and genes methylated in cancer cell lines and primary tumor tissue. A recent study combining data obtained from bacterial artificial chromosome comparative genome hybridization (BAC-CGH) array and RLGS demonstrated that genetic deletions and DNA methylation target different chromosomal regions (Zardo et al. 2002). Thus, it is very likely that the majority of methylated genes are still to be discovered, since past searches focused on genes frequently inactivated by genetic defects.

8
Number of Methylated CpG Islands in the Tumor Genome

Several important questions could be addressed using genome-wide searches for methylation. First, the extent of CpG island methylation in primary human malignancies was measured using RLGS. This two-dimensional gel electrophoresis displays about 1,500–2,000 unmethylated *Not*I restriction sites within a single gel. Comparison of RLGS profiles between normal and tumor allows the identification of those sequences that are lost (methylated) in the tumor profile. Using this assay, it was shown that on average 16 CpG islands out of a total of 1,184 analyzed, or 1.4%, were methylated in a set of 98 primary human malignancies composed of eight different tumor types (Costello et al. 2000). DMH analysis of primary breast carcinomas determined the average levels of CpG island methylation to be about 1%, confirming the range of methylation seen by RLGS (Yan et al. 2001). Based on this number and the 29,000 CpG islands identified by the human genome sequencing project (Venter et al. 2001), one can calculate that an average of 392 CpG islands are methylated in human malignancies. The range included tumors with no methylated CpG islands and others with frequencies as high as 8.3% in a patient with acute myeloid leukemia (Rush et al. 2001), 8.1% in a patient with chronic lymphocytic leukemia (Rush et al. 2004), 5.3% in a patient with non-small cell lung cancer (Dai et al. 2001), 2.1% in a patient's head and neck squamous cell carcinoma (Smiraglia et al. 2003), and 3.0% in a patient's testicular germ cell tumor (Smiraglia et al. 2002). Most human malignancies showed elevated levels of promoter methylation. Interesting, the seminomatous subgroup of testicular germ cell tumors showed only very little CpG island methylation with an average of 0.08%, whereas the nonseminomatous subgroup showed an average of 1.1% (Smiraglia et al. 2002). The frequencies of aberrant CpG island methylation are much higher in cancer cell lines where 5- to 93-fold increases in CpG island methylation have been reported (Smiraglia et al. 2001). The highest percentage of CpG island methylation was found in the leukemia cell line HL60, with 48% of all CpG islands methylated (Smiraglia et al. 2001).

These numbers were questioned using data obtained from MGR. In this approach, the colon cancer cell line RKO was treated with a low dose of 5-aza-2′-deoxycytidine and histone deacetylase inhibitor trichostatin A (TSA). Differences in gene expression with or without drug treatment were measured using complementary DNA (cDNA) microarrays following a cDNA subtraction step. Out of 10,841 gene sequences tested, 74 were upregulated with the treatment, suggesting that only roughly 0.7% of genes were epigenetically regulated (Suzuki et al. 2004). On the surface, these numbers may appear to contrast with the findings reported by RLGS and DMH where about 1.4% of CpG islands were demonstrated to be methylated in primary tumors and minimally 5% were methylated in cells lines (15% for colon cancer cell lines) (Costello et al. 2000; Smiraglia et al. 2001; Yan et al. 2001). However, it is inappropriate to try to directly compare these data sets. While the RLGS and DMH data are simply measuring the presence or absence of DNA methylation at a set of CpG islands, the MGR data are assessing the changes in gene expression following treatment with two drugs that have global effects on DNA methylation and histone acetylation. Treatment with 5-aza-2′-deoxycytidine and TSA might not lead to expression of a gene with a methylated CpG island for a number of possible reasons including:

1. The treatment is insufficient to appropriately alter the chromatin structure at a specific locus.
2. the cell no longer has the required set of transcription factors needed to drive expression of the gene, perhaps due to genetic damage.

3. the drugs may have induced the expression of a transcriptional repressor, or repressed the expression of a critical transcriptional activator for the gene.
4. the gene may not normally be expressed in the cell type.
5. the CpG island is involved with a function other than the control of expression of the nearby gene.

One question that needs to be addressed in this context is: Do all methylation events result in gene silencing? Given the large number of CpG islands shown to be methylated in cancer by RLGS and DMH data, it is unreasonable to expect that so many genes need be silenced in a cancer cell. The question then becomes: Which of these methylation events, or what proportion is associated with aberrant gene silencing? This question is usually addressed in a cell line system.

It is currently impossible to target a specific sequence for demethylation. The approach generally used is that cancer cell lines—in which the endogenous gene in question is hypermethylated and transcriptionally silenced—are treated with demethylating agents, 5-azacytidine or 5-aza-2′-deoxycytidine, and gene expression is tested by RT-PCR before and after treatment. If re-expression occurs, it is assumed that the candidate gene was silenced by DNA methylation.

Using this strategy, Costello et al. tested 16 methylated genes identified in a genome scan in glioblastomas and found six genes regulated by methylation (Costello et al. 2000). Similarly, Rush et al. tested five genes in acute myeloid leukemia cell lines and found restoration of gene expression in three (Rush et al. 2001), and in the chronic lymphocytic leukemia line WaC3CD5 five of five tested genes showed increased expression after treatment with 5-aza-2′-deoxycytidine (Rush et al. 2004). However, it is difficult to interpret the negative results from the Costello and Rush studies, and other similar studies. These data do not necessarily mean that those genes are not epigenetically regulated. In addition to problems of interpretation of negative 5-aza-2′-deoxycytidine data described above, studies on imprinted genes have shown us that methylated regions could potentially regulate several genes located within the imprinted gene cluster, and the genes could be at a distance exceeding 30 kb away from the methylated sequence, as for the mouse Rasgrf1 gene (Plass et al. 1996). In addition, the "true" promoter sequences may not be identified and a gene regulated by a certain methylation event could be missed. Therefore, it is very difficult to estimate how many CpG island hypermethylation events result in, or correlate with, gene silencing. Use of expression studies after 5-aza-2′-deoxycytidine and TSA treatment can help to shed light on this. However, while a positive result has relatively few interpretation problems, negative results have many; thus, the numbers provided by these types of drug-induced re-expression studies very likely represent a minimum estimate of the number of genes regulated by methylation.

9
Targets for Promoter Hypermethylation

RLGS data further demonstrated that patterns of CpG island methylation do not represent a random distribution of DNA methylation events over all CpG island sequences (Costello et al. 2000). These data were subsequently confirmed in an independent study utilizing a candidate gene approach and the MSP assay to detect promoter methylation (Esteller et al. 2001). Interestingly, about half of the CpG island sequences analyzed by RLGS were never methylated in any of the tumor samples ($n = 209$) studied, while other sequences are methylated at high frequencies, an observation not explainable if methylation targets were randomly selected (Plass and Smiraglia, unpublished data). Support for the non-random nature of the methylation patterns also comes from the observation that genes such as *GSTP1* and *MLH1* in prostate and colon cancers show very high frequencies of hypermethylation-associated silencing but few, if any, mutations in the sporadic cancers (Lee et al. 1994). Two possible explanations are currently discussed that may explain non-random patterns of DNA methylation and lead us to a proposed model of how aberrant DNA methylation patterns are established.

10
The Establishment of Tumor Type-Specific and Non-random Aberrant DNA Methylation

In our model (Fig. 1), we consider a genetic event, such as oncogene activation, that triggers the initiation of tumorigenesis and results in a deregulation of the DNA methylation machinery. Subsequently, we are considering three possible pathways (A–C in Fig. 1) that could generate non-random DNA methylation patterns. Our current knowledge may not be sufficient to decide if aberrant DNA methylation is caused by a genetic defect. One could assume that a genetic defect directly results in accelerated promoter methylation and other hypomethylation events by activating or disturbing the DNA methylation machinery. An alternative scenario would be that DNA methylation changes are an indirect result of accelerated growth of the tumor cells and the cells are not able to maintain the normal methylation patterns.

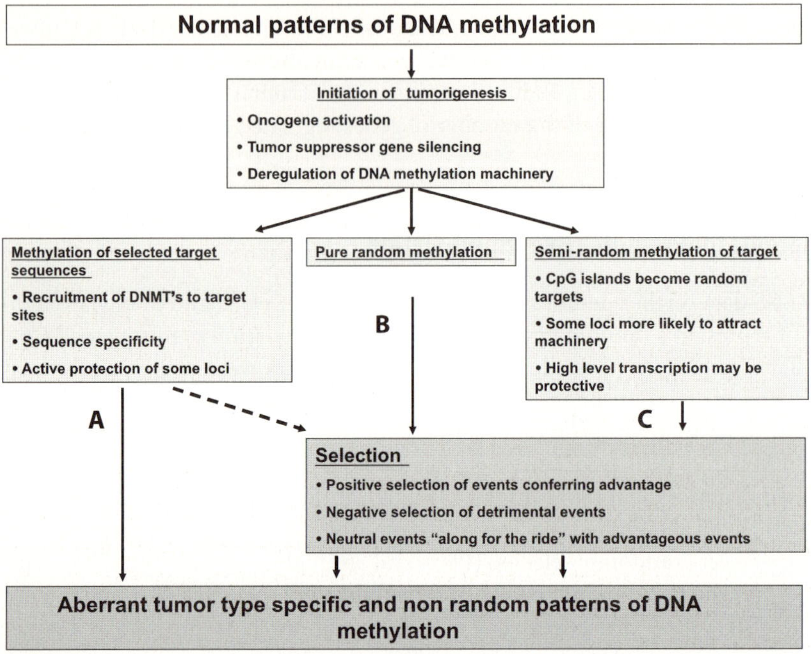

Fig. 1A–C A model explaining the establishment of tumor type-specific and non-random aberrant DNA methylation in human malignancies. See text for explanations

In pathway A, accelerated tumorigenesis would be mediated by intrinsic factors that regulate which sequences are "methylatable" and which ones are "unmethylatable" (Fig. 1). One possibility is that "unmethylatable" sequences are protected by the DNA binding factors preventing access of methyltransferases. It was proposed that transcription might in some way protect sequences from methylation (Clark and Melki 2002). Under this hypothesis, promoter sequences of housekeeping genes would be protected. Alternatively, methyltransferases may be recruited to certain target sequences by DNA binding factors. Support for this possibility comes from recent work on onco-fusion protein PML/RAR that binds DNMT1 and DNMT3A and recruits methyltransferase activity to RARβ promoter (Di Croce et al. 2002). Similar recruitment of DNMT1 by onco-fusion protein RUNX1/MTG8 has recently been reported. This fusion is a result of translocation, t(8;21)(q22;q22), a common abnormality in a cytogenetic subgroup of acute myeloid leukemia. The authors were able to show that in addition to the previously reported interaction with histone deacetylases (HDACs), RUNX1/MTG8 also recruits DNMT1 for gene repression (Liu et al. 2005). Similarly, oncogenic transcription factor

CMYC together with DNA-binding factor MIZ1 is able to target DNMT3a and redirects its activity to aberrant target sequences including p21Cip1 (Brenner et al. 2004). These various proposed mechanisms would allow for the establishment of non-random patterns of DNA methylation without further selection processes.

In pathway B, the establishment of non-random patterns of DNA methylation postulates an initially random event in which each CpG island has the same possibility to become methylated. Depending on the genes that would be influenced by this event, cells would either gain a growth advantage and thus conserve a methylation pattern, or the silencing of genes would be deleterious to the cell and a selective disadvantage would act against maintaining this pattern. Primary tumors represent the endpoint of a cellular evolutionary process; thus, it is difficult to determine the nature of non-random patterns and whether they arose out of selection or differential susceptibility to aberrant methylation. However, examples for "unmethylatable" sequences are RLGS fragments representing *CMYC*, *FOS*, *CDK6*, *MBD1*, and *SF3A1*, genes required for cell growth in any cell type and considered as housekeeping genes. It is likely that methylation of these genes is negatively selected against, and that this might explain why their methylation has not been observed. On the other hand, methylation of RLGS fragments representing tumor-suppressor genes such as *SOCS1* (Yoshikawa et al. 2001) and *BMP3B* (Dai et al. 2004) is frequently observed, which may suggest that when these loci are randomly methylated, those cells obtain a growth or survival advantage, thus perpetuating the event in the tumor. In the setting of a cell line, the drastic increase in CpG island hypermethylation might then be explained by (1) fewer genes critical to cell survival or growth given the vastly different in vitro environment, thus less negative selection against methylation, and (2) the energetic benefit of no longer expressing unneeded genes, thus more positive selective pressure for methylation.

Pathway C uses some of the concepts from both pathways A and B, where CpG island hypermethylation can be thought of as "semi-random." Under this scenario, we would propose that the process is random in that the aberrantly functioning methylation machinery does not purposefully target specific loci or types of loci (unlike the targeting by transcription factor DNA binding sites in pathway A); however, not every locus has the same chance of becoming hypermethylated. Certain loci may be highly susceptible to attracting the attention of an opportunistic aberrant methylation machinery, or others may be very resistant. For instance, genes that are highly transcribed may be far less likely to become hypermethylated than loci that are not transcribed or are being transcribed at a low level. In addition, certain sequence motifs might create secondary structures that are more or less attractive targets for aberrant methylation machinery. Furthermore, areas of DNA damage, such as double stranded breaks, might make for more attractive targets. Thus, each locus could have a level of susceptibility to aberrant hypermethylation that can be impacted either positively or negatively by many different factors including transcriptional activity, *cis* elements, and local DNA structure, and these factors may vary from cell type to cell type.

A recent study showed that certain sequences are more susceptible to become methylation targets (Feltus et al. 2003). In this study, overexpression of DNMT1 in SV40-transformed human fibroblasts resulted in consistent DNA methylation of a subset of 3.8% of CpG island sequences (of a total of 1,749 sequences) in different transformed lines, while other sequences where never methylated. Pattern recognition and supervised learning techniques identified sequence motifs that are more frequently associated with methylated than with unmethylated sequences. It is possible that the identified sequences are part of binding site motifs of transcription factors that interact with DNMTs or that these sequences are important for secondary structures in the chromosomes that supports DNMT activity. Using these sequence motifs it was possible to predict potential methylation targets with 82% accuracy. Since the data relied on an in vitro system to identify sequences preferentially methylated, it would be interesting to determine if these sequence motifs are also found in sequences aberrantly methylated in primary tumors and if there are differences between tumor types that may explain the tumor type-specific nature of aberrant DNA methylation patterns.

Beyond whether or not a locus is an initial target of hypermethylation is the question of whether or not the methylation event is perpetuated. This would then depend upon the consequence of the methylation event and the resulting selective pressure. Many events might have little or no effect because they occur where genes are already not expressed, the loss of expression of the gene is inconsequential, or perhaps the level of methylation at the particular locus is insufficient to have an effect. These events can be expected to have neutral selective pressure and may or may not be perpetuated, depending on other events in the cell. Methylation at other loci may impact the expression of important genes such that they give the cell a growth or survival advantage or disadvantage, and these will have positive or negative selective pressures, respectively. Thus, there could be an initially random process with each locus weighted either positively or negatively in terms of whether or not they would attract the aberrant methylation machinery, followed by selective screening of the consequences of these events, leading to a non-random output of aberrant hypermethylation that we have observed in cancers.

Acknowledgements The authors would like to thank Adam Karpf, Laura Smith, Romulo Martin Brena, and Shu-Hui Wang for critical discussions, and the citizens of Weissenburg for hosting the 2004 DNA Methylation meeting and providing a stimulating environment. This work was supported in part by grants CA93548 (CP) and PC040440 (DJS). C.P. is a Leukemia and Lymphoma Society of America Scholars.

References

Agathanggelou A, Honorio S, Macartney DP, Martinez A, Dallol A, Rader J, Fullwood P, Chauhan A, Walker R, Shaw JA, Hosoe S, Lerman MI, Minna JD, Maher ER, Latif F (2001) Methylation associated inactivation of RASSF1A from region 3p21.3 in lung, breast and ovarian tumours. Oncogene 20:1509–1518

Agirre X, Vizmanos JL, Calasanz MJ, Garcia-Delgado M, Larrayoz MJ, Novo FJ (2003) Methylation of CpG dinucleotides and/or CCWGG motifs at the promoter of TP53 correlates with decreased gene expression in a subset of acute lymphoblastic leukemia patients. Oncogene 22:1070–1072

Antequera F, Bird A (1993) CpG islands. Exs 64:169–185

Ast G (2004) How did alternative splicing evolve? Nat Rev Genet 5:773–782

Baylin SB, Herman JG, Graff JR, Vertino PM, Issa JP (1998) Alterations in DNA methylation: a fundamental aspect of neoplasia. Adv Cancer Res 72:141–196

Bernardino J, Roux C, Almeida A, Vogt N, Gibaud A, Gerbault-Seureau M, Magdelenat H, Bourgeois CA, Malfoy B, Dutrillaux B (1997) DNA hypomethylation in breast cancer: an independent parameter of tumor progression? Cancer Genet Cytogenet 97:83–89

Bird AP (1986) CpG-rich islands and the function of DNA methylation. Nature 321:209–213

Brenner C, Deplus R, Didelot C, Loriot A, Vire E, De Sme C, Gutierrez A, Danovi D, Bernard D, Boon T, Giuseppe Pelicci P, Amati B, Kouzarides T, de Launoit Y, Di Croce L, Fuks F (2004) Myc represses transcription through recruitment of DNA methyltransferase corepressor. EMBO J 24:336–346

Chen W, Cooper TK, Zahnow CA, Overholtzer M, Zhao Z, Ladanyi M, Karp JE, Gokgoz N, Wunder JS, Andrulis IL, Levine AJ, Mankowski JL, Baylin SB (2004) Epigenetic and genetic loss of Hic1 function accentuates the role of p53 in tumorigenesis. Cancer Cell 6:387–398

Clark SJ, Melki J (2002) DNA methylation and gene silencing in cancer: which is the guilty party? Oncogene 21:5380–5387

Clark SJ, Harrison J, Paul CL, Frommer M (1994) High sensitivity mapping of methylated cytosines. Nucleic Acids Res 22:2990–2997

Clark SJ, Harrison J, Frommer M (1995) CpNpG methylation in mammalian cells. Nat Genet 10:20–27

Cooper DN, Krawczak M (1989) Cytosine methylation and the fate of CpG dinucleotides in vertebrate genomes. Hum Genet 83:181–188

Costello JF, Plass C (2001) Methylation matters. J Med Genet 38:285–303

Costello JF, Fruhwald MC, Smiraglia DJ, Rush LJ, Robertson GP, Gao X, Wright FA, Feramisco JD, Peltomaki P, Lang JC, Schuller DE, Yu L, Bloomfield CD, Caligiuri MA, Yates A, Nishikawa R, Su Huang H, Petrelli NJ, Zhang X, O'Dorisio MS, Held WA, Cavenee WK, Plass C (2000) Aberrant CpG-island methylation has non-random and tumour-type-specific patterns. Nat Genet 24:132–138

Dai Z, Lakshmanan RR, Zhu WG, Smiraglia DJ, Rush LJ, Fruhwald MC, Brena RM, Li B, Wright FA, Ross P, Otterson GA, Plass C (2001) Global methylation profiling of lung cancer identifies novel methylated genes. Neoplasia 3:314–323

Dai Z, Popkie AP, Zhu WG, Timmers CD, Raval A, Tannehill-Gregg S, Morrison CD, Auer H, Kratzke RA, Niehans G, Amatschek S, Sommergruber W, Leone GW, Rosol T, Otterson GA, Plass C (2004) Bone morphogenetic protein 3B silencing in non-small-cell lung cancer. Oncogene 23:3521–3529

Dammann R, Li C, Yoon JH, Chin PL, Bates S, Pfeifer GP (2000) Epigenetic inactivation of a RAS association domain family protein from the lung tumour suppressor locus 3p21.3. Nat Genet 25:315–319

Del Senno L, Maestri I, Piva R, Hanau S, Reggiani A, Romano A, Russo G (1989) Differential hypomethylation of the c-myc protooncogene in bladder cancers at different stages and grades. J Urol 142:146–149

Di Croce L, Raker VA, Corsaro M, Fazi F, Fanelli M, Faretta M, Fuks F, Lo Coco F, Kouzarides T, Nervi C, Minucci S, Pelicci PG (2002) Methyltransferase recruitment and DNA hypermethylation of target promoters by an oncogenic transcription factor. Science 295:1079–1082

Doerfler W (1983) DNA methylation and gene activity. Annu Rev Biochem 52:93–124

Ehrlich M, Gama-Sosa MA, Huang LH, Midgett RM, Kuo KC, McCune RA, Gehrke C (1982) Amount and distribution of 5-methylcytosine in human DNA from different types of tissues of cells. Nucleic Acids Res 10:2709–2721

Esteller M, Corn PG, Baylin SB, Herman JG (2001) A gene hypermethylation profile of human cancer. Cancer Res 61:3225–3229

Fang JY, Zhu SS, Xiao SD, Jiang SJ, Shi Y, Chen XY, Zhou XM, Qian L (1996) Studies on the hypomethylation of c-myc, c-Ha-ras oncogenes and histopathological changes in human gastric carcinoma. J Gastroenterol Hepatol 11:1079–1082

Feinberg AP, Vogelstein B (1983) Hypomethylation distinguishes genes of some human cancers from their normal counterparts. Nature 301:89–92

Feltus FA, Lee EK, Costello JF, Plass C, Vertino PM (2003) Predicting aberrant CpG island methylation. Proc Natl Acad Sci USA 100:12253–12258

Feng Q, Zhang Y (2001) The MeCP1 complex represses transcription through preferential binding, remodeling, and deacetylating methylated nucleosomes. Genes Dev 15:827–832

Gama-Sosa MA, Slagel VA, Trewyn RW, Oxenhandler R, Kuo KC, Gehrke CW, Ehrlich M (1983) The 5-methylcytosine content of DNA from human tumors. Nucleic Acids Res 11:6883–6894

Goelz SE, Vogelstein B, Hamilton SR, Feinberg AP (1985) Hypomethylation of DNA from benign and malignant human colon neoplasms. Science 228:187–190

Gonzalgo ML, Liang G, Spruck CH 3rd, Zingg JM, Rideout WM 3rd, Jones PA (1997) Identification and characterization of differentially methylated regions of genomic DNA by methylation-sensitive arbitrarily primed PCR. Cancer Res 57:594–599

Goto T, Monk M (1998) Regulation of X-chromosome inactivation in development in mice and humans. Microbiol Mol Biol Rev 62:362–378

Hansen RS, Wijmenga C, Luo P, Stanek AM, Canfield TK, Weemaes CM, Gartler SM (1999) The DNMT3B DNA methyltransferase gene is mutated in the ICF immunodeficiency syndrome. Proc Natl Acad Sci U S A 96:14412–14417

Herman JG, Graff JR, Myohanen S, Nelkin BD, Baylin SB (1996) Methylation-specific PCR: a novel PCR assay for methylation status of CpG islands. Proc Natl Acad Sci USA 93:9821–9826

Hermann A, Gowher H, Jeltsch A (2004) Biochemistry and biology of mammalian DNA methyltransferases. Cell Mol Life Sci 61:2571–2587

Hogg RP, Honorio S, Martinez A, Agathanggelou A, Dallol A, Fullwood P, Weichselbaum R, Kuo MJ, Maher ER, Latif F (2002) Frequent 3p allele loss and epigenetic inactivation of the RASSF1A tumour suppressor gene from region 3p21.3 in head and neck squamous cell carcinoma. Eur J Cancer 38:1585–1592

International Human Genome Sequencing Consortium (2004) Finishing the euchromatic sequence of the human genome. Nature 431:931–945

Issa JP, Zehnbauer BA, Kaufmann SH, Biel MA, Baylin SB (1997) HIC1 hypermethylation is a late event in hematopoietic neoplasms. Cancer Res 57:1678–1681

Ji W, Hernandez R, Zhang XY, Qu GZ, Frady A, Varela M, Ehrlich M (1997) DNA demethylation and pericentromeric rearrangements of chromosome. Mutat Res 379:33–41

Jones PA, Baylin SB (2002) The fundamental role of epigenetic events in cancer. Nat Rev Genet 3:415–428

Kim YI, Giuliano A, Hatch KD, Schneider A, Nour MA, Dallal GE, Selhub J, Mason JB (1994) Global DNA hypomethylation increases progressively in cervical dysplasia and carcinoma. Cancer 74:893–899

Lee WH, Morton RA, Epstein JI, Brooks JD, Campbell PA, Bova GS, Hsieh WS, Isaacs WB, Nelson WG (1994) Cytidine methylation of regulatory sequences near the pi-class glutathione S-transferase gene accompanies human prostatic carcinogenesis. Proc Natl Acad Sci USA 91:11733–11737

Li E (2002) Chromatin modification and epigenetic reprogramming in mammalian development. Nat Rev Genet 3:662–673

Li E, Bestor TH, Jaenisch R (1992) Targeted mutation of the DNA methyltransferase gene results in embryonic lethality. Cell 69:915–926

Li E, Beard C, Jaenisch R (1993) Role for DNA methylation in genomic imprinting. Nature 366:362–365

Liu S, Shen C, Huynh L, Klisovic M, Rush L, Ford J, Yu J, Becknell B, Yu L, Liu C, Vukosavljevic T, Whitman S, Chang S, Byrd J, Perrotti D, Plass C, Marcucci G (2005) Interplay of RUNX1/MTG8 and DNA methyltransferase 1 in acute myeloid leukemia. Cancer Res 15:1277–1284

Lubbert M (2000) DNA methylation inhibitors in the treatment of leukemias, myelodysplastic syndromes and hemoglobinopathies: clinical results and possible mechanisms of action. Curr Top Microbiol Immunol 249:135–164

Miki Y, Nishisho I, Horii A, Miyoshi Y, Utsunomiya J, Kinzler KW, Vogelstein B, Nakamura Y (1992) Disruption of the APC gene by a retrotransposal insertion of L1 sequence in a colon cancer. Cancer Res 52:643–645

Momparler RL, Ayoub J (2001) Potential of 5-aza-2′-deoxycytidine (Decitabine) a potent inhibitor of DNA methylation for therapy of advanced non-small cell lung cancer. Lung Cancer 34 Suppl 4:111–115

Morse B, Rotherg PG, South VJ, Spandorfer JM, Astrin SM (1988) Insertional mutagenesis of the myc locus by a LINE-1 sequence in a human breast carcinoma. Nature 333:87–90

Narayan A, Ji W, Zhang XY, Marrogi A, Graff JR, Baylin SB, Ehrlich M (1998) Hypomethylation of pericentromeric DNA in breast adenocarcinomas. Int J Cancer 77:833–838

Ng HH, Zhang Y, Hendrich B, Johnson CA, Turner BM, Erdjument-Bromage H, Tempst P, Reinberg D, Bird A (1999) MBD2 is a transcriptional repressor belonging to the MeCP1 histone deacetylase complex. Nat Genet 23:58–61

Okano M, Bell DW, Haber DA, Li E (1999) DNA methyltransferases Dnmt3a and Dnmt3b are essential for de novo methylation and mammalian development. Cell 99:247–257

Oswald J, Engemann S, Lane N, Mayer W, Olek A, Fundele R, Dean W, Reik W, Walter J (2000) Active demethylation of the paternal genome in the mouse zygote. Curr Biol 10:475–478

Plass C, Shibata H, Kalcheva I, Mullins L, Kotelevtseva N, Mullins J, Kato R, Sasaki H, Hirotsune S, Okazaki Y, Held WA, Hayashizaki Y, Chapman VM (1996) Identification of Grf1 on mouse chromosome 9 as an imprinted gene by RLGS-M. Nat Genet 14:106–109

Plass C, Yu F, Yu L, Strout MP, El-Rifai W, Elonen E, Knuutila S, Marcucci G, Young DC, Held WA, Bloomfield CD, Caligiuri MA (1999) Restriction landmark genome scanning for aberrant methylation in primary refractory and relapsed acute myeloid leukemia; involvement of the WIT-1 gene. Oncogene 18:3159–3165

Prokhortchouk A, Hendrich B, Jorgensen H, Ruzov A, Wilm M, Georgiev G, Bird A, Prokhortchouk E (2001) The p120 catenin partner Kaiso is a DNA methylation-dependent transcriptional repressor. Genes Dev 15:1613–1618

Qu GZ, Grundy PE, Narayan A, Ehrlich M (1999) Frequent hypomethylation in Wilms tumors of pericentromeric DNA in chromosomes 1 and 16. Cancer Genet Cytogenet 109:34–39

Razin A, Cedar H (1994) DNA methylation and genomic imprinting. Cell 77:473–476

Rountree MR, Bachman KE, Baylin SB (2000) DNMT1 binds HDAC2 and a new co-repressor, DMAP1, to form a complex at replication foci. Nat Genet 25:269–277

Rush LJ, Dai Z, Smiraglia DJ, Gao X, Wright FA, Fruhwald M, Costello JF, Held WA, Yu L, Krahe R, Kolitz JE, Bloomfield CD, Caligiuri MA, Plass C (2001) Novel methylation targets in de novo acute myeloid leukemia with prevalence of chromosome 11 loci. Blood 97:3226–3233

Rush LJ, Raval A, Funchain P, Johnson AJ, Smith L, Lucas DM, Bembea M, Liu TH Heerema NA, Rassenti L, Liyanarachchi S, Davuluri R, Byrd JC, Plass C (2004) Epigenetic profiling in chronic lymphocytic leukemia reveals novel methylation targets. Cancer Res 64:2424–2433

Santos F, Hendrich B, Reik W, Dean W (2002) Dynamic reprogramming of DNA methylation in the early mouse embryo. Dev Biol 241:172–182

Silverman LR (2004) DNA methyltransferase inhibitors in myelodysplastic syndrome. Best Pract Res Clin Haematol 17:585–594

Smiragli DJ, Szymanska J, Kraggerud SM, Lothe RA, Peltomaki P, Plass C (2002) Distinct epigenetic phenotypes in seminomatous and nonseminomatous testicular germ cell tumors. Oncogene 21:3909–3916

Smiraglia DJ, Plass C (2002) The study of aberrant methylation in cancer via restriction landmark genomic scanning. Oncogene 21:5414–5426

Smiraglia DJ, Rush LJ, Fruhwald MC, Dai Z, Held WA, Costello JF, Lang JC, Eng C, Li B, Wright FA, Caligiuri MA, Plass C (2001) Excessive CpG island hypermethylation in cancer cell lines versus primary human malignancies. Hum Mol Genet 10:1413–1419

Smiraglia DJ, Smith LT, Lang JC, Rush LJ, Dai Z, Schuller DE, Plass C (2003) Differential targets of CpG island hypermethylation in primary and metastatic head and neck squamous cell carcinoma (HNSCC). J Med Genet 40:25–33

Song L, James SR, Kazim L, Karpf AR (2005) Specific method for the determination of genomic DNA methylation by liquid chromatography-electrospray ionization tandem mass spectrometry. Anal Chem 77:504–510

Suzuki H, Gabrielson E, Chen W, Anbazhagan R, Van Engeland M, Weijenberg MP, Herman JG, Baylin SB (2002) A genomic screen for genes upregulated by demethylation and histone deacetylase inhibition in human colorectal cancer. Nat Genet 31:141–149

Suzuki H, Watkins DN, Jair KW, Schuebel KE, Markowitz SD, Dong Chen W, Pretlow TP, Yang B, Akiyama Y, Van Engeland M, Toyota M, Tokino T, Hinoda Y, Imai K, Herman JG, Baylin SB (2004) Epigenetic inactivation of SFRP genes allows constitutive WNT signaling in colorectal cancer. Nat Genet 36:417–422

Tamaru H, Selker EU (2001) A histone H3 methyltransferase controls DNA methylation in Neurospora crassa. Nature 414:277–283

Toyota M, Ho C, Ahuja N, Jair KW, Li Q, Ohe-Toyota M, Baylin SB, Issa JP (1999) Identification of differentially methylated sequences in colorectal cancer by methylated CpG island amplification. Cancer Res 59:2307–2312

Vachtenheim J, Horakova I, Novotna H (1994) Hypomethylation of CCGG sites in the 3′ region of H-ras protooncogene is frequent and is associated with H-ras allele loss in non-small cell lung cancer. Cancer Res 54:1145–1148

Venter JC, Adams MD, Myers EW, Li PW, Mural RJ, Sutton GG, Smith HO, Yandell M, Evans CA, Holt RA, Gocayne JD, Amanatides P, Ballew RM, Huson DH, Wortman JR, Zhang Q, Kodira CD, Zheng XH, Chen L, Skupski M, Subramanian G, Thomas PD, Zhang J, Gabor Miklos GL, Nelson C, Broder S, Clark AG, Nadeau J, McKusick VA, Zinder N, Levine AJ, Roberts RJ, Simon M, Slayman C, Hunkapiller M, Bolanos R, Delcher A, Dew I, Fasulo D, Flanigan M, Florea L, Halpern A, Hannenhalli S, Kravitz S, Levy S, Mobarry C, Reinert K, Remington K, Abu-Threideh J, Beasley E, Biddick K, Bonazzi V, Brandon R, Cargill M, Chandramouliswaran I, Charlab R, Chaturvedi K, Deng Z, Francesco VD, Dunn P, Eilbeck K, Evangelista C, Gabrielian AE, Gan W, Ge W, Gong F, Gu Z, Guan P, Heiman TJ, Higgins ME, Ji RR, Ke Z, Ketchum KA, Lai Z, Lei Y, Li Z, Li J, Liang Y, Lin X, Lu F, Merkulov GV, Milshina N, Moore HM, Naik AK, Narayan VA, Neelam B, Nusskern D, Rusch DB, Salzberg S, Shao W, Shue B, Sun J, Wang ZY, Wang A, Wang X, Wang J, Wei MH, Wides R (2001) The sequence of the human genome. Science 5507:1304–1351

Widschwendter M, Jiang G, Woods C, Muller HM, Fiegl H, Goebel G, Marth C, Muller-Holzner E, Zeimet AG, Laird PW, Ehrlich M (2004) DNA hypomethylation and ovarian cancer biology. Cancer Res 64:4472–4480

Xiong Z, Laird PW (1997) COBRA: a sensitive and quantitative DNA methylation assay. Nucleic Acids Res 25:2532–2534

Xu GL, Bestor TH, Bourc'his D, Hsieh CL, Tommerup N, Bugge M, Hulten M, Qu X, Russo JJ, Viegas-Pequignot E (1999) Chromosome instability and immunodeficiency syndrome caused by mutations in a DNA methyltransferase gene. Nature 402:187–191

Yan PS, Chen CM, Shi H, Rahmatpanah F, Wei SH, Caldwell CW, Huang TH (2001) Dissecting complex epigenetic alterations in breast cancer using CpG island microarrays. Cancer Res 61:8375–8380

Yoshikawa H, Matsubara K, Qian GS, Jackson P, Groopman JD, Manning JE, Harris CC, Herman JG (2001) SOCS-1, a negative regulator of the JAK/STAT pathway, is silenced by methylation in human hepatocellular carcinoma and shows growth-suppression activity. Nat Genet 28:29–35

Zardo G, Tiirikainen MI, Hong C, Misra A, Feuerstein BG, Volik S, Collins CC, Lamborn KR, Bollen A, Pinkel D, Albertson DG, Costello JF (2002) Integrated genomic and epigenomic analyses pinpoint biallelic gene inactivation in tumors. Nat Genet 32:453–458

Zhang L, Eugeni EE, Parthun MR, Freitas MA (2003) Identification of novel histone post-translational modifications by peptide mass fingerprinting. Chromosoma 112:77–86

CTMI (2006) 310:199–210

Decreased Fidelity in Replicating DNA Methylation Patterns in Cancer Cells Leads to Dense Methylation of a CpG Island

N. Watanabe · E. Okochi-Takada · Y. Yagi · J.-I. Furuta · T. Ushijima (✉)

Carcinogenesis Division, National Cancer Center Research Institute,
5-1-1 Tsukiji, Chuo-ku, 104-0045 Tokyo, Japan
tushijim@ncc.go.jp

Abstract Cancer cells that have a large number of aberrantly methylated CpG islands (CGIs) are known to have CpG island methylator phenotype (CIMP), and decreased fidelity in replicating methylation patters has been analyzed as an underlying mechanism. First we developed a method to analyze the number of errors in replicating CpG methylation patterns in a defined period. A single cell was expanded into 10^6 cells, and the number of errors during the culture was measured by counting the deviation from the original methylation patterns. It was shown that methylated status of a CpG site was more stably inherited than unmethylated status, suggesting that the genome is constantly exposed to *de novo* methylation. Promoter CGIs showed higher fidelities than CGIs outside promoter regions. We then analyzed error rates in two gastric cancer cell lines without CIMP and two with CIMP for five promoter CGIs. Two CIMP(−) cell lines showed error rates smaller than 1.0×10^{-3} errors per site per generation (99.90%–100% fidelity) for all the five CGIs. In contrast, AGS cells showed significantly elevated error rates, mainly due to increased *de novo* methylation, in three CGIs (1.6- to 3.2-fold), and KATOIII cells showed a significantly elevated error rate in one CGI (2.2-fold). Presence of densely methylated DNA molecules was observed only in KATOIII and AGS. These data demonstrated that some cancer cells have decreased fidelity in replicating CpG methylation patterns that underlie CIMP.

1
Introduction

DNA methylation is inherited upon cell division, and methylation of CpG islands (CGIs) in gene promoter regions is known to suppress the expression of the downstream genes (Jones and Baylin 2002; Bird 2002). Aberrant methylation of promoter CGIs of tumor-suppressor genes is known to be deeply involved in carcinogenesis. At the same time, it has recently been recognized that various CGIs, not only those of tumor-suppressor genes, are methylated in cancer cells (Costello et al. 2000; Sato et al. 2003; Ushijima 2005). Some cancers are known to have methylation of multiple CGIs, and this phenotype was designated as CGI methylator phenotype (CIMP) (Toyota et al. 1999; Issa 2004). When a cancer has a CIMP, it has been proposed that a number of important genes are inactivated due to methylation of promoter CGIs, and that this will have a significant impact on the behavior of cancers with CIMP. In fact, our recent study demonstrated that neuroblastomas with CIMP are associated with a significantly and markedly lower probability of survival (Abe et al. 2005).

On the other hand, Yamashita et al.could not find the presence of a distinct phenotype with methylation of multiple CGIs, based on their methylation analysis of seven CGIs of known tumor-related genes and 30 *Not*I sites randomly selected from the genome (Yamashita et al. 2003). This raised several issues that we have to consider when analyzing the presence of CIMP. First, as described in the first report by Toyota et al. (1999), selection of appropriate CGIs is important. Methylation of appropriate CGIs should not cause selection of cells with their methylation, because selection can cause an apparent increase of cells with methylation. Also, appropriate CGIs should not be methylated in non-cancerous tissues, since CIMP refers to abnormal cellular capacity to induce methylation of CGIs, and not to age-dependent methylation (Issa et al. 1994, 2001). Second, analysis of appropriate regions within a CGI is necessary. Methylation statuses within a CGI are not homogeneous (Ushijima 2005). A totally different methylation profile can be obtained when a core region within a CGI is analyzed and when non-core regions within a CGI are analyzed. Third, and most importantly, analysis on the dynamic speed of occurrence (rate) of methylation of CGIs or CpG sites in a defined period of time is necessary (Ushijima and Okochi-Takada 2005). Most studies so far have analyzed the number of aberrantly methylated CGIs, which is dependent upon multiple factors, including the rate of occurrence of methylation, the number of past events of clonal selection, and the number of methylated CGIs in the precursor cell.

To analyze possibly increased rates of occurrence of methylation in cancer cells, information on the rate in normal cells is indispensable. However, analysis of the rates of methylation errors has been limited. The methylated status of an exogenously introduced DNA was maintained with fidelity of 94% per generation per site by Southern blot analysis (Wigler et al. 1981). Pfeifer et al. developed the ligation-mediated PCR (LMPCR) method, and analyzed the efficiency of maintenance methylation (E_m) and that of *de novo* methylation (E_d) separately, using CpG sites within a CGI in the 5′ region of the *PGK1* gene on the inactive X chromosome (Pfeifer et al. 1990a, b). They observed an E_m of 98.8%–99.9% per site per generation and E_d of 5%. E_d of 5% corresponds to a fidelity of 95% in keeping the unmethylated status of a CpG site.

Recent advancements in bisulfite sequencing methods have enabled researchers to analyze methylation status at the nucleotide level (Clark et al. 1994). Taking advantage of bisulfite sequencing, we decided to observe a large number of CpG sites within a CGI, and measure the fidelity in maintaining their methylated or unmethylated status, and then to analyze changes in the fidelity in cancer cells.

2 Fidelity in Normal Mammary Epithelial Cells

Before analyzing fidelity in cancer cells, we had to establish a system in which we could measure the number of errors in replicating methylated or unmethylated statuses of individual CpG sites in a defined number of cell divisions (Ushijima et al. 2003). For this purpose, we seeded a single human mammary epithelial cell (HMEC) in a well of a 96-well plate, and expanded it up to 10^6 cells (Fig. 1A). From the actual count of the number of cells at harvest and the number of cells lost during two transfers, we calculated the actual number of cell divisions during the culture. Using DNA extracted from the final 10^6 cells, methylation statuses of individual CpG sites were examined. To exclude artifacts due to insufficient bisulfite treatment, unconversion rates were measured using unmethylated control DNA, and were confirmed to be small enough compared with error rates.

Methylation patterns of the differentially methylated region (DMR) of *H19* was initially examined, since distinction of maternal and paternal alleles was possible by a polymorphism and also by the overall methylation statuses (Fig. 1B). All of the unmethylated DNA molecules (molecules 1–9 in Fig. 1) had similar methylation patterns and the T polymorphism, while all the methylated DNA molecules (molecules 10–12 in Fig. 1) had similar methylation patterns and the G polymorphism. This showed that the number of errors in replicating methylation patterns was not too large during the expansion from 1 to 10^6 cells, that the methylation patterns of the two alleles in the original single cell can be inferred (molecules 1–7 for the unmethylated allele, and

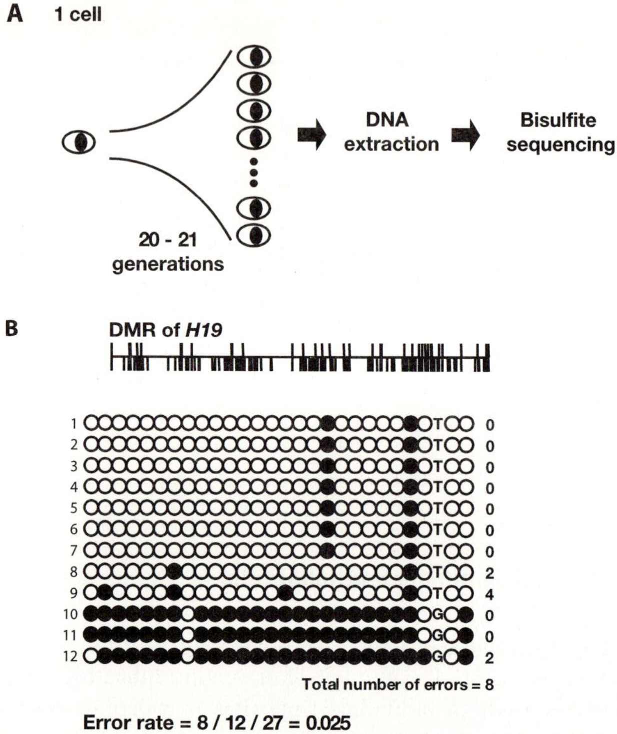

Fig. 1A, B The method to measure the fidelity in replicating methylation patterns. A A single cell was expanded to 10^6 cells, and methylation patterns in the final cell population were analyzed. **B** Example of an analysis. Twelve DNA molecules were analyzed for methylation patterns of the *H19* DMR, and deviation from the inferred original methylation patterns was calculated (the numbers of errors are shown to the *right* of clones). *Open and closed circles*: unmethylated and methylated CpG sites, respectively. T or G is a reported polymorphism. Based on the total number of CpG sites analyzed and the observed number of errors, the error rate in the defined period was measured. Six independent cultures were analyzed for each region

molecules 10 and 11 for the methylated allele), and that the number of errors in replicating methylation patterns (shown in the right of each molecule) can be measured. To obtain an accurate number of errors, six independent cultures were analyzed, and the average number of errors was calculated. Possible errors due to erroneous selection of the original methylation patterns were examined by selecting different patterns as the original methylation patterns (permutation test), and we confirmed that selection errors do not cause significant changes in the error rates.

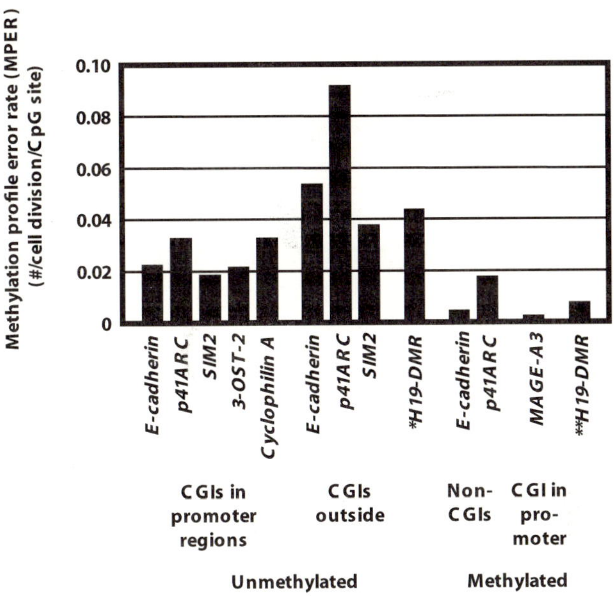

Fig. 2 Error rates in various regions of the genome in normal human mammary epithelial cells. The numbers of errors per cell division per CpG site are shown. Unmethylated regions showed higher error rates than methylated regions. This was true even when the unmethylated allele (marked with *) and methylated allele (shown by **) of *H19* were compared. CGIs in promoter regions showed lower error rates than CGIs outside promoter regions

The analysis was expanded to five CGIs in promoter regions, three CGIs outside promoter regions, CpG sites outside CGIs (non-CGIs), and a normally methylated CGI in a promoter region (Fig. 2). When unmethylated regions and methylated regions were compared, it was clear that error rates were higher in unmethylated regions. Even limited to DMR of *H19*, the unmethylated allele showed a higher rate of errors. This showed that keeping the unmethylated status of CpG sites is much more prone to errors than keeping the methylated status of CpG sites. This finding was reasonably explained using the assumption that the genome is constantly exposed to pressure of *de novo* methylation, which is in good accordance with a pioneering finding (Pfeifer et al. 1990b).

When CGIs in promoter regions and CGIs outside promoter regions were compared, the former had lower error rates. Since methylation of promoter CGIs leads to silencing of the downstream genes and is potentially harmful to a cell, it appeared that CGIs in promoter regions were protected from *de novo* methylation in a safer manner than CGIs outside.

The measurement system does not take account of errors in the very early stages of culture (founder errors), and cannot make clear distinction between a failure in maintaining methylated status and that in maintaining unmethylated status. However, the effect of founder errors was considered very small because the variation among six independent experiments was reasonably small. Since error rates in unmethylated regions and methylated regions were clearly different, distinction of the two types of errors is important, and the development of a new system that can distinguish them is necessary.

3
Decreased Fidelity in Gastric Cancer Cells

Since the system seemed to be working, we shifted to analysis of cancer cells (Ushijima et al. 2005). For this purpose, we chose two gastric cancer cell lines without CIMP (HSC39 and HSC57) and two with CIMP (KATOIII and AGS) (Kaneda et al. 2002). The fidelity was analyzed for five promoter CGIs of five genes: *bA305P22.2.3* (*bA305P*), *FLJ32130*, a homolog of *RIKEN2210016F16* (*RIKEN2210016*; currently *C9orf64*), *E-cadherin*, and *cyclophilin A*. Since cancer cells might have aneuploidy of the genes analyzed, the copy numbers were analyzed in all the four gastric cancer cell lines by fluorescence in situ hybridization (FISH) and Southern blot analysis. For each CGI, three times as many clones (or more), vs the number of alleles, were analyzed. As was the case in normal mammary epithelial cells, the six experiments were repeated. As a result, we sequenced 1,495 clones.

Gastric cancer cell lines without CIMP (HSC39 and HSC57) showed error rates smaller than 0.02 errors/CpG site per observed generation for all the five CGIs (Fig. 3). This corresponded to fidelities of 99.90%–100%. In contrast, KATOIII and AGS showed significantly elevated error rates, mainly due to increased *de novo* methylation, in one CGI (2.2-fold) and in three CGIs (1.6- to 3.2-fold), respectively. This showed that the two gastric cancer cell lines with CIMP had decreased fidelity in replicating methylation patterns that produced scattered methylation of a CGI (Fig. 4B). Interestingly, the decreased fidelity was prominent in specific CGIs, such as promoter CGIs of *bA305P* and *RIKEN2210016*.

4
Decreased Fidelity and Induction of Dense Methylation

The next question was whether or not the scattered methylation induced by the decreased fidelity really leads to induction of methylation of an entire CGI

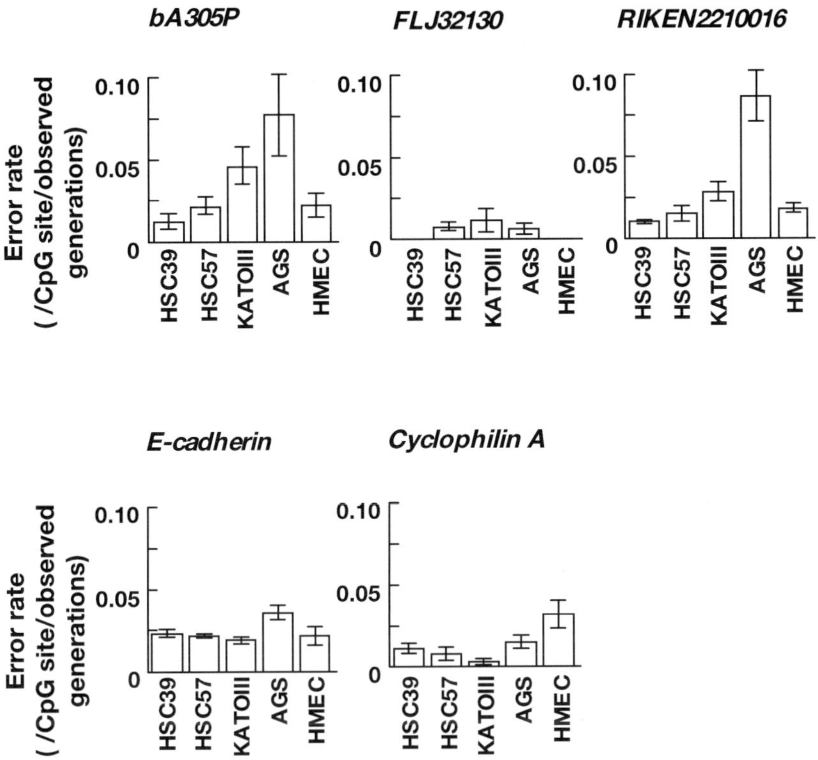

Fig. 3 Error rates of two gastric cancer cell lines without CIMP (HSC39 and HSC57) and two with CIMP (KATOIII and AGS) in five promoter CGIs. The numbers of errors per CpG sites in 21.6–23.1 generations are shown (note that the unit is different from Fig. 2). KATOIII showed increased error rates in *bA305P*, and AGS showed increased error rates in *bA305P*, *RIKEN2210016*, and *E-cadherin*

(dense methylation; Fig. 4E). As for the role of scattered methylation, Song et al. reported that both "seeds of methylation," which they created by *Hpa*II methylase, and decreased gene expression were important for induction of dense methylation of a promoter CGI (Song et al. 2002). The finding was further confirmed in the authors' following report (Stirzaker et al. 2004). Encouraged by these reports, we decided to detect densely methylated DNA molecules by selective amplification of such molecules by methylation-specific PCR (MSP) (Herman et al. 1996). MSP is known to be capable of detecting a small number of methylated DNA molecules embedded in an excess amount of unmethylated DNA molecules.

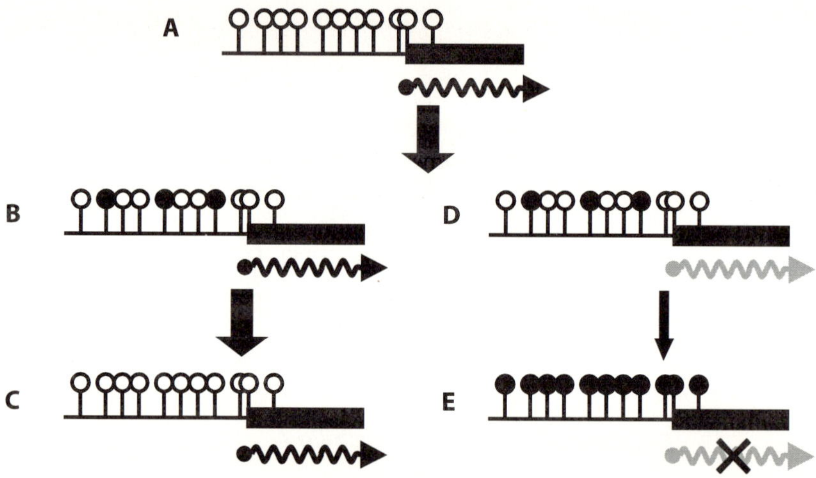

Fig. 4A–E A scheme how scattered methylation leads to dense methylation. CGIs in promoter regions are generally kept unmethylated (**A**). In cells with decreased fidelity, scattered methylation is constantly produced (**B** and **D**). In the majority of cells, the scattered methylation is erased by unknown mechanisms (**C**). On the other hand, dense methylation is induced in a minor fraction (**E**). For the induction, there is a possibility that low gene expression levels are involved

For each culture, four aliquots of bisulfite-modified DNA were amplified using primers specific to methylated *RIKEN2210016* promoter CGI. Amplification of possibly densely methylated DNA molecules was stochastically detected in KATOIII and AGS, but never in HSC39 and HSC57. The amplified DNA fragments were sequenced, and all the CpG sites between the MSP primers were shown to be methylated. The stochastic amplification suggested only one or no densely methylated DNA molecule was present in the template bisulfite-modified DNA solution. By calculating the copy number of template DNA molecule based on the efficiency of bisulfite modification, it was suggested that 1 of 250–380 DNA molecules was densely methylated in the *RIKEN2210016* promoter CGI in KATOIII and AGS.

Based on the above data, the model for occurrence of dense methylation in cell lines with CIMP is shown in Fig. 4. Due to the decreased fidelity, methylation of scattered CpG sites takes place (Fig. 4B or 4D). In most cases, the scattered methylation is erased and unmethylated status of a CGI is maintained (Fig. 4C). However, as a rare event, dense methylation of the entire CGI is induced (Fig. 4E). Since the chance that most CpG sites in a CGI are simultaneously methylated is extremely low, there should be a mechanism that induces the dense methylation when the number of seeds is high enough.

We consider that one possible determinant between erasure and induction of dense methylation is the gene expression level in a cell. Gene expression levels normally fluctuate from one cell to another, even among a homogeneous cell population. Based on the previous findings that low gene expression levels are important for induction of dense methylation (De Smet et al. 2004; Song et al. 2002; Stirzaker et al. 2004), there is a possibility that a cell with a lower expression level might undergo dense methylation.

Once dense methylation of a promoter CGI of a gene is induced, the downstream gene is silenced. The fate of a cell with the dense methylation is considered dependent upon the function of the gene silenced. If the silencing brings a growth advantage to the cell, the population will increase rapidly. If the silencing does not, the population will stay as small.

5
Molecular Basis for CIMP, and Variation of the Fidelity Among CGIs

The molecular mechanisms for the decreased fidelity in KATOIII and AGS are important. Since the decrease is mainly due to increased *de novo* methylation, increased activity of *de novo* methyltransferase and impaired function of protection mechanisms of CGIs from *de novo* methylation were theoretically postulated. Therefore, we analyzed messenger RNA (mRNA) expression levels of *DNA methyltransferase 1* (*DNMT1*), *DNMT3A*, and *DNMT3B* in the four cell lines. Although *DNMT1* and *DNMT3A* did not show any concordant expression levels, *DNMT3B* was expressed four and eight times as high as HSC39 in KATOIII and AGS, respectively (Ushijima et al. 2005). The role of high *DNMT3B* expression in the decreased fidelity needs to be further studied.

The decreased fidelity in replicating methylation patterns was prominent in some CGIs. The most probable factor was gene expression levels, but no clear association between low expression levels and high error rates was observed (Ushijima et al. 2005). Then, to explore a possible involvement of chromatin structure in the different susceptibility of CGIs, we performed chromatin-immunoprecipitation analysis. The numbers of DNA molecules bound to histone H3 acetylated (AcH3) and those bound to histone H3 dimethylated at lysine 9 (MetH3K9) were quantitatively measured, and normalized to the number of input DNA molecules (Fig. 5). Although a large amount of MetH3K9-bound DNA was observed at the *bA305P* CGI in AGS, where a high error rate was observed, other CGIs where high error rates were observed did not have high amounts of MetH3K9-bound DNA. Therefore, little molecular explanation as to why some CGIs show higher susceptibility to decreased fidelity is available so far.

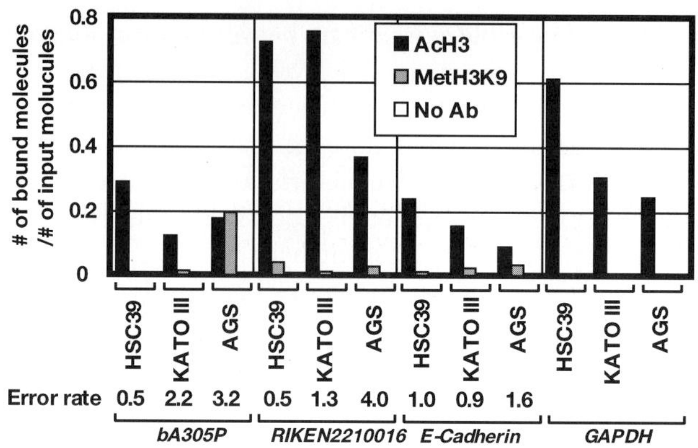

Fig. 5 Histone modifications in the five promoter CGIs analyzed for error rates. The number of DNA molecules bound to acetylated histone H3 (AcH3) and that to histone H3 dimethylated at lysine 9 (MetH3K9) were measured by real-time PCR, and were normalized to the number of molecules in input DNA. There was no clear association between histone modification and increased error rates

6 Epilogue

The fact that some cancer cells have impaired capacity in replicating CpG methylation patterns was demonstrated. The impairment was due to increased *de novo* methylation that scattered within a CGI. Although the frequency was low, the scattered methylation led to dense methylation of the entire CGI. Along with nice pioneering studies by the Clark laboratory (Song et al. 2002; Stirzaker et al. 2004), our study showed the important role of "seeds of methylation" in induction of dense methylation. The decreased fidelity was thus considered as one of the mechanisms for CIMP. However, CIMP is a complex phenotype with many different mechanisms, and the decreased fidelity cannot be applied to all cancers with CIMP.

The absolute value of fidelity measured by our system incorporates possible "error repair" mechanisms during the culture. The presence of such mechanisms is strongly suggested, since methylation statuses of many CpG sites within a CGI are generally unified. Since such error repair mechanisms are functioning even during the culture from a single cell to 10^6 cells, there is a possibility that fidelity measured by our system is higher than that measured without the influence of such an error repair system. Actually, an innovative technique recently reported by Laird et al., hairpin-bisulfite PCR, measures errors at one replication, and fidelities lower than those measured here were reported (Laird et al. 2004; Riggs and Xiong 2004). Caution seems to be necessary when we compare absolute values of fidelity measured by different systems.

Efforts to clarify the regulatory mechanisms of epigenetic fidelity now seem to be worth being invested.

Acknowledgements The authors are grateful to Drs. W. Doerfler and P. Boehm for the wonderful meeting in Weissenburg and their dedicated editing of this issue; and to Drs. M. Abe and T. Niwa for critical reading of the manuscript. The studies summarized here were supported by the Third-Term Comprehensive Cancer Control Strategy from the Ministry of Health, Labor, and Welfare, Japan. N.W. is a recipient of Research Resident Fellowship from the Foundation of Promotion of Cancer Research.

References

Abe M, Ohira M, Kaneda A, Yagi Y, Yamamoto S, Kitano Y, Takato T, Nakagawara A, Ushijima T (2005) CpG island methylator phenotype is a strong determinant of poor prognosis in neuroblastomas. Cancer Res 65:828–834

Bird A (2002) DNA methylation patterns and epigenetic memory. Genes Dev 16:6–21

Clark SJ, Harrison J, Paul CL, Frommer M (1994) High sensitivity mapping of methylated cytosines. Nucleic Acids Res 22:2990–2997

Costello JF, Fruhwald MC, Smiraglia DJ, Rush LJ, Robertson GP, Gao X, Wright FA, Feramisco JD, Peltomaki P, Lang JC, Schuller DE, Yu L, Bloomfield CD, Caligiuri MA, Yates A, Nishikawa R, Su Huang H, Petrelli NJ, Zhang X, O'Dorisio MS, Held WA, Cavenee WK, Plass C (2000) Aberrant CpG-island methylation has non-random and tumour-type-specific patterns. Nat Genet 24:132–138

De Smet C, Loriot A, Boon T (2004) Promoter-dependent mechanism leading to selective hypomethylation within the 5′ region of gene MAGE-A1 in tumor cells. Mol Cell Biol 24:4781–4790

Herman JG, Graff JR, Myohanen S, Nelkin BD, Baylin SB (1996) Methylation-specific PCR: a novel PCR assay for methylation status of CpG islands. Proc Natl Acad Sci USA 93:9821–9826

Issa JP (2004) CpG island methylator phenotype in cancer. Nat Rev Cancer 4:988–993

Issa JP, Ottaviano YL, Celano P, Hamilton SR, Davidson NE, Baylin SB (1994) Methylation of the oestrogen receptor CpG island links ageing and neoplasia in human colon. Nat Genet 7:536–540

Issa JP, Ahuja N, Toyota M, Bronner MP, Brentnall TA (2001) Accelerated age-related CpG island methylation in ulcerative colitis. Cancer Res 61:3573–3577

Jones PA, Baylin SB (2002) The fundamental role of epigenetic events in cancer. Nat Rev Genet 3:415–428

Kaneda A, Kaminishi M, Yanagihara K, Sugimura T, Ushijima T (2002) Identification of silencing of nine genes in human gastric cancers. Cancer Res 62:6645–6650

Laird CD, Pleasant ND, Clark AD, Sneeden JL, Hassan KM, Manley NC, Vary JC Jr, Morgan T, Hansen RS, Stoger R (2004) Hairpin-bisulfite PCR: assessing epigenetic methylation patterns on complementary strands of individual DNA molecules. Proc Natl Acad Sci USA 101:204–209

Pfeifer GP, Tanguay RL, Steigerwald SD, Riggs AD (1990a) In vivo footprint and methylation analysis by PCR-aided genomic sequencing: comparison of active and inactive X chromosomal DNA at the CpG island and promoter of human PGK-1. Genes Dev 4:1277–1287

Pfeifer GP, Steigerwald SD, Hansen RS, Gartler SM, Riggs AD (1990b) Polymerase chain reaction-aided genomic sequencing of an X chromosome-linked CpG island: methylation patterns suggest clonal inheritance, CpG site autonomy, and an explanation of activity state stability. Proc Natl Acad Sci USA 87:8252–8256

Riggs AD, Xiong Z (2004) Methylation and epigenetic fidelity. Proc Natl Acad Sci USA 101:4–5

Sato N, Fukushima N, Maitra A, Matsubayashi H, Yeo CJ, Cameron JL, Hruban RH, Goggins M (2003) Discovery of novel targets for aberrant methylation in pancreatic carcinoma using high-throughput microarrays. Cancer Res 63:3735–3742

Song JZ, Stirzaker C, Harrison J, Melki JR, Clark SJ (2002) Hypermethylation trigger of the glutathione-S-transferase gene (GSTP1) in prostate cancer cells. Oncogene 21:1048–1061

Stirzaker C, Song JZ, Davidson B, Clark SJ (2004) Transcriptional gene silencing promotes DNA hypermethylation through a sequential change in chromatin modifications in cancer cells. Cancer Res 64:3871–3877

Toyota M, Ahuja N, Ohe-Toyota M, Herman JG, Baylin SB, Issa JP (1999) CpG island methylator phenotype in colorectal cancer. Proc Natl Acad Sci USA 96:8681–8686

Ushijima T (2005) Detection and interpretation of altered methylation in cancer cells. Nat Rev Cancer 5:223–231

Ushijima T, Okochi-Takada E (2005) Aberrant methylations in cancer cells: where do they come from? Cancer Sci 96:206–211

Ushijima T, Watanabe N, Okochi E, Kaneda A, Sugimura T, Miyamoto K (2003) Fidelity of the methylation pattern and its variation in the genome. Genome Res 13:868–874

Ushijima T, Watanabe N, Shimizu K, Miyamoto K, Sugimura T, Kaneda A (2005) Decreased fidelity in replicating CpG methylation patterns in cancer cells. Cancer Res 65:11–17

Wigler M, Levy D, Perucho M (1981) The somatic replication of DNA methylation. Cell 24:33–40

Yamashita K, Dai T, Dai Y, Yamamoto F, Perucho M (2003) Genetics supersedes epigenetics in colon cancer phenotype. Cancer Cell 4:121–131

CTMI (2006) 310:211–250

Methylation of Endogenous Human Retroelements in Health and Disease

W. A. Schulz[1] (✉) · C. Steinhoff[2] · A. R. Florl[1]

[1]Urologische Klinik, Heinrich Heine Universität,
Moorenstr. 5, 40225 Düsseldorf, Germany
wolfgang.schulz@uni-duesseldorf.de

[2]Computational Molecular Biology, Max Planck Institute for Molecular Genetics,
Berlin, Germany

Abstract Retroelements constitute approximately 45% of the human genome. Long interspersed nuclear element (LINE) autonomous retrotransposons are predominantly represented by LINE-1, nonautonomous small interspersed nuclear elements (SINEs) are primarily represented by ALUs, and LTR retrotransposons by several families of human endogenous retroviruses (HERVs). The vast majority of LINE and HERV elements are densely methylated in normal somatic cells and contained in inactive chromatin. Methylation and chromatin structure together ensure a stable equilibrium between retroelements and their host. Hypomethylation and expression in developing germ cells opens a "window of opportunity" for retrotransposition and recombination that contribute to human evolution, but also inherited disease. In somatic cells, the presence of retroelements may be exploited to organize the genome into active and inactive regions, to separate domains and functional regions within one chromatin domain, to suppress transcriptional noise, and to regulate transcript stability. Retroelements, particularly ALUs, may also fulfill physiological roles during responses to stress and infections. Reactivation and hypomethylation of LINEs and HERVs may be important in the pathophysiology of cancer and various autoimmune diseases, contributing to chromosomal instability and chronically aberrant immune responses. The emerging insights into the pathophysiological importance of endogenous retroelements accentuate the gaps in our knowledge of how these elements are controlled in normal developing and mature cells.

1 Introduction

The draft sequence published in early 2001 (Lander et al. 2001) yielded the first realistic estimate of the number and distribution of single-copy genes in the human genome. It also put the repetitive sequences contained in the euchromatic regions of the genome into the limelight. While it had long been known that centromeric and pericentromeric heterochromatin consisted overwhelmingly of satellite tandem repeats, with some interspersed retrotransposon and single copy sequences, the analysis of the draft genome sequence revealed that roughly 45% of the euchromatic regions is also made up of repeats. This is likely an underestimate, because evolutionarily older sequences accumulate mutations in a random fashion. Therefore, they gradually lose their similarity to each other, disappearing as seemingly unique sequences into the genome background.

The predominant classes of interspersed sequences in the human genome are all retrotransposons. According to the 2001 analysis, which has been confirmed overall by the 2004 update (International Human Genome Sequencing Consortium 2004), small interspersed nuclear elements (SINEs) account for 13%, long interspersed nuclear elements (LINEs) for 20%, and long-terminal repeat (LTR) retroposons for 8%, respectively, of the sequenced genome. All DNA transposons together amount to less than 3%. SINEs are overwhelmingly represented by ALU sequences. More than one million ALUs from more than 20 subfamilies are present, which add up to roughly 10% of the genome in spite of their small size. In addition, several subfamilies of the evolutionarily older mammalian-wide interspersed repeats (MIRs) contribute. Long interspersed nuclear elements are dominated in a similar way by the LINE-1 family. Its proportion in the total genome may approach 20%, because LINE-1 elements are also relatively frequent in heterochromatin. In addition, LINE-2 and LINE-3 subfamilies can be distinguished. The LTR class is more diversified, consisting of several families and subfamilies of (human) endogenous retroviruses (HERVs) plus "mammalian apparent LTR-retrotransposons" (MaLRs).

In this review, the focus will be on the major families in the SINE and LINE classes, i.e., ALUs and LINE-1s, and on HERVs, since these are by far the best studied and seem most relevant in the context of human genetic and acquired diseases.

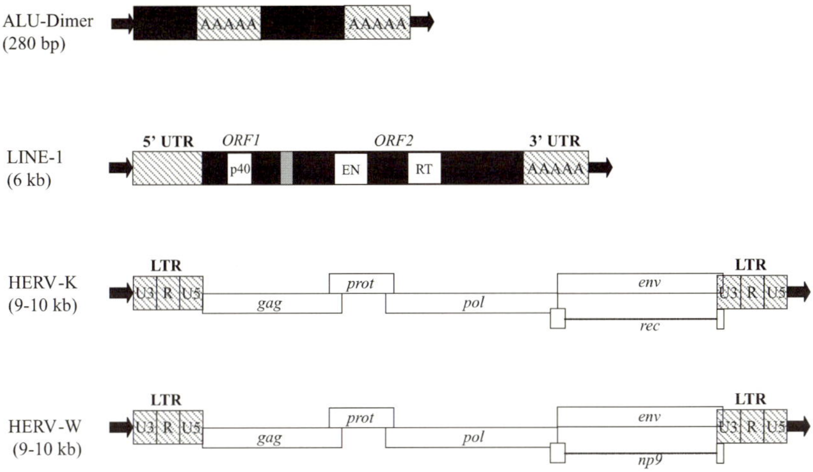

Fig. 1 The most important retroelements in the human genome. Schematic representation of ALU, LINE-1, HERV-K, and HERV-W. The lengths of the different regions are not to scale. Poly-A runs can differ in length. *Arrows* indicate repeat sequences present at many, but not all, insertion sites. *ORF1* and *ORF2*, open reading frame; *EN*, endonuclease activity; *RT*, reverse transcriptase

Intact HERV sequences (Fig. 1) possess the canonical structure of retroproviruses (Lower 1999; Nelson et al. 2003; Bannert and Kurth 2004). Two approximately 1-kb LTRs flank gag, pol, and env genes. Like most retroviruses, HERVs encode auxiliary proteins that regulate RNA processing, designated Rec and Np9. Very few HERVs in the human genome, however, have remained intact. Almost all are mutated, deleted, and truncated, and most often, the intermediary coding sequencing have been deleted by recombination between LTRs. Therefore, solitary LTRs are much more frequent than proviruses and in evolutionary terms, HERVs "tether on the brink of extinction" (Lander et al. 2001). The more surprising is the emerging evidence implicating these "vestiges of evolution" (Doerfler 1991) in the pathophysiology of several human diseases.

LINEs and SINEs are LTR-less retrotransposons. Intact LINE-1 elements (Ostertag and Kazazian 2001) are approximately 6 kb long, with two open reading frames, ORF1 and ORF2, and a terminal poly-A sequence (Fig. 1). They are autonomous but, unlike retroviruses, they are restricted to their cell of origin. ORF1 encodes a 40-kDa RNA-binding protein that appears to coordinate the transport and translation of the retrotransposon RNA. ORF2 encodes a 150-kDa protein that yields an endonuclease and a reverse transcriptase. Transcription occurs from an internal promoter located at the

5′-end of intact elements, which can be functionally divided into two parts (Swergold 1990; Hata and Sasaki 1997; Steinhoff and Schulz 2003; Lavie et al. 2004). The first approximately 200 bp are important for basic transcription, while the following 300–400-bp increase promoter strength and are necessary for ubiquitous activity. While the promoter contains sequences resembling A and B boxes typical of RNA polymerase (Pol) III promoters, most evidence now indicates that transcription is really performed by Pol II, albeit in an unusual fashion. Transcription factors acting at the promoter likely include Yin–Yang (YY)1, runt-related transcription factor (Runx), and SOX proteins (Becker et al. 1993; Kurose et al. 1995; Tchénio et al. 2000; Yang et al. 2003), but overall the mechanisms controlling LINE-1 transcription remain poorly defined. The involvement of YY1 is intriguing, since this protein has also been implicated at the HERV LTR (Knössl et al. 1999). Moreover, this transcription factor has been named Yin–Yang 1 for good reason, since it can function as an activator or a repressor.

The mechanisms involved in LINE retrotransposition have been clarified during the last few years, although not all aspects are understood (Kazazian 2004). The endonuclease induces single-strand breaks at AT-rich DNA target regions, preferably at consensus TTTT/A sites. These properties explain the tendency of LINE-1s to integrate into AT-rich heterochromatin and G-bands, and in particular their tendency to form clusters in the genome, since LINEs contain consensus sites. Following endonuclease action, the poly-A tail of the LINE RNA pairs with oligo-dT sequences in the target DNA, which serve as primers for the reverse transcriptase. Reverse transcription yields a branched DNA structure, which is resolved in a poorly understood fashion, presumably by cellular DNA repair systems. Two issues in this process are worth extra mention. First, as DNA synthesis starts from the very 3′-end of the sequence, it is evident why the majority of LINEs in the human genome are truncated at the 5′-end. Second, the retrotransposition mechanism involves an obligatory recombination and two DNA single-strand breaks in relatively close proximity, which effectively behave as a double-strand break. Therefore, the retrotransposition process carries a risk of eliciting chromosome breaks, deletions, translocations, and recombinations (Symer et al. 2002). In the human genome, less than 100 elements are thought to be intact and potentially active (Kazazian 2004). However, a larger number of elements retain intact features and could serve as a source of LINE-encoded proteins (Fig. 2).

ALU elements (Schmid 1998; Batzer and Deininger 2002) consist of two similar repeats of an approximately 150-bp sequence presumably derived from 7SL RNA, each followed by a more or less long and conserved poly-A sequence (Fig. 1). Several subfamilies can be distinguished, which differ in

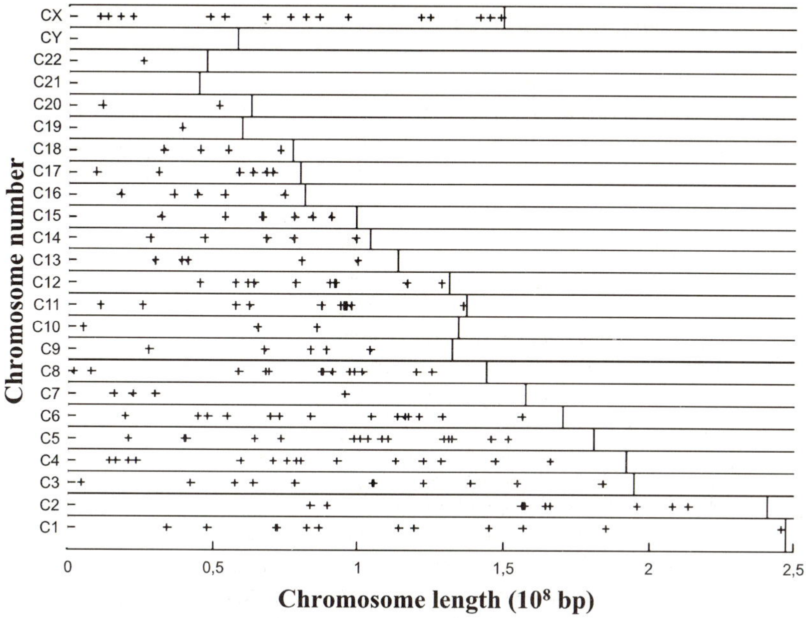

Fig. 2 Distribution of full-length LINE-1 sequences in the human genome. Full-length LINE-1 sequences are represented by *crosses* indicating their position along each chromosome. Note their overrepresentation on the X chromosome. On all other chromosomes, their prevalence does not significantly deviate from the expected frequency, if chromosome size is taken into account. The figure was obtained using a novel method that involves initially searching separately for the three conserved segments, promoter, ORF1, and ORF2, and then identifying elements with correct distances between the segments. Interestingly, not all full-length elements are members of the L1 Ta subfamily. More details on the distribution of LINE sequences and a description of the method used can be found at http://edoc.mpg.de (ID:20927.0)

sequence and GC content. It is thought that during a given evolutionary period only one subfamily is active with regard to retrotransposition, and that only a small number of elements serve as "masters" from which all others are derived. Accordingly, younger elements have recognizable internal Pol III promoters with A and B boxes. ALUs are obviously too small to encode enzymes for retrotransposition and are accordingly non-autonomous. There is now convincing experimental evidence that they use the enzymatic machinery provided by LINE-1 (Kajikawa and Okada 2002; Hagan et al. 2003). Thus, active ALU retrotransposition is dependent on LINE-1 expression. Similar examples of co-evolved pairs of SINEs and LINEs exist in other organisms.

2
Methylation of Retroelements

In somatic tissues and mature germ cells, retrotransposition and even transcriptional activity of retroelements is largely suppressed. Only retroelement sequences located in gene introns or in untranslated regions (UTRs) are transcribed, particular ALUs, which are present in roughly 75% of human genes. A window for transcriptional activity opens during germ cell development. Therefore, expression of retroelement proteins and transposition events are normally restricted to developing germ cells. Expression is also found in placental tissue, where DNA is hypomethylated overall (Li 2002). There are probably several mechanisms that prohibit retroelement activity in somatic cells. DNA methylation is crucial, as detailed below, and synergizes with tight packaging into chromatin (Kondo and Issa 2003). This is evident for retroelements like LINEs located in constitutive and facultative heterochromatin, but it also occurs in a more localized fashion. There is evidence for additional mechanisms preventing retrotransposition, even if intact elements are transcribed. LINE-1 RNA is apparently unstable and poorly translated (Han et al. 2004). ALU elements appear to activate stress responses (Schmid 1998). Moreover, ALU-containing RNAs are subject to editing (Athanasiadis et al. 2004; Kim et al. 2004; Levanon et al. 2004). Retrotransposition yields DNA strand-breaks and unusual DNA structures that ought to activate checkpoint signaling e.g., through TP53. Obviously, in meiotic germ cells, these checkpoint responses must be desensitized. Accordingly, germ cell tumors are among the few cancer types in which TP53 mutations are rare (Looijenga and Oosterhuis 2003) and which express high levels of retroelements (Table 1). Re-expression of retroelements is also found in some cancers of somatic cell origin with deficient checkpoint signaling and mutant TP53, but overall at lower levels (Table 1). Interestingly, TP53 represses transcription by Pol III (White 2004), especially of ALUs (Chesnokov et al. 1996) and severe global hypomethylation activates apoptosis via TP53, as well as through other mechanisms (Jackson-Grusby et al. 2001)

DNA methylation is central to the control of retroelements in the human genome, in human ontogeny and phylogeny (Yoder et al. 1997). Most retroelements are densely methylated in somatic cells. LINEs are most intensely studied. Analyses by Southern blot hybridization reveal intense methylation across all elements and specifically at the CpG-rich 5′-ends of intact sequences (Florl et al. 1999). Techniques that allow the investigation of individual sites, such as ligation-mediated PCR and bisulfite-PCR, indicate that a limited number of individual sites are unmethylated (Florl et al. 1999; Chalitchagorn et al. 2004; Yang et al. 2004). One study using bisulfite sequencing has reported

Table 1 Altered methylation and expression of retroelements in human cancers

Element	Change	Cancer	Remarks	Study
ALU	Hypomethylation	Germ cell cancers		Rubin et al. 1994
HERV	Hypomethylation expression	Testicular germ cell cancer		Götzinger et al. 1996
	Hypomethylation	Bladder cancer	Expression only in teratocarcinoma	Florl et al. 1999
	Expression	Breast cancer	Env and auxiliary	Wang-Johanning et al. 2001
	Expression	Prostate cancer	Auxiliary	Wang-Johanning et al. 2003
	Hypomethylation	Ovarian carcinoma		Menendez et al. 2004
LINE-1	Expression	Teratocarcinoma	Cell lines	Skowronski et al. 1988; Bratthauer and Fanning 1992
	Hypomethylation	Various	Cell lines	Dante et al. 1992
	Hypomethylation	Various	Cell lines	Alves et al. 1996
	Hypomethylation expression	Bladder cancer	Expression stronger in teratocarcinomas	Jürgens et al. 1996; Florl et al. 1999
	No hypomethylation	Renal carcinoma	Tissues only	Florl et al. 1999
	Hypomethylation	Prostate cancer	Increases with stage and metastasis	Santourlidis et al. 1999; Schulz et al. 2002; Florl et al. 2004
	Hypomethylation	Liver carcinoma		Takai et al. 2000
	Hypomethylation expression	Liver carcinoma	Hypomethylation, but not expression cancer-specific	Lin et al. 2001
	Hypomethylation	Various cancers	Differences between cancer types	Chalitchagorn et al. 2004
	Hypomethylation	Colon cancer	Begins in preneoplastic mucosa	Suter et al. 2004
	Hypomethylation	Gastric cancer	Correlates with overall hypomethylation	Kaneda et al. 2004
	Hypomethylation	Ovarian carcinoma		Menendez et al. 2004

hemimethylation of the 5′-region (Woodcock et al. 1997). Methylation of ALUs is more difficult to study because of their variability. Reports using Southern-blot analysis and a range of PCR-based techniques concur to indicate a more heterogeneous pattern of methylation (Schmid 1998; Yang et al. 2004). Hence, while the majority of elements are densely methylated, a substantial fraction is undermethylated. A recent study estimates this fraction to be 10%–15%, albeit in cell lines. The methylation of HERVs is not well investigated. The few published studies suggest that proviruses and solitary LTRs are densely methylated under normal physiological circumstances, except in developing germ cells and in the placenta (Bannert and Kurth 2004).

Both ALU and LINE promoters contain CpG sites in the promoter region that are largely methylated in normal cells. Promoter methylation would be expected to suppress transcription. In ALUs, crucial CpG sites are localized in the essential A and B boxes and their methylation prevents binding of Pol III cofactors (Liu and Schmid 1993; Kochanek et al. 1995). Repression of the internal LINE-1 promoter by methylation appears to be mediated by the methylcytosine-binding proteins MeCP2 and MBD2 (Yu et al. 2001; Steinhoff and Schulz 2003). Moreover, methylation of individual CpG sites may directly interfere with binding of transcriptional activators. The effects of methylation on HERV regulation have not been studied in detail, but some transcription factors eliciting expression in germ cells have been defined (Knössl et al. 1999; Schön et al. 2001). HERV expression has also been reported in some carcinomas with hypomethylated genomes, but at relatively low levels (Florl et al. 1999; Armbruester et al. 2002; Wang-Johanning 2001, 2003). This may reflect a requirement for germ-cell specific transcription factors in addition to hypomethylation. In contrast, transcription factors for ALU and LINE promoters appear to be ubiquitous. In summary, although more studies are required for all classes of endogenous retroelements, it is safe to conclude that their transcriptional activity is limited in somatic cells by DNA methylation.

A second mechanism involving methylation is thought to operate in the course of evolution. Methylated sites in the human genome may mutate at a higher rate, since deamination of methylcytosine yields thymidine, which is less obviously alien to DNA than uracil derived from cytosine by the same reaction. This process is generally made responsible for the peculiar distribution of CpG sites in the human genome. According to this explanation, CpG sites in methylated sequences including retroelements become gradually depleted by mutation, while sequences that are never methylated are retained as CpG islands. Indeed, older classes of ALUs contain fewer CpGs (Schmid 1998; Batzer and Deininger 2002). Newer findings suggest that this traditional explanation on the role of methylation in retroelement evolution may need some modifications. Thus, ALU sequences may mutate at non-CpG sites at similar rates (Xing et al. 2004). Moreover, a specific repair system for G–T mismatches efficiently prevents mutations at methylated CpGs (Hardeland et al. 2001).

Unfortunately, the mechanisms mediating retroelement methylation and chromatin packaging are still essentially unknown. At later stages of germ cell development, most retroelements become methylated (Li 2002), although some ALUs are exempt (Schmid 1998). The DNA methyltransferase Dnmt3L has been found to be essential for methylation and silencing of L1 Line sequences during male germ cell development in mice (Bourc'his and Bestor 2004). In the early embryo, genome-wide demethylation occurs that is followed by another round of methylation around the time of gastrulation in the actual fetus. CpG islands and selected other genes are exempt from these changes. In contrast, extra-fetal tissues become strongly hypomethylated overall. The DNA methyltransferases Dnmt1, Dnmt3a, and Dnmt3b are essential for establishment and/or maintenance of genome-wide methylation patterns, although to different extents (Bestor 2000; Li 2002). On a note of caution, these processes are much better described in the mouse than in man, for understandable reasons.

This description of development changes suggests that methylation is established as a sort of default state, i.e., all sequences become methylated unless specific factors prohibit methylation, e.g., at CpG islands and imprinted genes. It is thought that genes lacking CpG islands might also become initially methylated, but when strong transcriptional activators overcome repression and reactivate them at later stages of development, their regulatory regions become demethylated again. According to this explanation, retroelements are methylated in adult somatic cells, because DNA methylation established in the fetus is perpetuated by maintenance methylation in association with methylcytosine-binding proteins, histone deacetylases, and histone methylases. Transcriptional activators interacting with LINE-1 promoters and ALU promoters would not be strong enough to overcome the silencing effects of DNA methylation and chromatin structure.

This hypothesis is plausible, the more so as ALU and LINE-1 promoters are ubiquitously active, if they are unmethylated (see above). Moreover, ALUs re-expression during cell stress (see the following section) appears to depend on chromatin remodeling (Kim et al. 2001). Nevertheless, it remains possible that methylation of retroelements during fetal development may involve additional factors to ensure efficient recognition and silencing. One hint in this direction is the requirement for Dnmt3L for L1 methylation in mouse germ cells, and another one is the requirement of DNMT3B for the methylation of specific repeat sequences during human development. Several proteins beyond DNMTs have also been implicated. Defects in the α-thalassemia/mental retardation syndrome (ATRX) protein cause demethylation of a specific range of sequences in affected children (Gibbons et al. 2000). In the mouse, knock-out of the Lsh helicase leads to widespread hypomethylation, which preferentially affects repetitive elements (Huang et al. 2004). Its human homologs SMARCA6 and SMARCA4 are altered in some cancers (Fukuoka et al. 2004; Hendricks et al. 2004; Medina et al. 2004; Yano et al. 2004), but the relationship to DNA methylation has not been investigated. Importantly, transfection experiments suggest that de novo methylation may not be a default mechanism, but preferentially recognizes certain sequences, especially SINEs (Hasse and Schulz 1994; Turker 2002). In plants, the establishment of silenced chromatin has been shown to be directed by double-stranded (ds)RNA (Lippman et al. 2004). Because of the high propensity of ALU RNA to form double-stranded structures and of the presence of both sense and antisense transcripts of repetitive retroelements in the nucleus, it is tempting to speculate that some sort of dsRNA-directed mechanism may also be involved in retroelement silencing in humans. The results of ongoing studies are therefore eagerly awaited.

3 Functions of Retroelements in the Human Genome

Retroelements are largely inactive in normal somatic cells, and most of their activity is restricted to specific phases during the development of germ cells and to the placenta (Yoder et al. 1997; Kazazian 2004). Hypomethylation in germ cells opens a "window of opportunity" during which transposition of active elements and recombination between active, but also transcriptionally inactive, elements can take place. The changes arising in the genome can be passed on in the germ-line, unless they are lethal for the germ cell or the individual. In this fashion, retroelements have contributed to human evolution and continue to do so. This aspect has been repeatedly reviewed (Tomilin 1999; Shaffer and Lupski 2000; Ostertag and Kazazian 2001; Batzer and Deininger 2002; Ovchinnikov et al. 2002; Bannert and Kurth 2004; Kazazian 2004) and will not be treated in detail here.

Retrotransposition and recombination involving retroelements in germ cells also contribute to human inherited disease. Some events are passed on within families or in specific populations, while others cause disease in individuals, e.g., cancer resulting from translocations or gene disruption in the germ cell of a parent or during early development. These aspects, too, have been thoroughly summarized by others (Schaffer and Lupski 2000; Ovchinnikov et al. 2002; Bannert and Kurth 2004).

A further aspect of this "window of opportunity" concerns the influence of retroelements on gene regulation and has received less attention. In a considerable number of instances, transcriptional regulatory sequences from retroelements, typically from HERV LTRs, but also from LINEs, appear to have been co-opted for gene regulation (Tomilin 1999; van de Lagemaat 2003; Bannert and Kurth 2004; Kazazian 2004). In some cases, these sequences have changed almost beyond recognition and behave like host sequences with regard to methylation and activity. Specifically, they remain active in somatic cells. For instance, the enhancer of the apolipoprotein A gene is derived from a LINE-1 sequence (Yang et al. 1998). In other cases, sequences used for gene regulation still behave like other retroelements and become inactivated in somatic cells, typically associated with DNA methylation. This effectively restricts gene activity to developing germ cells, the placenta, and perhaps early embryonic development. Thus, placenta-specific expression of the growth factor pleiotropin is attributable to a germ-line insertion by a member of the HERV family (Schulte et al. 1996). The protein syncytin, which is essential for the development of multinucleated syncytiotrophoblasts in the placenta, is an env protein from an endogenous retrovirus (Mi et al. 2000). Pleiotropin and syncytin are crucial for the particular structure of the human placenta optimized for a prolonged gestational period and nutrition of a big fetus. In summary, the cooption of retroelement regulatory sequences allows targeting of transcriptional activity to specific phases in development. Conversely, in somatic cells, retroelements located at appropriate sites can function as silencers (Hewitt et al. 1995).

Importantly, testis and placenta are immunoprivileged tissues. Therefore, proteins encoded by the LINE ORFs and by HERVs are normally not presented as antigens to the immune system. Therefore, when peptides from these proteins are presented on the MHC of somatic cells, they tend to elicit immune responses. In theory, this presentation could lead to the elimination of cells with active retroelement expression and serve as a protective mechanism against cells with abnormal DNA methylation. This mechanism is hypothetical in normal cells. However, a process involving recognition of retroelement sequences, proteins, or both is strongly implicated in the pathogenesis of autoimmune disease (see Sect. 6).

While the functions of retroelements in the germ cell lineage are now relatively well understood, it is unclear to which extent they contribute to the physiology of somatic lineages. In fact, active as well as inactive retroelements might serve a variety of purposes in somatic cells. Perhaps the least controversial issue concerns the formation of heterochromatin. In some organisms, heterochromatin consists of retrotransposon sequences. These are often highly methylated and always densely packed into chromatin. It is now thought likely that heterochromatin formation in such species is directed by dsRNA resulting from the transcription of retroelements in opposite directions (Lippman et al.

2004). While in human cells only a fraction of centromeric heterochromatin consists of retrotransposons, especially LINEs (Laurent et al. 1997), they certainly contribute to its formation. LINE sequences are also overrepresented in the late-replicating G-bands of human chromosomes (Lander et al. 2001). It is plausible that their presence is responsible for the more heterochromatic character of these parts of the genome. Along the same line of argument, it has been proposed that LINE sequences may act as "way stations" during X chromosome inactivation (Bailey et al. 2000; Hansen 2003) and LINE sequences are associated with nuclear matrix attachment regions (Khodarev et al. 2000).

The association of retroelements, especially LINEs and HERVs, with inactive chromatin could also be exploited at more local levels of genome organization. Clusters of retroelement sequences could serve as boundaries between active gene domains preventing transcriptional regulatory elements of one gene from interfering with the regulation of others. This function would e.g., explain why HOX clusters, which require long-range interactions for proper regulation, are almost free of retroelement sequences (Lander et al. 2001).

In addition, methylated retroelement clusters forming inactive chromatin could serve as barriers during DNA repair and recombination. Evidence for this idea comes from a study of deletions at 9p21 in squamous cell carcinoma cells (Raschke et al. 2005). Deletion ends were found to be preferentially located in clusters of repeat sequences (LTR and LINEs) flanking the *CDKN2A* locus. The breakpoint junctions showed hallmarks of DNA double-strand break repair by non-homologous end-joining (NHEJ). A plausible explanation for this finding is that the NHEJ protein complex processing the damaged DNA ends is slowed down by denser chromatin at retroelement sequences surrounding the locus, thereby limiting the size of the ensuing deletions. Conceivably, clusters of densely methylated retroelements tightly packed into chromatin might affect DNA repair by homologous recombination in a similar manner (Fig. 3). During homologous recombination repair of DNA double-strand breaks, single-strand ends are generated that invade the homologous dsDNA and are used to prime DNA synthesis. In effect, this mode of repair then leads to the formation of a Holliday junction, which can migrate along the two helices. Both the extension of DNA synthesis and the migration of the Holliday junction might be limited by repeat clusters, thereby curbing the extent of the recombination. Conceivably, in cancer cells with hypomethylated genomes, these barrier functions might be less effective and the length of the recombinated sequence might increase.

Many retroelements are located in genome regions that are not or only weakly transcribed in normal somatic cells, but many are located within transcriptional units. Theoretically, depending on their orientation, they might disturb transcription (Whitelaw and Martin 2001) by interference (antisense)

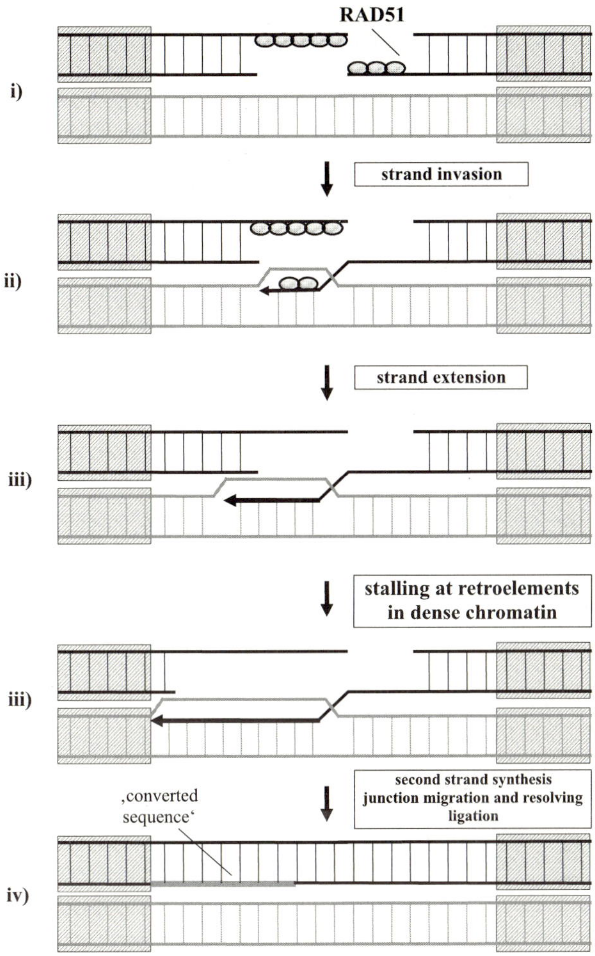

Fig. 3 Limitation of homologous recombination by retroelement methylation in somatic cells. A sketch of DNA double-strand break repair by homologous recombination in human cells. Ladders represent DNA helices, the *gray rectangles* to the *left* and *right* are schematic representations of denser chromatin resulting from the presence of methylated retroelements. For simplicity, only the limitation of DNA repair synthesis is shown. Presumably, tightly packed chromatin would also hinder repair synthesis of the second strand and migration of the Holliday junction

or competition (sense). There is little evidence for such effects in either normal somatic human cells or, more surprisingly, cancer cells. Likewise, the poly-adenylation sites in intragenic retroelements seem to be ignored by the transcriptional machinery, at least in most cases. In contrast, the presence of LINEs and ALUs in transcribed sequences may have substantial effects on RNA stability. Recently, LINE-1 transcripts were reported to be quite unstable in human cells. If this observation can be generalized, the presence of LINE sequences in many introns might make functional sense by destabilizing unspliced genomic transcripts (Han et al. 2004). Likewise, the presence of ALU elements in the 3′-UTR of many genes may have substantial effects on the stability of their transcripts. Due to their tandem structure, ALUs form very stable dsRNA. Therefore, mRNAs containing ALU sequences should be good substrates for Dicer-type RNases and accordingly tend to be unstable. The positioning of ALU sequences in the 3′-UTR of many genes may even provide a means to regulate mRNA stability through editing of the ALU sequences in the transcripts. Editing adenines to inosines prevents dsRNA formation and would be expected to increase the stability of the transcripts (Athanasiadis et al. 2004; Kim et al. 2004; Levanon et al. 2004).

Older ALU elements and MIR family SINEs may overall behave in similar ways as the majority of LINEs and HERVs. A number of diverse observations indicate that evolutionarily younger ALU subfamilies may fulfill additional, specific functions in the human genome. The first clue comes from their peculiar distribution in the human genome. Throughout the genome, the presence of ALUs is highly significantly correlated with gene and GC content (Grover et al. 2004). Specifically, many genes contain one or several ALUs in close proximity 5′ to their CpG islands (Lander et al. 2001). It is currently not understood what this non-random arrangement signifies. Conceivably, it contributes to gene regulation, perhaps by providing a mark for the edge of basal promoters. Moreover, ALUs are densely methylated in normal cells and in transfection experiments, ALUs and similar elements are preferentially recognized as "centers of de novo methylation" from which methylation spreads into surrounding DNA (Turker 2002). Hence, the presence of ALUs upstream of active genes and of retroelements in their bodies may establish a generally methylated state that "highlights" the unmethylated CpG island where transcriptional activators and the Pol II complex bind. In this fashion, retroelements could help to facilitate recognition of gene regulatory regions and to suppress spurious initiation of transcription.

At some loci, ALUs located upstream of CpG islands may be involved in gene regulation. An example is provided by the *KIR* gene cluster at 19q13, which originated through repeated duplication and recombination (Wilson et al. 2000). The *KIR* genes encode surface proteins that determine the specificity of natural killer cells. In each clone, a subset of the genes is selected for expression, contributing to diversity together with genetic variation through highly variable haplotypes. The genes not selected for expression become

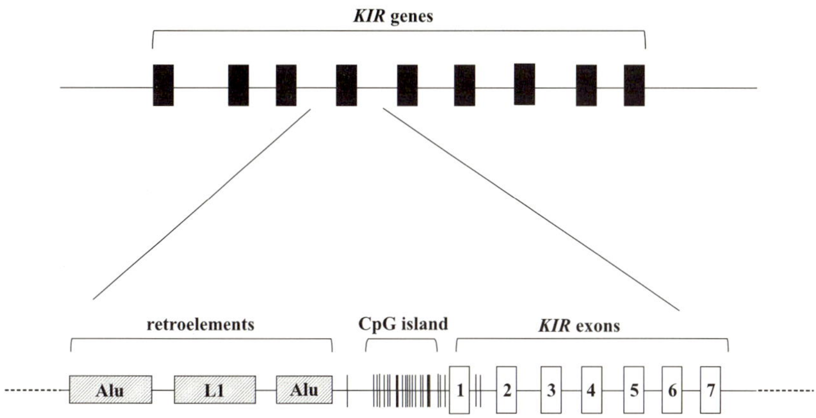

Fig. 4 Retroelements in the regulation of *KIR* gene selection. Schematic representation of the organization of an invididual gene within the *KIR* gene cluster. See Wilson et al. (2000) for the detailed structure of the cluster

methylated and silenced. Intriguingly, each *KIR* gene retains retroelements, typically ALU of the S family or LINE fragments (or both) upstream of its CpG island (Fig. 4). It is tempting to speculate that in the non-selected genes methylation spreads from these repeat sequences into the adjacent promoter to lock in the silenced state (Santourlidis et al. 2002). Such a function of ALU sequences is also consistent with their absence from many imprinted genes (Greally 2002).

A caricature of this mechanism may act during carcinogenesis. Many genes hypermethylated in cancer contain ALUs upstream of their CpG islands (Graff et al. 1997). In some cases where the mechanisms have been more closely investigated, methylation appears to spread from upstream ALUs into the promoter region (Song et al. 2002; Levine et al. 2003; Stirzaker et al. 2004).

While under normal circumstances ALU sequences are only represented in the cytoplasm as parts of protein-coding mRNAs, bona fide ALU transcripts are found in the cytoplasm during various types of cellular stress, including heat-shock and inhibition of translation (Schmid 1998). Cytotoxic chemotherapy may elicit a similar response (Hagan and Rudin 2002). The ALU transcripts represent more than 100 different elements. Monomeric transcripts from the upstream half predominate. This is one of several arguments suggesting that regulatory sites upstream of the individual ALUs determine the set of elements that is induced (Vorce et al. 1994; Li and Schmid 2001). The induced ALU RNAs appear to exert several effects, including regulation of translation, inhibition of RNA-dependent protein kinase (PKR), and perhaps activation of toll-like receptors (Schmid 1998; Goldberg et al. 2000; Rubin et al. 2002; Sivori et al. 2004).

It is ALUs that have been predominantly implicated as mediators of cellular stress responses, but some evidence, not least from human autoimmune diseases (see Sect. 6), implicates HERV and LINE sequences as well. If active LINE retrotransposons were indeed induced, some types of cellular stress might provide an opportunity for retrotransposition. This has been experimentally shown for an ALU introduced into mouse cells (Hagan et al. 2003). While the use of mouse cells in this experiment is understandable, because ALU retrotranspositions are hard to detect against a background of more than a million similar sequences, it creates some uncertainty, because mouse L1 retrotransposons are much less well controlled than human LINEs (Ostertag and Kazazian 2001). The same caveat applies to other examples of retroelements activated in rodent cells. For instance, anoxia activates VL30 elements in rodent fibroblasts, which elicit a number of responses including secretion of metalloproteinases and increased chromosomal instability (Estes et al. 1995). Unquestionably, if similar processes occur in humans too, they would contribute to inflammation and to tumor progression during hypoxia and chemotherapy. Whether they do happen in vivo is uncertain, but further studies are clearly worthwhile.

In summary, there is evidence that the presence of retroelements in the human genome is exploited to organize the genome into active and inactive regions, to separate domains and functional regions within one chromatin domain, to direct transcription and regulate transcript stability, and to respond to cellular stress. Accordingly, changes in DNA methylation and chromatin structure at retroelement sequences carry the potential to influence a variety of cellular functions. It is currently not known to which extent methylation changes in retroelements modulate gene expression in physiological states. More clearly, such changes contribute to the pathophysiology of a range of human diseases.

4
Endogenous Retroelements in Acquired Human Diseases

It is now well documented how retroelements cause inherited human diseases by transpositions and recombinations in germ cells. Very likely, they also contribute to acquired diseases, but neither the extent nor the mechanisms of their involvement are adequately understood. Many studies suggest a role for retroelements in human cancer that may be intimately related to alterations of DNA methylation in this group of diseases. By now, diseases in which retroelements are implicated include a much wider range of afflictions, such as atherosclerosis (Hiltunen and Yla-Herttuala 2003), schizophrenia (Shastry 2002; Kan et al. 2004), and even aging in general (Fuke et al. 2004). A particularly fast-growing volume of evidence links retroelements to inflammatory and autoimmune diseases, specifically lupus erythematosus, rheumatoid arthritis, juvenile diabetes, and multiple sclerosis. With some justification, atherosclerosis could also be counted among these diseases (as among tumors). In the present article, we will focus on cancer and autoimmune diseases, since these are best studied.

5 Retroelement Methylation and Expression in Human Cancer

Alterations of DNA methylation accompany carcinogenesis in many human tissues, and DNA methylation is now regarded as one of several mechanisms driving cancer development and progression (Jones and Baylin 2002). In the genome of cancer cells, DNA methylation is often severely disturbed. CpG islands surrounding the transcriptional start regions of individual genes become hypermethylated. This aberrant methylation is typically associated with gene silencing or a promoter switch. Hypermethylation affects only a few genes in some cancers, but hundreds in others. In spite of local hypermethylation, the overall methylcytosine content in the DNA of cancer cells is often decreased as a consequence of hypomethylation of repeat and unique sequences, and specifically of retroelements (Ehrlich 2002). This phenomenon is commonly denominated "genome-wide" or "global" hypomethylation.

Although genome-wide hypomethylation in human cancers as such was first described 20 years ago, it is still insufficiently understood. Even the description of hypomethylation changes in human cancers is fragmentary. For instance, HERV methylation has been studied in only a few cancers (Table 1). The data suggest that endogenous retroviruses are affected by genome-wide hypomethylation in parallel to LINE-1 sequences (Florl et al. 1999). In selected cancers, the sequences may be almost completely unmethylated. As one might expect, HERV sequences are strongly hypomethylated in testicular cancers (Götzinger et al. 1996) in accordance with their origin from germ cell precursors with hypomethylated genomes (Schulz 1998; Smiraglia et al. 2002; Looijenga and Oosterhuis 2003). Accordingly, expressed sequences derived from HERVs are found in germ cell cancers, and antibodies directed against HERV-encoded proteins are found in the blood of patients (Götzinger et al. 1996). They can be used to monitor the course of the disease during chemotherapy (Kleiman et al. 2004). In cancers of somatic cell origin, bona fide transcripts for env and the auxiliary proteins have been reported, especially in breast cancers (Wang-Johanning et al. 2001; Armbruester et al. 2002). Some results suggest that expression is found in a wider range of cancers and even normal tissues (Sugimoto et al. 2001; Stauffer et al. 2004; Yi et al. 2004). These data need further verification to exclude artifacts from genomic DNA and unspliced transcripts. Moreover, the somewhat surprising findings that different transcripts from different subfamilies may be expressed in a cancer type-specific fashion call for a closer analysis of the mechanisms involved.

Overall ALU methylation in cancers is also insufficiently studied (Table 1), even though individual elements located at CpG islands have been investigated in detail by bisulfite sequencing. This state of things may partly reflect that studying methylation of ALU sequences is tedious due to their heterogeneity and their tendency to form stable secondary structures. Like HERVs, many ALUs are clearly hypomethylated in germ cell cancers (Rubin et al. 1994). There is less evidence that extensive ALU hypomethylation occurs in cancers of somatic cell origin, but it is unclear whether this is due to publication bias against negative findings or to a lack of studies. ALUs located near CpG islands are typically methylated in normal cells and remain so when aberrant CpG island hypermethylation develops. As speculated above, these ALUs may in fact represent the centers from which methylation spreads into downstream promoter sequences (Turker 2002). Detailed analyses of hypermethylation in the *glutathione S-transferase pi* (*GSTP1*) and *target of methylation-induced silencing* (*TMS1*) genes suggest that hypermethylation gradually extends from upstream ALUs into their CpG islands (Song et al. 2002; Levine et al. 2003; Stirzaker et al. 2004). In the *GSTP1* gene, an AT-rich boundary separates an upstream ALU from the CpG island, whose function may break down during prostate carcinogenesis. Similarly, in breast cancers, hypermethylation appears to gradually "creep in" from the 5′-direction into the *TMS1* promoter CpG island. Such findings indicate that at least ALUs adjacent to hypermethylated CpG islands do not become significantly hypomethylated in human cancers. Obviously, ALUs located away from genes may behave differently.

Hypomethylation of LINE-1 sequences is better studied and has been detected in many human cancers (Table 1). In general, it appears to parallel overall hypomethylation (Kaneda et al. 2004). LINE-1 methylation has traditionally been investigated by digestion of DNA with methylation-sensitive restriction enzymes followed by Southern blotting or ligation-mediated PCR (Florl et al 1999). More recently, bisulfite-based PCR methods have been developed to estimate the status of LINE-1 methylation (Chalitchagorn et al. 2004; Yang et al. 2004). An important conclusion from such studies is that LINE-1 hypomethylation is not uniform in different cancers in two respects. First, different extents of hypomethylation are found in cancers of the same type. These differences persist in cancer cell lines and are therefore not only due to differences in the proportion of tumor cells in the tissue samples. Second, cancer types differ with regard to the stage at which hypomethylation appears. For instance, it is an early event in colon, gastric, and bladder cancers, but sets in at a later stage in prostate carcinomas. Primary renal

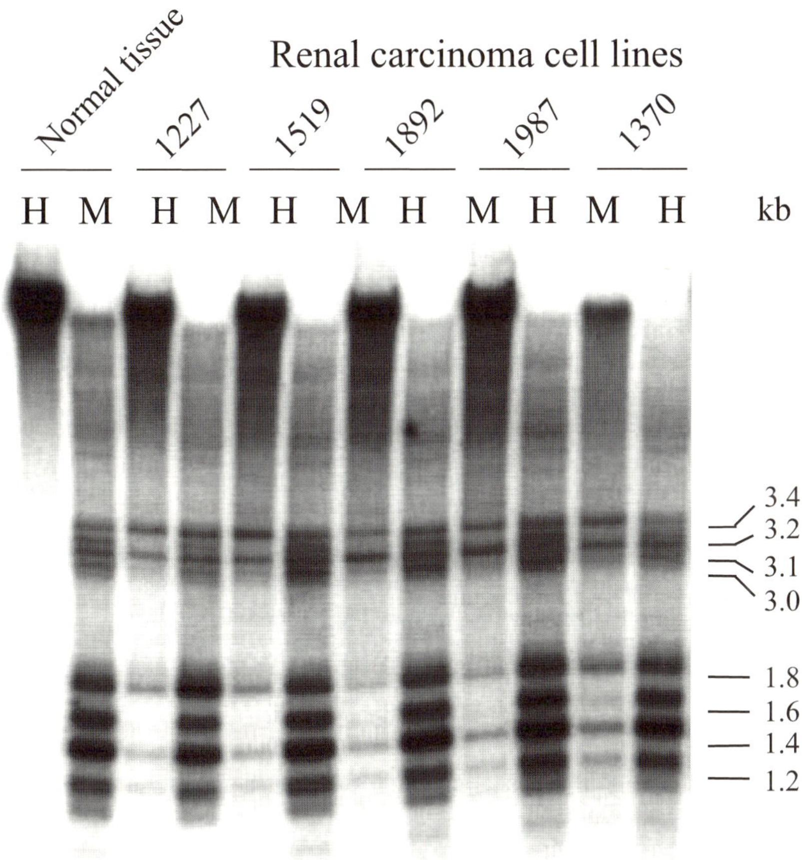

Fig. 5 Hypomethylation of LINE-1 sequences in renal carcinoma cell lines. Analysis of LINE-1 hypomethylation in five renal carcinoma cell lines by Southern blot analysis. DNA was cut either with the methylation-sensitive restriction enzyme *Hpa*II (*H*) or its methylation-insensitive isoschizomer *Msp*I, run on an agarose gel, blotted and hybridized to a LINE-1 probe (see Florl et al. 1999 for details). All cell lines show substantial hypomethylation, in contrast to normal and cancerous renal tissues

carcinomas typically lack LINE-1 hypomethylation, although it appears in cell lines (Fig. 5). Germ cell cancers have generally hypomethylated genomes. Accordingly, expression of full-length LINE-1 sequences is by far strongest in teratocarcinoma cell lines, while weaker expression is also observed in carcinoma cells exhibiting hypomethylation (Florl et al. 1999).

The causes for the differences between different cancer types remain unknown, like the causes for global hypomethylation in general. Several plausible hypotheses have been put forward, but at present none can be considered proven (Ehrlich 2002; Hoffmann and Schulz 2005). DNA hypomethylation may be a consequence of (1) *S*-adenosylmethionine deficiency in replicating cancer cells, (2) misregulation of DNA methyltransferases or presumptive DNA demethylases, or (3) both. Perhaps most likely, it could be associated with the general reorganization of chromatin structure in aneuploid cancer cells that disturbs the compartmentation of the genome (Ferreira et al. 2001; Geiman and Robertson 2002; Hoffmann and Schulz 2005). Indeed, in addition to DNA methylation itself, a variety of chromatin regulator proteins have been reported to be aberrantly expressed or even to be mutated in human cancers (Ferreira et al. 2001; Geiman and Robertson 2002; Muegge et al. 2003; Lund and van Lohuizen 2004; Hoffmann and Schulz 2005).

In this context, the question is often posed whether altered methylation of retroelements in cancer cells is a cause or consequence of the alterations in DNA methylation and chromatin structure (or both). This question relates to that discussed above: How are retroelements recognized when they become inactivated during embryonic development? Since intact LINE-1 and ALU retrotransposons contain ubiquitously active promoters, their inactivation cannot be simply a consequence of transcriptional inactivity, but must be actively established. Most researchers agree that this could occur as a one-time event during fetal development (Li 2002). It is less clear whether maintenance of the inactive state solely relies on the combined action of DNA methylation and an inactive chromatin structure or whether it is aided by a more specific repression mechanism. This distinction becomes important when considering hypomethylation of intact LINE-1 retrotransposons in cancer cells (Fig. 6). If a specific mechanism were involved, hypomethylation could be initiated by its failure. No global repressor of retroelements is established, but certain proteins possess some of the necessary properties, e.g., Lsh alias SMARCA6 (Huang et al. 2004). Alternatively, if silencing of retroelements in somatic cells relied exclusively on the maintenance of an inactive state, random genome-wide hypomethylation could initiate demethylation of LINE-1 sequences. Partial demethylation could then allow recognition by ubiquitous transcriptional activators promoting further hypomethylation, to the point of re-activation. In the case of HERVs, the issue is more complex, since their

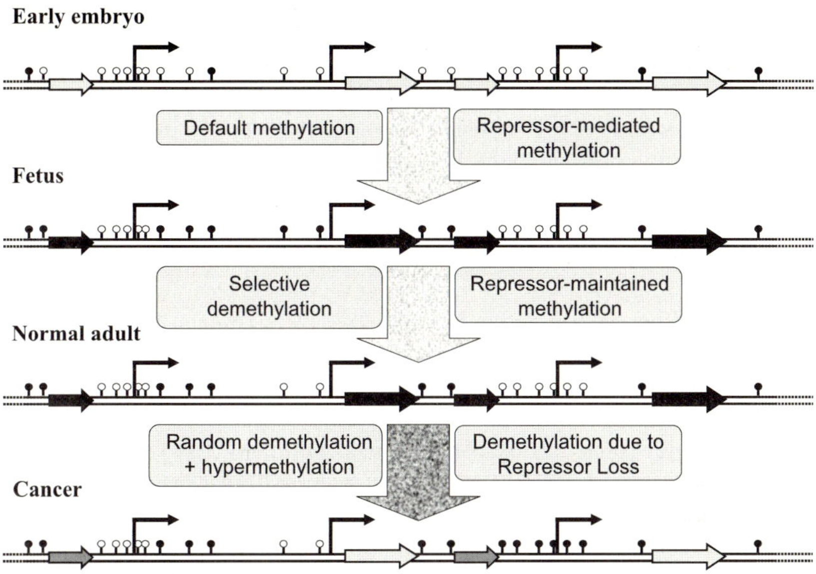

Fig. 6 Default methylation vs specific repression of retroelements in development and cancer. Two alternative explanations for methylation in development and cancer are depicted (*left* and *right* of the *large vertical arrows*). Three genes are shown schematically; that in the *center* does not possess a CpG island. Retroelements are symbolized by *horizontal arrows*, their methylation state by *shades of gray*; *short arrows* are ALUs, *longer arrows* are LINEs. *Black or white circles* denote methylated CpG sites in single-copy sequences

LTR promoters appear to be more cell-type specific. Of course, incomplete retroelements in the genome are most likely to be indeed methylated during embryogenesis as part of a global, unspecific methylation process and to become hypomethylated in cancer cells by random demethylation.

In many, albeit not all, cancer types, the extent of global hypomethylation is associated with disease progression and specifically with metastasis (Ehrlich 2002; Hoffmann and Schulz 2005). Since global hypomethylation not only affects retroelements, but also other repeat sequences such as classical CpG-rich satellites and selected single-copy genes, hypomethylation of retroelements is not necessarily the only or even most important factor responsible for this association. For instance, hypomethylation of juxtacentromeric satellite sequences has been linked to particular chromosomal alterations in specific cancers (Ji et al. 1997; Qu et al. 2000; Wong et al. 2001; Widschwendter et al. 2004). Hypomethylation of tandem repeat satellite sequences is thought to cause decondensation of pericentromeric chromatin and an increased propensity for chromosomal breaks and rearrangements in this region. In a similar fashion, hypomethylation of retroelement sequences dispersed in the genome could facilitate illegitimate recombination. Indeed, ALU sequences appear to be overrepresented near the breakpoints of translocations in hematological cancers (Kolomietz et al. 2002). LINE sequences may promote formation of double minute circular chromosomes in cancer cells (Jones and Potter 1995; Huang et al. 1998). They are also enriched at the ends of 3p14.1 and 9p21 deletions in carcinomas (Mimori et al. 1999; Florl and Schulz 2003; Raschke et al. 2005).

These findings are suggestive, but there are several caveats. First, far too few chromosomal breakpoints have been investigated, especially in carcinomas. Second, it is not known whether hypomethylation of repeat sequences really preceded the chromosomal alterations in any of the cases where they were associated with translocations or deletions. Third, ALU sequences contain a core sequence (Rudiger et al. 1995; Jeffs et al. 1998) that may preferentially bind proteins mediating homologous recombinations, thereby necessitating their location near breakpoint junctions. More generally, recombinations arising during homologous repair of DNA double-strand breaks may preferentially employ longer repeat sequences in the genome, which are essentially all retroelements, as a target for invasion by the processed single-strand from the damaged DNA helix. This would likewise explain the presence of repeat sequences, particular LINEs, at sites of illegitimate recombinations. In addition, as discussed in Sect. 3, the preferential location of deletion endpoints at retroelement clusters could also be due to stalling of DNA repair and homologous recombination at retrotransposon sequences densely packed into chromatin (Fig. 3). Paradoxically, retroelement hypomethylation could therefore diminish the tendency of breakpoints to be located at retroelement sequences, while causing an increase in the size of deleted and recombinated sequences.

The effects of retroelement re-expression in cancer cells are also inadequately understood. In the germ-line, retrotransposition events take place quite regularly, at an estimated rate of 1 event per 100 births (Kazazian 2004). Re-expression of retroelements in cancer cells might therefore be expected to result at least occasionally in retrotranspositions leading to tumor suppressor gene disruption or oncogene activation. Since such events ought to be strongly selected for during cancer growth, one would expect to find a considerable number of retrotranspositions in cancer cells. In fact, very few reports on such events have appeared in the literature, and it is not even certain whether those observed really originated in somatic cells. Since this lack of reports can hardly be caused by a publication bias, an explanation for the enigma is required. One possible explanation is that LINE-1 transcripts are unstable

(Han et al. 2004). However, in germ-cell cancers, LINE-1 transcripts are abundant (Bratthauer and Fanning 1992; Skowronski et al. 1998; Florl et al. 1999), but retrotransposition events have neither been reported. Hence, transcript instability may be only part of the explanation. Another interesting possibility is that retrotransposition events become initiated quite frequently, but are regularly aborted due to deficiencies in DNA repair and recombination in cancer cells. This speculative hypothesis is in accord with the preferential location of deletion ends in carcinomas at LINE-1 clusters. It would also explain why no retroelements have been found to insert at sites of DNA double-strand break repair in cancer cells, although they do so in model experiments (Morrish et al. 2002). If the hypothesis is correct, retroelements could contribute directly to chromosomal instability in cancer cells by creating DNA breaks, even though retrotransposition as such may be rare.

Another gap in our knowledge concerns the proteins encoded by HERVs and LINEs. Reverse transcriptases and endonucleases from both classes of elements, and the auxiliary Rec and Np9 proteins of HERVs could have profound effects on genomic stability and on gene expression (Bannert and Kurth 2004). These proteins are expressed in germ cell cancers, but definitive data on somatic cell cancers are missing. In addition, as discussed in Sect. 6, there is evidence from human autoimmune diseases that retroelement proteins can be recognized as antigens by the immune system. Many cancers express proteins that are otherwise only present in fetal tissues and germ cells (Zendman et al. 2003). Oncofetal or cancer-testis antigens are known to elicit immune responses directed at the cancer. Interestingly, several genes encoding such antigens have been shown to be controlled by DNA methylation and, accordingly, re-expression to be associated with promoter hypomethylation. The best-characterized group of such genes encodes melanoma-associated antigen (MAGE)-A cancer-testis antigens (de Smet et al. 2004). Conceivably, retroelement proteins may also behave as cancer-testis antigens. If so, antibodies directed against HERV proteins in testicular cancer patients (Götzinger et al. 1996) are the tip of the iceberg, and a wider range of immune responses may be elicited by retroelement proteins (Schiavetti et al. 2002). Finally, as discussed already, ALUs and other retroelements are induced by cellular stress in human cells (Schmid 1998). Global hypomethylation would be expected to significantly enhance this induction process. Therefore, induction of retroelement expression is expected to occur in hypoxic tumor cells and during chemotherapy (Hagan and Rudin 2002). The obvious question yet to be answered is whether these changes are strong enough to significantly alter the properties of tumor cells.

6 Retroelement Methylation and Expression in Human Autoimmune Diseases

Chronic lupus erythematosus is an autoimmune disease in which autoantibodies develop against cytosolic and nuclear self-antigens including histone epitopes and even DNA. The symptoms are very heterogeneous and variable, ranging from mild skin rashes to lethal kidney failure and brain damage. It is usually treated by unspecific immune-suppressive drugs such as glucocorticoids, with varying success. The causes of the disease are multifactorial. Genetic predispositions are evidently involved, but the disease is usually precipitated by infectious diseases, medical drugs, or a combination of the two factors (Richardson 2003; Januchowski et al. 2004).

According to current understanding, a crucial component of the disease is an unspecific activation of $CD4^+$ T lymphocytes. These "helper" cells become active independent of presented antigens and acquire properties of cytotoxic T cells such as the expression of perforin. A comparable loss of specificity can be induced experimentally by inhibitors of DNMTs or by inhibitors of canonical mitogen-activated protein kinase (MAPK) signaling which prevent the induction of DNMT1 that is normally associated with increased DNA synthesis and proliferation during T cell activation (Oelke et al. 2004). Accordingly, several genes induced in this condition, including *LFA-1A* (*Leukocyte-adhesion glycoprotein subunit alpha*) and *PRF1* (*perforin*), have been shown to become hypomethylated (Kaplan et al. 2004). In addition, HERV transcription becomes induced, apparently as a consequence of genome-wide hypomethylation (Okada et al. 2002; Ogasawara et al. 2003). As HERV proteins can elicit the production of autoantibodies when presented to the immune system (Götzinger et al. 1996; Herve et al. 2002), the provirus sequences may contribute to the production of the signals that lead to misdirection of the immune response in this disease.

Several drugs that can precipitate lupus erythematosus have meanwhile been shown to interfere with DNA methylation (Richardson 2003). Most prominently, procainamide is a relatively efficacious inhibitor of the major maintenance methyltransferase, DNMT1 (Villar-Garea et al. 2003). It is therefore plausible that such drugs may act by eliciting DNA hypomethylation in a sensitized immune system. Therefore, on a note of caution, novel drugs blocking DNMT1 that are being developed for cancer treatment (Szyf 2005) should be closely monitored for their effects on the immune system. In fact, the specificity of $CD4^+$ cells is by far not the only property in the immune system that is fixed by DNA methylation (Fitzpatrick and Wilson 2003; Teitell and Richardson 2003). Another example is provided by KIR proteins on natural killer (NK) cells discussed in Sect. 3.

A similar, but far from complete line of evidence links hypomethylation endogenous retroelements to rheumatoid arthritis (Seemayer et al. 2001). In this chronic progressive disease, synovial lining cells proliferate, while joint cell linings and cartilage are destroyed by an uncontrolled inflammatory reaction that involves various immune cells, including autoreactive T cells and activated synovial e activation of the fibroblasts in this disease resembles in many respects that in the stroma of malignant tumors, with enhanced proliferation and migration, and the secretion of cytokines, chemokines, and proteases. In particular, there is evidence that the fibroblast genomes are hypomethylated. Accordingly, promoters were found to be hypomethylated and full-length LINE-1 sequences to be expressed (Neidhart et al. 2000). Overexpression of the p40 ORF2 protein has been suggested to activate stress-induced protein kinases (Kuchen et al. 2004). The primary cause of rheumatoid arthritis is unknown. Speculations invoke an aberrant reaction to infection by an as-yet-undefined virus. In this context, hypomethylation and expression of endogenous retroelements are considered to potentially serve as an amplification step in the pathogenesis of the disease (Seemayer et al. 2001).

A very specific hypothesis has been proposed how volved in the pathogenesis of type I (juvenile) diabetes et al. 2001). In this autoimmune disease affecting children and adolescents, insulin-secreting pancreatic β-cells are destroyed by autoreactive cytotoxic T cells. Susceptibility to the disease is inherited, and is very strongly associated with specific major histocompatibility (MHC isotypes, including Vβ7. Autoimmunity is likely precipitated by common viral childhood infections to which predisposed children overreact. Interestingly, susceptibility to the disease is also highly significantly associated with a polymorphism in the HERV-K18 provirus located in the CD48 gene (Marguerat et al. 2004). Expression of the provirus and T cells reactive to an antigen encoded by HERV-K18 has been reported to be detectable in tissue during early phases of the disease (Conrad et al. 1997). Thus, in susceptible individuals, the initial viral infection may elicit the expression of the HERV sequences, which perpetuates the immune response. It has been proposed that the expression of the endogenous antigen may be induced by 5; in response to the initial viral infection (Stauffer et al. 2001). According to this hypothesis, the predisposition toward the disease arises by the interaction of antigens from a particular endogenous retrovirus with a specific MHC isoform that causes the immune reaction to its encoded antigen to get out of control.

A fourth variation on the same theme emerges from investigations of multiple sclerosis. This autoimmune disease, in which activated immune and glial cells destroy Schwann cells providing the myelin sheaths for peripheral axons, has long been speculated to be precipitated by viral infections. Most commonly, herpes viruses such as HHV-6 and HSV-1 have been implicated, and again, genetic predisposition, including linkage to particular HLA haplotypes, is thought to be involved. In the late 1990s, at least two HERV members, a HERV-W (designated MSRV in this context) and a HERV-H (alias RGH), were found to be activated in the affected tissues (Perron et al. 1997; Monteyne et al. 1998; Fujinami and Libbey 1999). Several mechanisms for their activation have been proposed, including direct transactivation by herpes viruses and indirect transcriptional stimulation as a consequence of cytokine action enhanced by the primary infection or the secondary immune response (Clerici et al. 1999; Perron et al. 2000; Serra et al. 2003; Brudek et al. 2004). It is not known for certain whether activation is associated with hypomethylation of HERV provirus sequences. The protein products of the HERVs are thought to contribute to the pathophysiology of the disease as autoantigens, but also directly as cytotoxic agents, especially the HERV-W env protein syncytin (Antony et al. 2004). Intriguingly, as in lupus patients, MSRV activation has also been observed in B cells isolated from multiple sclerosis patients (Perron et al. 1997).

In spite of considerable gaps in our knowledge, if considered together, the results from several autoimmune diseases suggest a common pathogenetic mechanism in which endogenous retroelements are crucially involved (Fig. 7). In the first phase of the disease, certain infections, particularly by viruses or drugs (or both) interfering with determinants of the immune system fixed by DNA methylation may elicit the expression of endogenous retroelements, prominently HERVs. Open questions are whether retroelement expression is generally associated with or even caused by hypomethylation and whether it is induced directly through transactivation by exogenous viruses or by cytokines secreted in response to viral infections. In each case, however, in a second phase of the disease, the expression of endogenous retroelements appears to contribute to a sustained inflammation and the development of an autoreactive immune response in susceptible individuals. Disease susceptibility may depend on the relationship between polymorphic retroelement antigens and the immune repertoire. In addition, intracellular stress responses elicited by retroelement expression may enhance cell activation and disturb recognition by the immune system. Specific retroelement products may also be cytotoxic. In the third chronic phase of disease, tissue destruction by immune and inflammatory cells leads to the repeated presentation of retroelement proteins and hypomethylated DNA sequences from cell debris, perpetuating a vicious cycle. On a note of caution, HERV env proteins have on one hand been reported to behave as "superantigens" (Sutkowski et al. 2001), but also postulated to exhibit immunosuppressive properties (Larsson and Andersson 1998).

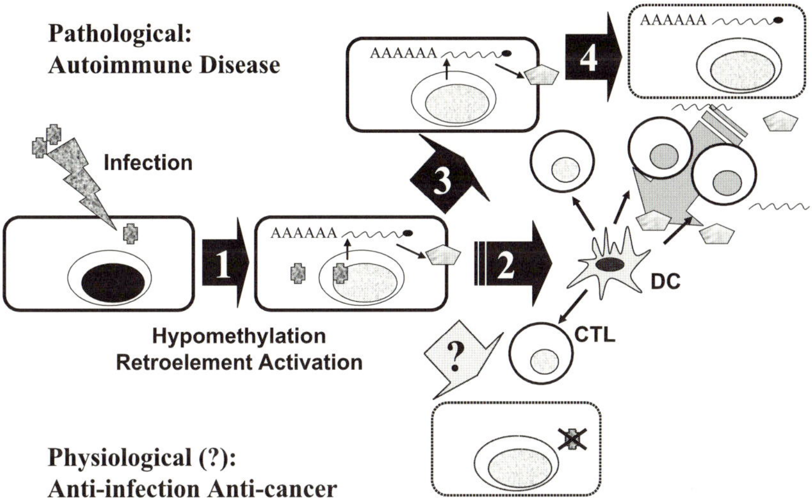

Fig. 7 Hypothetical role of retroelement activation in the pathogenesis of autoimmune diseases and as part of a physiological defense mechanism. It is postulated that certain viral infections (particles entering at *lightning symbol*) may lead to genome-wide hypomethylation resulting in the expression of retroelement RNA and proteins (*pentangles*). These may enhance recognition of altered cells by dendritic cells (*DC*) that in turn activate cytotoxic T cells (*CTL*). In a normal immune response, these would remove the infected cell, terminating the response. In a pathological response, overactivity of the immune system, perhaps assisted by persistent hypomethylation, would lead to a vicious cycle manifesting as autoimmune disease

7
Conclusion: Open Questions

We still know very little about the half of our genome that is composed of repetitive sequences. Retroelements form a major part of it. The phase of rapid expansion of endogenous retroviruses and retrotransposons lies in the distant past of human evolution, fortunately, one would assume. Meanwhile, humans and their genomic parasites have reached a sort of truce, and some elements should rather be considered symbionts. In ways that require further detailed investigation, they may contribute to the organization and regulation of the function of the human genome, in developing germ cells and the placenta, where they are mainly expressed, but also in somatic cells. Moreover, the very presence of unstable genome components may allow the continued evolution of humans. The balance between beneficial and adverse effects of retroelements is crucially dependent on their containment by DNA methylation and chromatin structure. Again, important details of the mechanisms involved are unknown, particularly how the vast majority of retroelements are silenced during germ cell development and once more in the fetus, following their hypomethylation in the early embryo. Several newer findings converge to indicate that default methylation may not be the complete answer to the question, but too few specifics are known.

The benefits derived from the presence and perhaps the activity of retroelements in the human genome remain to be better defined. Are LINEs responsible for the heterochromatic character of G-bands? To which extent does this depend on their methylation and to which extent on particular chromatin proteins? In which fashion do ALU elements so peculiarly located near active genes influence their regulation? Do retroelements really function as boundaries and, if so, how dynamic are these during development, cell differentiation, and in pathological states?

The importance of containment of endogenous retroelements by DNA methylation becomes most evident when this process is disturbed. Hypomethylation and activation of retroelements in cancer has been considered for a long time, but even the description of the changes is incomplete, and our understanding of its consequences is wanting. Like the mechanisms establishing methylation patterns during development, those deficient in cancer cells remain to be defined. The completion of the human genome sequence has provided an important tool to follow up these questions. Experiments in cell lines and in mouse models have yielded important insights, but it remains to be seen to which extent they can be extended to human cells in vivo. The almost complete absence of reported transpositions in cancer cells is a case in point. Some newer approaches are promising. Therefore, the structure of the LINE-1 endonuclease has been solved (Weichenrieder et al. 2004) and suitable antibodies for detection of LINE-encoded proteins have been developed (Ergün et al. 2004).

There is suggestive evidence implicating endogenous retroelements in human autoimmune diseases, even if the initial stimuli may be exogenous agents. As proposed above, the outlines of a common mechanism in these diseases may be emerging, although, evidently, many details remain to be clarified. Again, the growing knowledge on which retroelements are present in the human genome, how they are controlled, and which proteins they encode, may help to address the mechanisms by which they are involved in these diseases. Obviously, further diseases—most prominently atherosclerosis and perhaps even human ageing—could be investigated in this context.

An interesting speculation is that the mechanism apparent in autoimmune diseases may really be a physiological one going astray (Fig. 7). This is suggested by the various findings indicating that retroelements, particularly ALU sequences, are induced by cellular stress (see Sect. 3), as would occur during infection or at certain phases of cancer development. ALU induction has been experimentally found to impinge on protein synthesis, transcriptional regulation, and genomic stability intracellularly. If ALUs are not the only retroelements induced under stress conditions, as has been suggested by several studies, stress responses could lead to the presentation of endogenous retroelement antigens and also elicit immune responses. One wonders whether this reaction could represent an attempt to alert the immune system to the presence of cells with altered methylation. Indeed, methylation changes have been invoked as part of the cellular reaction toward viral infections (Müller et al. 2001; Muegge et al. 2003) and they are certainly an important component in the formation of cancer. It is tempting to speculate that hypomethylation of repeat sequences might occur as a consequence of viral infection or of hyperproliferation without adequate maintenance methylation and lead to increased expression of retroelement RNAs that activate intracellular defense systems and the immune system through presentation of antigenic retroelement proteins. According to this view, DNA hypomethylation observed in autoimmune diseases with overshooting immune reactions and persisting in cancers in spite of immune responses might reflect pathological extensions of a physiological protective response.

This speculation illustrates well the limits of our understanding of the retroelements in our genomes. We will have to fill many gaps in our knowledge before we can refute many ideas on their function and control, or confirm them.

Acknowledgements Work in our lab is supported by the Deutsche Forschungsgemeinschaft and the Deutsche Krebshilfe.

References

Alves G, Tatro A, Fanning T (1996) Differential methylation of human LINE-1 retrotransposons in malignant cells. Gene 176:39–44

Antony JM, van Marle G, Opii W, Butterfield DA, Mallet F, Yong VW, Wallace JL, Deacon RM, Warren K, Power C (2004) Human endogenous retrovirus glycoprotein-mediated induction of redox reactants causes oligodendrocyte death and demyelination. Nat Neurosci 7:1088–1095

Armbruester V, Sauter M, Krautkraemer E, Meese E, Kleiman A, Best B, Roemer K, Mueller-Lantzsch N (2002) A novel gene from the human endogenous retrovirus K expressed in transformed cells. Clin Cancer Res 8:1800–1807

Athanasiadis A, Rich A, Maas S (2004) Widespread A-to-I RNA editing of Alu-containing mRNAs in the human transcriptome. PLoS Biol 2:e391

Bailey JA, Carrel L, Chakravarti A, Eichler EE (2000) Molecular evidence for a relationship between LINE-1 elements and X chromosome inactivation: the Lyon repeat hypothesis. Proc Natl Acad Sci U S A 97:6634–6639

Bannert N, Kurth R (2004) Retroelements and the human genome: new perspectives on an old relation. Proc Natl Acad Sci U S A 101 Suppl 2:14572–14579

Batzer MA, Deininger PL (2002) Alu repeats and human genomic diversity. Nat Rev Genet 3:370–379

Becker KG, Swergold GD, Ozato K, Thayer RE (1993) Binding of the ubiquitous nuclear transcription factor YY1 to a cis regulatory sequence in the human LINE-1 transposable element. Hum Mol Genet 2:1697–1702

Bestor TH (2000) The DNA methyltransferases of mammals. Hum Mol Genet 9:2395–2402

Bourc'his D, Bestor TH (2004) Meiotic catastrophe and retrotransposon reactivation in male germ cells lacking Dnmt3L. Nature 431:96–99

Bratthauer GL, Fanning TG (1992) Active LINE-1 retrotransposons in human testicular cancer. Oncogene 7:507–510

Brudek T, Christensen T, Hansen HJ, Bobecka J, Moller-Larsen A (2004) Simultaneous presence of endogenous retrovirus and herpes virus antigens has profound effect on cell-mediated immune responses: implications for multiple sclerosis. AIDS Res Hum Retroviruses 20:415–423

Chalitchagorn K, Shuangshoti S, Hourpai N, Kongruttanachok N, Tangkijvanich P, Thong-Ngam D, Voravud N, Sriuranpong V, Mutirangura A (2004) Distinctive pattern of LINE-1 methylation level in normal tissues and the association with carcinogenesis. Oncogene 23:8841–8846

Chesnokov I, Chu WM, Botchan MR, Schmid CW (1996) p53 inhibits RNA polymerase III-directed transcription in a promoter-dependent manner. Mol Cell Biol 16:7084–7088

Clerici M, Fusi ML, Caputo D, Guerini FR, Trabattoni D, Salvaggio A, Cazzullo CL, Arienti D, Villa ML, Urnovitz HB, Ferrante P (1999) Immune responses to antigens of human endogenous retroviruses in patients with acute or stable multiple sclerosis. J Neuroimmunol 99:173–182

Conrad B, Weissmahr RN, Boni J, Arcari R, Schupbach J, Mach B (1997) A human endogenous retroviral superantigen as candidate autoimmune gene in type I diabetes. Cell 90:303–313

Dante R, Dante-Paire J, Rigal D, Roizes G (1992) Methylation patterns of long interspersed repeated DNA and alphoid repetitive DNA from human cell lines and tumors. Anticancer Res 12:559–563

De Smet C, Loriot A, Boon T (2004) Promoter-dependent mechanism leading to selective hypomethylation within the 5′ region of gene MAGE-A1 in tumor cells. Mol Cell Biol 24:4781–4790

Dennis K, Fan T, Geiman T, Yan Q, Muegge K (2001) Lsh, a member of the SNF2 family, is required for genome-wide methylation. Genes Dev 15:2940–2944

Doerfler W (1991) Patterns of DNA methylation—evolutionary vestiges of foreign DNA inactivation as a host defense mechanism. A proposal. Biol Chem Hoppe Seyler 372:557–564

Eden A, Gaudet F, Waghmare A, Jaenisch R (2003) Chromosomal instability and tumors promoted by DNA hypomethylation. Science 300:455

Ehrlich M (2002) DNA methylation in cancer: too much, but also too little. Oncogene 21:5400–5413

Ehrlich M, Jiang G, Fiala E, Dome JS, Yu MC, Long TI, Youn B, Sohn OS, Widschwendter M, Tomlinson GE, Chintagumpala M, Champagne M, Parham D, Liang G, Malik K, Laird PW (2002) Hypomethylation and hypermethylation of DNA in Wilms tumors. Oncogene 21:6694–6702

Ergün S, Buschmann C, Heukeshoven J, Dammann K, Schnieders F, Lauke H, Chalajour F, Kilic N, Stratling WH, Schumann GG (2004) Cell type-specific expression of LINE-1 open reading frames 1 and 2 in fetal and adult human tissues. J Biol Chem 279:27753–27763

Estes SD, Stoler DL, Anderson GR (1995) Anoxic induction of a sarcoma virus-related VL30 retrotransposon is mediated by a cis-acting element which binds hypoxia-inducible factor 1 and an anoxia-inducible factor. J Virol 69:6335–6341

Ferreira R, Naguibneva I, Pritchard LL, Ait-Si-Ali S, Harel-Bellan A (2001) The Rb/chromatin connection and epigenetic control: opinion. Oncogene 20:3128–3133

Fitzpatrick DR, Wilson CB (2003) Methylation and demethylation in the regulation of genes, cells, and responses in the immune system. Clin Immunol 109:37–45

Florl AR, Schulz WA (2003) Peculiar structure and location of 9p21 homozygous deletion breakpoints in human cancer cells. Genes Chromosomes Cancer 37:141–148

Florl AR, Lower R, Schmitz-Drager BJ, Schulz WA (1999) DNA methylation and expression of LINE-1 and HERV-K provirus sequences in urothelial and renal cell carcinomas. Br J Cancer 80:1312–1321

Florl AR, Steinhoff C, Muller M, Seifert HH, Hader C, Engers R, Ackermann R, Schulz WA (2004) Coordinate hypermethylation at specific genes in prostate carcinoma precedes LINE-1 hypomethylation. Br J Cancer 91:985–994

Fujinami RS, Libbey JE (1999) Endogenous retroviruses: are they the cause of multiple sclerosis? Trends Microbiol 7:263–264

Fuke C, Shimabukuro M, Petronis A, Sugimoto J, Oda T, Miura K, Miyazaki T, Ogura C, Okazaki Y, Jinno Y (2004) Age related changes in 5-methylcytosine content in human peripheral leukocytes and placentas: an HPLC-based study. Ann Hum Genet 68:196–204

Fukuoka J, Fujii T, Shih JH, Dracheva T, Meerzaman D, Player A, Hong K, Settnek S, Gupta A, Buetow K, Hewitt S, Travis WD, Jen J (2004) Chromatin remodeling factors and BRM/BRG1 expression as prognostic indicators in non-small cell lung cancer. Clin Cancer Res 10:4314–4324

Geiman TM, Robertson KD (2002) Chromatin remodeling, histone modifications, and DNA methylation—how does it all fit together? J Cell Biochem 87:117–125

Gibbons RJ, McDowell TL, Raman S, O'Rourke DM, Garrick D, Ayyub H, Higgs DR (2000) Mutations in ATRX, encoding a SWI/SNF-like protein, cause diverse changes in the pattern of DNA methylation. Nat Genet 24:368–371

Goldberg B, Urnovitz HB, Stricker RB (2000) Beyond danger: unmethylated CpG dinucleotides and the immunopathogenesis of disease. Immunol Lett 73:13–18

Götzinger N, Sauter M, Roemer K, Mueller-Lantzsch N (1996) Regulation of human endogenous retrovirus-K Gag expression in teratocarcinoma cell lines and human tumours. J Gen Virol 77:2983–2990

Graff JR, Herman JG, Myohanen S, Baylin SB, Vertino PM (1997) Mapping patterns of CpG island methylation in normal and neoplastic cells implicates both upstream and downstream regions in de novo methylation. J Biol Chem 272:22322–22329

Greally JM (2002) Short interspersed transposable elements (SINEs) are excluded from imprinted regions in the human genome. Proc Natl Acad Sci U S A 99:327–332
Grover D, Mukerji M, Bhatnagar P, Kannan K, Brahmachari SK (2004) Alu repeat analysis in the complete human genome: trends and variations with respect to genomic composition. Bioinformatics 20:813–817
Hagan CR, Rudin CM (2002) Mobile genetic element activation and genotoxic cancer therapy: potential clinical implications. Am J Pharmacogenomics 2:25–35
Hagan CR, Sheffield RF, Rudin CM (2003) Human Alu element retrotransposition induced by genotoxic stress. Nat Genet 35:219–220
Han JS, Szak ST, Boeke JD (2004) Transcriptional disruption by the L1 retrotransposon and implications for mammalian transcriptomes. Nature 429:268–274
Hansen RS (2003) X inactivation-specific methylation of LINE-1 elements by DNMT3B: implications for the Lyon repeat hypothesis. Hum Mol Genet 12:2559–2567
Hardeland U, Bentele M, Lettieri T, Steinacher R, Jiricny J, Schar P (2001) Thymine DNA glycosylase. Prog Nucleic Acid Res Mol Biol 68:235–253
Hasse A, Schulz WA (1994) Enhancement of reporter gene de novo methylation by DNA fragments from the alpha-fetoprotein control region. J Biol Chem 269:1821–1826
Hata K, Sakaki Y (1997) Identification of critical CpG sites for repression of L1 transcription by DNA methylation. Gene 189:227–234
Hendricks KB, Shanahan F, Lees E (2004) Role for BRG1 in cell cycle control and tumor suppression. Mol Cell Biol 24:362–376
Herve CA, Lugli EB, Brand A, Griffiths DJ, Venables PJ (2002) Autoantibodies to human endogenous retrovirus-K are frequently detected in health and disease and react with multiple epitopes. Clin Exp Immunol 128:75–82
Hewitt SM, Fraizer GC, Saunders GF (1995) Transcriptional silencer of the Wilms' tumor gene WT1 contains an Alu repeat. J Biol Chem 270:17908–17912
Hiltunen MO, Yla-Herttuala S (2003) DNA methylation, smooth muscle cells, and atherogenesis. Arterioscler Thromb Vasc Biol 23:1750–1753
Hoffmann MJ, Schulz WA (2005) Causes and consequences of DNA hypomethylation in human cancer. Biochem Cell Biol 83:296–321
Huang H, Qian J, Proffit J, Wilber K, Jenkins R, Smith DI (1998) FRA7G extends over a broad region: coincidence of human endogenous retroviral sequences (HERV-H) and small polydispersed circular DNAs (spcDNA) and fragile sites. Oncogene 16:2311–2319
Huang J, Fan T, Yan Q, Zhu H, Fox S, Issaq HJ, Best L, Gangi L, Munroe D, Muegge K (2004) Lsh, an epigenetic guardian of repetitive elements. Nucleic Acids Res 32:5019–5028
International Human Genome Sequencing Consortium (2004) Finishing the euchromatic sequence of the human genome. Nature 431:931–945
Jackson-Grusby L, Beard C, Possemato R, Tudor M, Fambrough D, Csankovszki G, Dausman J, Lee P, Wilson C, Lander E, Jaenisch R (2001) Loss of genomic methylation causes p53-dependent apoptosis and epigenetic deregulation. Nat Genet 27:31–39
Januchowski R, Prokop J, Jagodzinski PP (2004) Role of epigenetic DNA alterations in the pathogenesis of systemic lupus erythematosus. J Appl Genet 45:237–248
Jeffs AR, Benjes SM, Smith TL, Sowerby SJ, Morris CM (1998) The BCR gene recombines preferentially with Alu elements in complex BCR-ABL translocations of chronic myeloid leukaemia. Hum Mol Genet 7:767–776

Ji W, Hernandez R, Zhang XY, Qu GZ, Frady A, Varela M, Ehrlich M (1997) DNA demethylation and pericentromeric rearrangements of chromosome 1. Mutat Res 379:33–41

Jones PA, Baylin SB (2002) The fundamental role of epigenetic events in cancer. Nat Rev Genet 3:415–428

Jones RS, Potter SS (1985) L1 sequences in HeLa extrachromosomal circular DNA: evidence for circularization by homologous recombination. Proc Natl Acad Sci U S A 82:1989–1993

Jürgens B, Schmitz-Drager BJ, Schulz WA (1996) Hypomethylation of L1 LINE sequences prevailing in human urothelial carcinoma. Cancer Res 56:5698–5703

Kajikawa M, Okada N (2002) LINEs mobilize SINEs in the eel through a shared 3′ sequence. Cell 111:433–444

Kan PX, Popendikyte V, Kaminsky ZA, Yolken RH, Petronis A (2004) Epigenetic studies of genomic retroelements in major psychosis. Schizophr Res 67:95–106

Kaneda A, Tsukamoto T, Takamura-Enya T, Watanabe N, Kaminishi M, Sugimura T, Tatematsu M, Ushijima T (2004) Frequent hypomethylation in multiple promoter CpG islands is associated with global hypomethylation, but not with frequent promoter hypermethylation. Cancer Sci 95:58–64

Kaplan MJ, Lu Q, Wu A, Attwood J, Richardson B (2004) Demethylation of promoter regulatory elements contributes to perforin overexpression in CD4+ lupus T cells. J Immunol 172:3652–3661

Kazazian HH Jr (2004) Mobile elements: drivers of genome evolution. Science 303:1626–1632

Khodarev NN, Bennett T, Shearing N, Sokolova I, Koudelik J, Walter S, Villalobos M, Vaughan AT (2000) LINE L1 retrotransposable element is targeted during the initial stages of apoptotic DNA fragmentation. J Cell Biochem 79:486–495

Kim C, Rubin CM, Schmid CW (2001) Genome-wide chromatin remodeling modulates the Alu heat shock response. Gene 276:127–133

Kim DD, Kim TT, Walsh T, Kobayashi Y, Matise TC, Buyske S, Gabriel A (2004) Widespread RNA editing of embedded alu elements in the human transcriptome. Genome Res 14:1719–1725

Kleiman A, Senyuta N, Tryakin A, Sauter M, Karseladze A, Tjulandin S, Gurtsevitch V, Mueller-Lantzsch N (2004) HERV-K(HML-2) GAG/ENV antibodies as indicator for therapy effect in patients with germ cell tumors. Int J Cancer 110:459–461

Knössl M, Lower R, Lower J (1999) Expression of the human endogenous retrovirus HTDV/HERV-K is enhanced by cellular transcription factor YY1. J Virol 73:1254–1261

Kochanek S, Renz D, Doerfler W (1995) Transcriptional silencing of human Alu sequences and inhibition of protein binding in the box B regulatory elements by 5′-CG-3′ methylation. FEBS Lett 360:115–120

Kolomietz E, Meyn MS, Pandita A, Squire JA (2002) The role of Alu repeat clusters as mediators of recurrent chromosomal aberrations in tumors. Genes Chromosomes Cancer 35:97–112

Kondo Y, Issa JP (2003) Enrichment for histone H3 lysine 9 methylation at Alu repeats in human cells. J Biol Chem 278:27658–27662

Kuchen S, Seemayer CA, Rethage J, von Knoch R, Kuenzler P, Beat AM, Gay RE, Gay S, Neidhart M (2004) The L1 retroelement-related p40 protein induces p38delta MAP kinase. Autoimmunity 37:57–65

Kurose K, Hata K, Hattori M, Sakaki Y (1995) RNA polymerase III dependence of the human L1 promoter and possible participation of the RNA polymerase II factor YY1 in the RNA polymerase III transcription system. Nucleic Acids Res 23:3704–3709

Lander ES, Linton LM, Birren B, Nusbaum C, Zody MC, Baldwin J, Devon K, Dewar K, Doyle M, FitzHugh W, Funke R, Gage D, Harris K, Heaford A, Howland J, Kann L, Lehoczky J, LeVine R, McEwan P, McKernan K, Meldrim J, Mesirov JP, Miranda C, Morris W, Naylor J, Raymond C, Rosetti M, Santos R, Sheridan A, Sougnez C, Stange-Thomann N, Stojanovic N, Subramanian A, Wyman D, Rogers J, Sulston J, Ainscough R, Beck S, Bentley D, Burton J, Clee C, Carter N, Coulson A, Deadman R, Deloukas P, Dunham A, Dunham I, Durbin R, French L, Grafham D, Gregory S, Hubbard T, Humphray S, Hunt A, Jones M, Lloyd C, McMurray A, Matthews L, Mercer S, Milne S, Mullikin JC, Mungall A, Plumb R, Ross M, Shownkeen R, Sims S, Waterston RH, Wilson RK, Hillier LW, McPherson JD, Marra MA, Mardis ER, Fulton LA, Chinwalla AT, Pepin KH, Gish WR, Chissoe SL, Wendl MC, Delehaunty KD, Miner TL, Delehaunty A, Kramer JB, Cook LL, Fulton RS, Johnson DL, Minx PJ, Clifton SW, Hawkins T, Branscomb E, Predki P, Richardson P, Wenning S, Slezak T, Doggett N, Cheng JF, Olsen A, Lucas S, Elkin C, Uberbacher E, Frazier M, Gibbs RA, Muzny DM, Scherer SE, Bouck JB, Sodergren EJ, Worley KC, Rives CM, Gorrell JH, Metzker ML, Naylor SL, Kucherlapati RS, Nelson DL, Weinstock GM, Sakaki Y, Fujiyama A, Hattori M, Yada T, Toyoda A, Itoh T, Kawagoe C, Watanabe H, Totoki Y, Taylor T, Weissenbach J, Heilig R, Saurin W, Artiguenave F, Brottier P, Bruls T, Pelletier E, Robert C, Wincker P, Smith DR, Doucette-Stamm L, Rubenfield M, Weinstock K, Lee HM, Dubois J, Rosenthal A, Platzer M, Nyakatura G, Taudien S, Rump A, Yang H, Yu J, Wang J, Huang G, Gu J, Hood L, Rowen L, Madan A, Qin S, Davis RW, Federspiel NA, Abola AP, Proctor MJ, Myers RM, Schmutz J, Dickson M, Grimwood J, Cox DR, Olson MV, Kaul R, Raymond C, Shimizu N, Kawasaki K, Minoshima S, Evans GA, Athanasiou M, Schultz R, Roe BA, Chen F, Pan H, Ramser J, Lehrach H, Reinhardt R, McCombie WR, de la BM, Dedhia N, Blocker H, Hornischer K, Nordsiek G, Agarwala R, Aravind L, Bailey JA, Bateman A, Batzoglou S, Birney E, Bork P, Brown DG, Burge CB, Cerutti L, Chen HC, Church D, Clamp M, Copley RR, Doerks T, Eddy SR, Eichler EE, Furey TS, Galagan J, Gilbert JG, Harmon C, Hayashizaki Y, Haussler D, Hermjakob H, Hokamp K, Jang W, Johnson LS, Jones TA, Kasif S, Kaspryzk A, Kennedy S, Kent WJ, Kitts P, Koonin EV, Korf I, Kulp D, Lancet D, Lowe TM, McLysaght A, Mikkelsen T, Moran JV, Mulder N, Pollara VJ, Ponting CP, Schuler G, Schultz J, Slater G, Smit AF, Stupka E, Szustakowski J, Thierry-Mieg D, Thierry-Mieg J, Wagner L, Wallis J, Wheeler R, Williams A, Wolf YI, Wolfe KH, Yang SP, Yeh RF, Collins F, Guyer MS, Peterson J, Felsenfeld A, Wetterstrand KA, Patrinos A, Morgan MJ, de Jong P, Catanese JJ, Osoegawa K, Shizuya H, Choi S, Chen YJ (2001) Initial sequencing and analysis of the human genome. Nature 409:860–921

Larsson E, Andersson G (1998) Beneficial role of human endogenous retroviruses: facts and hypotheses. Scand J Immunol 48:329–338

Laurent AM, Puechberty J, Prades C, Gimenez S, Roizes G (1997) Site-specific retrotransposition of L1 elements within human alphoid satellite sequences. Genomics 46:127–132

Lavie L, Maldener E, Brouha B, Meese EU, Mayer J (2004) The human L1 promoter: variable transcription initiation sites and a major impact of upstream flanking sequence on promoter activity. Genome Res 14:2253–2260

Levanon EY, Eisenberg E, Yelin R, Nemzer S, Hallegger M, Shemesh R, Fligelman ZY, Shoshan A, Pollock SR, Sztybel D, Olshansky M, Rechavi G, Jantsch MF (2004) Systematic identification of abundant A-to-I editing sites in the human transcriptome. Nat Biotechnol 22:1001–1005

Levine JJ, Stimson-Crider KM, Vertino PM (2003) Effects of methylation on expression of TMS1/ASC in human breast cancer cells. Oncogene 22:3475–3488

Li E (2002) Chromatin modification and epigenetic reprogramming in mammalian development. Nat Rev Genet 3:662–673

Li TH, Schmid CW (2001) Differential stress induction of individual Alu loci: implications for transcription and retrotransposition. Gene 276:135–141

Lin CH, Hsieh SY, Sheen IS, Lee WC, Chen TC, Shyu WC, Liaw YF (2001) Genome-wide hypomethylation in hepatocellular carcinogenesis. Cancer Res 61:4238–4243

Lippman Z, Gendrel AV, Black M, Vaughn MW, Dedhia N, McCombie WR, Lavine K, Mittal V, May B, Kasschau KD, Carrington JC, Doerge RW, Colot V, Martienssen R (2004) Role of transposable elements in heterochromatin and epigenetic control. Nature 430:471–476

Liu WM, Schmid CW (1993) Proposed roles for DNA methylation in Alu transcriptional repression and mutational inactivation. Nucleic Acids Res 21:1351–1359

Looijenga LH, Oosterhuis JW (2002) Pathobiology of testicular germ cell tumors: views and news. Anal Quant Cytol Histol 24:263–279

Lower R (1999) The pathogenic potential of endogenous retroviruses: facts and fantasies. Trends Microbiol 7:350–356

Lund AH, van Lohuizen M (2004) Epigenetics and cancer. Genes Dev 18:2315–2335

Marguerat S, Wang WY, Todd JA, Conrad B (2004) Association of human endogenous retrovirus K-18 polymorphisms with type 1 diabetes. Diabetes 53:852–854

Medina PP, Carretero J, Fraga MF, Esteller M, Sidransky D, Sanchez-Cespedes M (2004) Genetic and epigenetic screening for gene alterations of the chromatin-remodeling factor, SMARCA4/BRG1, in lung tumors. Genes Chromosomes Cancer 41:170–177

Menendez L, Benigno BB, McDonald JF (2004) L1 and HERV-W retrotransposons are hypomethylated in human ovarian carcinomas. Mol Cancer 3:12

Mi S, Lee X, Li X, Veldman GM, Finnerty H, Racie L, LaVallie E, Tang XY, Edouard P, Howes S, Keith JC Jr, McCoy JM (2000) Syncytin is a captive retroviral envelope protein involved in human placental morphogenesis. Nature 403:785–789

Mimori K, Druck T, Inoue H, Alder H, Berk L, Mori M, Huebner K, Croce CM (1999) Cancer-specific chromosome alterations in the constitutive fragile region FRA3B. Proc Natl Acad Sci U S A 96:7456–7461

Monteyne P, Bureau JF, Brahic M (1998) Viruses and multiple sclerosis. Curr Opin Neurol 11:287–291

Morrish TA, Gilbert N, Myers JS, Vincent BJ, Stamato TD, Taccioli GE, Batzer MA, Moran JV (2002) DNA repair mediated by endonuclease-independent LINE-1 retrotransposition. Nat Genet 31:159–165

Muegge K, Young H, Ruscetti F, Mikovits J (2003) Epigenetic control during lymphoid development and immune responses: aberrant regulation, viruses, and cancer. Ann N Y Acad Sci 983:55–70

Muller K, Heller H, Doerfler W (2001) Foreign DNA integration. Genome-wide perturbations of methylation and transcription in the recipient genomes. J Biol Chem 276:14271–14278

Neidhart M, Rethage J, Kuchen S, Kunzler P, Crowl RM, Billingham ME, Gay RE, Gay S (2000) Retrotransposable L1 elements expressed in rheumatoid arthritis synovial tissue: association with genomic DNA hypomethylation and influence on gene expression. Arthritis Rheum 43:2634–2647

Nelson PN, Carnegie PR, Martin J, Davari EH, Hooley P, Roden D, Rowland-Jones S, Warren P, Astley J, Murray PG (2003) Demystified. Human endogenous retroviruses. Mol Pathol 56:11–18

Oelke K, Lu Q, Richardson D, Wu A, Deng C, Hanash S, Richardson B (2004) Overexpression of CD70 and overstimulation of IgG synthesis by lupus T cells and T cells treated with DNA methylation inhibitors. Arthritis Rheum 50:1850–1860

Ogasawara H, Okada M, Kaneko H, Hishikawa T, Sekigawa I, Hashimoto H (2003) Possible role of DNA hypomethylation in the induction of SLE: relationship to the transcription of human endogenous retroviruses. Clin Exp Rheumatol 21:733–738

Okada M, Ogasawara H, Kaneko H, Hishikawa T, Sekigawa I, Hashimoto H, Maruyama N, Kaneko Y, Yamamoto N (2002) Role of DNA methylation in transcription of human endogenous retrovirus in the pathogenesis of systemic lupus erythematosus. J Rheumatol 29:1678–1682

Ostertag EM, Kazazian HH Jr (2001) Biology of mammalian L1 retrotransposons. Annu Rev Genet 35:501–538

Ovchinnikov I, Rubin A, Swergold GD (2002) Tracing the LINEs of human evolution. Proc Natl Acad Sci U S A 99:10522–10527

Perron H, Garson JA, Bedin F, Beseme F, Paranhos-Baccala G, Komurian-Pradel F, Mallet F, Tuke PW, Voisset C, Blond JL, Lalande B, Seigneurin JM, Mandrand B (1997) Molecular identification of a novel retrovirus repeatedly isolated from patients with multiple sclerosis. The Collaborative Research Group on Multiple Sclerosis. Proc Natl Acad Sci U S A 94:7583–7588

Perron H, Perin JP, Rieger F, Alliel PM (2000) Particle-associated retroviral RNA and tandem RGH/HERV-W copies on human chromosome 7q: possible components of a 'chain-reaction' triggered by infectious agents in multiple sclerosis? J Neurovirol 6 Suppl 2:S67–S75

Qu GZ, Grundy PE, Narayan A, Ehrlich M (1999) Frequent hypomethylation in Wilms tumors of pericentromeric DNA in chromosomes 1 and 16. Cancer Genet Cytogenet 109:34–39

Raschke S, Balz V, Efferth T, Schulz WA, Florl AR (2005) Homozygous deletions of CDKN2A caused by alternative mechanisms in various human cancer cell lines. Genes Chromosomes Cancer 42:58–67

Richardson B (2003) DNA methylation and autoimmune disease. Clin Immunol 109:72–79

Rubin CM, VandeVoort CA, Teplitz RL, Schmid CW (1994) Alu repeated DNAs are differentially methylated in primate germ cells. Nucleic Acids Res 22:5121–5127
Rubin CM, Kimura RH, Schmid CW (2002) Selective stimulation of translational expression by Alu RNA. Nucleic Acids Res 30:3253–3261
Rudiger NS, Gregersen N, Kielland-Brandt MC (1995) One short well conserved region of Alu-sequences is involved in human gene rearrangements and has homology with prokaryotic chi. Nucleic Acids Res 23:256–260
Santourlidis S, Florl A, Ackermann R, Wirtz HC, Schulz WA (1999) High frequency of alterations in DNA methylation in adenocarcinoma of the prostate. Prostate 39:166–174
Santourlidis S, Trompeter HI, Weinhold S, Eisermann B, Meyer KL, Wernet P, Uhrberg M (2002) Crucial role of DNA methylation in determination of clonally distributed killer cell Ig-like receptor expression patterns in NK cells. J Immunol 169:4253–4261
Schiavetti F, Thonnard J, Colau D, Boon T, Coulie PG (2002) A human endogenous retroviral sequence encoding an antigen recognized on melanoma by cytolytic T lymphocytes. Cancer Res 62:5510–5516
Schmid CW (1998) Does SINE evolution preclude Alu function? Nucleic Acids Res 26:4541–4550
Schön U, Seifarth W, Baust C, Hohenadl C, Erfle V, Leib-Mosch C (2001) Cell type-specific expression and promoter activity of human endogenous retroviral long terminal repeats. Virology 279:280–291
Schulte AM, Lai S, Kurtz A, Czubayko F, Riegel AT, Wellstein A (1996) Human trophoblast and choriocarcinoma expression of the growth factor pleiotrophin attributable to germ-line insertion of an endogenous retrovirus. Proc Natl Acad Sci U S A 93:14759–14764
Schulz WA (1998) DNA methylation in urological malignancies (review). Int J Oncol 13:151–167
Schulz WA, Elo JP, Florl AR, Pennanen S, Santourlidis S, Engers R, Buchardt M, Seifert HH, Visakorpi T (2002) Genomewide DNA hypomethylation is associated with alterations on chromosome 8 in prostate carcinoma. Genes Chromosomes Cancer 35:58–65
Seemayer CA, Distler O, Kuchen S, Muller-Ladner U, Michel BA, Neidhart M, Gay RE, Gay S (2001) Die Rheumatoide Arthritis: Neuentwicklungen in der Pathogenese unter besonderer Berücksichtigung der synovialen Fibroblasten. Z Rheumatol 60:309–318
Serra C, Mameli G, Arru G, Sotgiu S, Rosati G, Dolei A (2003) In vitro modulation of the multiple sclerosis (MS)-associated retrovirus by cytokines: implications for MS pathogenesis. J Neurovirol 9:637–643
Shaffer LG, Lupski JR (2000) Molecular mechanisms for constitutional chromosomal rearrangements in humans. Annu Rev Genet 34:297–329
Shastry BS (2002) Schizophrenia: a genetic perspective (review). Int J Mol Med 9:207–212
Sivori S, Falco M, Della CM, Carlomagno S, Vitale M, Moretta L, Moretta A (2004) CpG and double-stranded RNA trigger human NK cells by Toll-like receptors: induction of cytokine release and cytotoxicity against tumors and dendritic cells. Proc Natl Acad Sci U S A 101:10116–10121

Skowronski J, Fanning TG, Singer MF (1988) Unit-length line-1 transcripts in human teratocarcinoma cells. Mol Cell Biol 8:1385–1397

Smiraglia DJ, Szymanska J, Kraggerud SM, Lothe RA, Peltomaki P, Plass C (2002) Distinct epigenetic phenotypes in seminomatous and nonseminomatous testicular germ cell tumors. Oncogene 21:3909–3916

Song JZ, Stirzaker C, Harrison J, Melki JR, Clark SJ (2002) Hypermethylation trigger of the glutathione-S-transferase gene (GSTP1) in prostate cancer cells. Oncogene 21:1048–1061

Stauffer Y, Marguerat S, Meylan F, Ucla C, Sutkowski N, Huber B, Pelet T, Conrad B (2001) Interferon-alpha-induced endogenous superantigen. A model linking environment and autoimmunity. Immunity 15:591–601

Stauffer Y, Theiler G, Sperisen P, Lebedev Y, Jongeneel CV (2004) Digital expression profiles of human endogenous retroviral families in normal and cancerous tissues. Cancer Immun 4:2

Steinhoff C, Schulz WA (2003) Transcriptional regulation of the human LINE-1 retrotransposon L1.2B. Mol Genet Genomics 270:394–402

Stirzaker C, Song JZ, Davidson B, Clark SJ (2004) Transcriptional gene silencing promotes DNA hypermethylation through a sequential change in chromatin modifications in cancer cells. Cancer Res 64:3871–3877

Sugimoto J, Matsuura N, Kinjo Y, Takasu N, Oda T, Jinno Y (2001) Transcriptionally active HERV-K genes: identification, isolation, and chromosomal mapping. Genomics 72:137–144

Suter CM, Martin DI, Ward RL (2004) Hypomethylation of L1 retrotransposons in colorectal cancer and adjacent normal tissue. Int J Colorectal Dis 19:95–101

Sutkowski N, Conrad B, Thorley-Lawson DA, Huber BT (2001) Epstein-Barr virus transactivates the human endogenous retrovirus HERV-K18 that encodes a superantigen. Immunity 15:579–589

Swergold GD (1990) Identification, characterization, and cell specificity of a human LINE-1 promoter. Mol Cell Biol 10:6718–6729

Symer DE, Connelly C, Szak ST, Caputo EM, Cost GJ, Parmigiani G, Boeke JD (2002) Human l1 retrotransposition is associated with genetic instability in vivo. Cell 110:327–338

Szyf M (2003) DNA methylation and cancer therapy. Drug Resist Updat 6:341–353

Szyf M (2005) DNA methylation and cancer therapy. Landes Bioscience, Georgetown, pp 1–239

Takai D, Yagi Y, Habib N, Sugimura T, Ushijima T (2000) Hypomethylation of LINE1 retrotransposon in human hepatocellular carcinomas, but not in surrounding liver cirrhosis. Jpn J Clin Oncol 30:306–309

Tchénio T, Casella JF, Heidmann T (2000) Members of the SRY family regulate the human LINE retrotransposons. Nucleic Acids Res 28:411–415

Teitell M, Richardson B (2003) DNA methylation in the immune system. Clin Immunol 109:2–5

Tomilin NV (1999) Control of genes by mammalian retroposons. Int Rev Cytol 186:1–48

Tuck-Muller CM, Narayan A, Tsien F, Smeets DF, Sawyer J, Fiala ES, Sohn OS, Ehrlich M (2000) DNA hypomethylation and unusual chromosome instability in cell lines from ICF syndrome patients. Cytogenet Cell Genet 89:121–128

Turker MS (2002) Gene silencing in mammalian cells and the spread of DNA methylation. Oncogene 21:5388–5393

van de Lagemaat LN, Landry JR, Mager DL, Medstrand P (2003) Transposable elements in mammals promote regulatory variation and diversification of genes with specialized functions. Trends Genet 19:530–536

Villar-Garea A, Fraga MF, Espada J, Esteller M (2003) Procaine is a DNA-demethylating agent with growth-inhibitory effects in human cancer cells. Cancer Res 63:4984–4989

Vorce RL, Lee B, Howard BH (1994) Methylation- and mutation-dependent stimulation of Alu transcription in vitro. Biochem Biophys Res Commun 203:845–851

Wang-Johanning F, Frost AR, Johanning GL, Khazaeli MB, LoBuglio AF, Shaw DR, Strong TV (2001) Expression of human endogenous retrovirus k envelope transcripts in human breast cancer. Clin Cancer Res 7:1553–1560

Wang-Johanning F, Frost AR, Jian B, Azerou R, Lu DW, Chen DT, Johanning GL (2003) Detecting the expression of human endogenous retrovirus E envelope transcripts in human prostate adenocarcinoma. Cancer 98:187–197

Weichenrieder O, Repanas K, Perrakis A (2004) Crystal structure of the targeting endonuclease of the human LINE-1 retrotransposon. Structure (Camb) 12:975–986

White RJ (2004) RNA polymerase III transcription and cancer. Oncogene 23:3208–3216

Whitelaw E, Martin DI (2001) Retrotransposons as epigenetic mediators of phenotypic variation in mammals. Nat Genet 27:361–365

Widschwendter M, Jiang G, Woods C, Muller HM, Fiegl H, Goebel G, Marth C, Muller-Holzner E, Zeimet AG, Laird PW, Ehrlich M (2004) DNA hypomethylation and ovarian cancer biology. Cancer Res 64:4472–4480

Wilson MJ, Torkar M, Haude A, Milne S, Jones T, Sheer D, Beck S, Trowsdale J (2000) Plasticity in the organization and sequences of human KIR/ILT gene families. Proc Natl Acad Sci U S A 97:4778–4783

Wong N, Lam WC, Lai PB, Pang E, Lau WY, Johnson PJ (2001) Hypomethylation of chromosome 1 heterochromatin DNA correlates with q-arm copy gain in human hepatocellular carcinoma. Am J Pathol 159:465–471

Woodcock DM, Lawler CB, Linsenmeyer ME, Doherty JP, Warren WD (1997) Asymmetric methylation in the hypermethylated CpG promoter region of the human L1 retrotransposon. J Biol Chem 272:7810–7816

Xing J, Hedges DJ, Han K, Wang H, Cordaux R, Batzer MA (2004) Alu element mutation spectra: molecular clocks and the effect of DNA methylation. J Mol Biol 344:675–682

Yang AS, Estecio MR, Doshi K, Kondo Y, Tajara EH, Issa JP (2004) A simple method for estimating global DNA methylation using bisulfite PCR of repetitive DNA elements. Nucleic Acids Res 32:e38

Yang N, Zhang L, Zhang Y, Kazazian HH Jr (2003) An important role for RUNX3 in human L1 transcription and retrotransposition. Nucleic Acids Res 31:4929–4940

Yang Z, Boffelli D, Boonmark N, Schwartz K, Lawn R (1998) Apolipoprotein(a) gene enhancer resides within a LINE element. J Biol Chem 273:891–897

Yano M, Ouchida M, Shigematsu H, Tanaka N, Ichimura K, Kobayashi K, Inaki Y, Toyooka S, Tsukuda K, Shimizu N, Shimizu K (2004) Tumor-specific exon creation of the HELLS/SMARCA6 gene in non-small cell lung cancer. Int J Cancer 112:8–13

Yi JM, Kim HM, Kim HS (2004) Expression of the human endogenous retrovirus HERV-W family in various human tissues and cancer cells. J Gen Virol 85:1203–1210

Yoder JA, Walsh CP, Bestor TH (1997) Cytosine methylation and the ecology of intragenomic parasites. Trends Genet 13:335–340

Yu F, Zingler N, Schumann G, Stratling WH (2001) Methyl-CpG-binding protein 2 represses LINE-1 expression and retrotransposition but not Alu transcription. Nucleic Acids Res 29:4493–4501

Zendman AJ, Ruiter DJ, Van Muijen GN (2003) Cancer/testis-associated genes: identification, expression profile, and putative function. J Cell Physiol 194:272–288

CTMI (2006) 310:251–274

Cancer-Linked DNA Hypomethylation and Its Relationship to Hypermethylation

M. Ehrlich (✉)

Human Genetics Program, Department of Biochemistry, and Tulane Cancer Center, Tulane Medical School, 1430 Tulane Ave., New Orleans, LA 70112, USA
ehrlich@tulane.edu

Abstract It is not surprising that cancer, a kind of derangement of development, hijacks DNA methylation, which is necessary for normal mammalian embryogenesis. Both decreases and increases in DNA methylation are a frequent characteristic of a wide variety of cancers. There is often more hypomethylation than hypermethylation of DNA during carcinogenesis, leading to a net decrease in the genomic 5-methylcytosine content. Although the exact methylation changes between different cancers of the same type are not the same, there are cancer type-specific differences in the frequency of hypermethylation or hypomethylation of certain genomic sequences. These opposite types of DNA methylation changes appear to be mostly independent of one another, although they may arise because of a similar abnormality leading to long-lasting epigenetic instability in cancers. Both tandem and interspersed DNA repeats often exhibit cancer-associated hypomethylation. However, one of these repeated sequences

(NBL2) displayed predominant increases in methylation in some ovarian carcinomas and Wilms tumors and decreases in others. Furthermore, decreases and increases in CpG methylation can be interspersed within a small subregion of the 1.4-kb repeat unit of these tandem arrays. While the transcription-silencing role of DNA hypermethylation at promoters of many tumor-suppressor genes is clear, the biological effects of cancer-linked hypomethylation of genomic DNA are less well understood. Evidence suggests that DNA hypomethylation functions in direct or indirect control of transcription and in destabilizing chromosomal integrity. Recent studies of cancer-linked DNA hypomethylation indicate that changes to DNA methylation during tumorigenesis and tumor progression have a previously underestimated plasticity and dynamic nature.

This chapter is dedicated to Charles W. Gehrke, a careful and path-setting analytical biochemist, whose collaboration brought my lab into the study of cancer epigenetics.

1 Overview

Epigenetic changes in cancer, both in DNA (cytosine methylation) and chromatin (including histone modification and chromatin remodeling), are critically involved in carcinogenesis (Baylin and Herman 2000; Cairns 2001; Ehrlich 2000, 2002; Klochendler-Yeivin et al. 2002). Both increases in methylation in some portions of the genome and decreases in others are characteristic of cancer, although they seem to be independent changes. Furthermore, a subregion of the genome can be subject to overall increases in methylation in some cancers and decreases in other tumors of the same type. After giving some background on mammalian DNA epigenetics, this review will summarize the nature and biological significance of hypomethylation of DNA in cancer, which is less well understood than is cancer-linked hypermethylation of DNA. Lastly, it will explore the possible relationships between these opposite types of tumor-associated methylation changes.

2 Background: Tissue-Specific DNA Methylation

Some of the early research on the species-specific distribution of 5-methylcytosine (m^5C) in DNA in non-human vertebrate tissues was begun by Boris Vanyushin and colleagues (Vanyushin et al. 1970, 1973; Romanov and

Vanyushin 1981). Our lab subsequently confirmed the tissue-specificity of genomic m^5C levels in animals (Gama-Sosa et al. 1983a) and demonstrated such tissue-specificity also for human specimens (Ehrlich et al. 1982). The idea that cancer represents a special kind of derangement of differentiation led us to look for cancer-specific aberrations in DNA methylation, first at the level of the genomic m^5C content determined by high-performance liquid chromatography (HPLC) (Gama-Sosa et al. 1983b; Ehrlich 2002), as described in Sect. 5.

3
Background: Vertebrate DNA Methylation and Gene Expression

In 1975, critical reviews of vertebrate DNA methylation by Holliday and Pugh (1975) and Riggs (1975) advanced our understanding of vertebrate DNA methylation with their hypotheses about maintenance vs de novo methylation and the involvement of this methylation in differentiation and X chromosome inactivation. In the late 1970s and early 1980s, there was an initial flurry of activity in which associations of differential promoter or gene methylation with tissue-specific repression were found (Ehrlich and Wang 1981; Cooper 1983; Riggs and Jones 1983; Bird 1984). During that time, Walter Doerfler's laboratory pioneered important studies of the effects of adenoviral infection on DNA methylation and the spreading of methylation along the DNA (Doerfler 1984). Although many examples of inverse associations between gene expression and promoter methylation for endogenous vertebrate genes and infecting viral genes were discovered, some genes display tissue-specific differences in DNA methylation that do not correlate with expression (Ehrlich and Wang 1981; Cooper 1983; Riggs and Jones 1983; Bird 1984; Doerfler 1984). Even more genes, especially constitutively expressed ones with CpG-rich promoters and 5′ gene regions (5′ CpG islands), have little or no methylation in these regions in a wide variety of tissues.

Vertebrate DNA methylation at transcription control regions appears to often modulate gene expression or help maintain an already established inactive state, rather than simply acting as an on–off switch. However, most methylation of vertebrate genomes is not in such transcription control elements (Romanov and Vanyushin 1981; Gama-Sosa et al. 1983a), and methylation of transcription control elements does not always control gene expression in vivo. Moreover, the inverse correlations between expression and methylation that are seen for many gene regulatory regions (Heard et al. 1997; Attwood et al. 2002; Plass and Soloway 2002) could be consequences of changes in gene expression rather than regulators of changes in expression. Nonetheless, an overview of results from various experimental approaches using diverse genes convincingly demonstrates the biological importance of vertebrate DNA methylation to gene expression during normal development (Bruniquel and Schwartz 2003; Ehrlich 2003; Makar et al. 2003; Martinowich et al. 2003; Cusack et al. 2004; Geyer et al. 2004; Heard 2004; Iwano et al. 2004; Namihira et al. 2004; Nishino et al. 2004; Strathdee et al. 2004; Tagoh et al. 2004; Weaver et al. 2004).

4
Background: Vertebrate DNA Methylation and Development

A problematic aspect of DNA methylation research during the 1980s and 1990s was the mistaken notion (Anonymous 1985) that methylation of the vertebrate genome is probably of little consequence to vertebrate development simply because *Drosophila* had not been found to have DNA methylation. The reasoning was that *Drosophila*, like vertebrates, is a higher eukaryote with complicated development pathways. So if *Drosophila* could accomplish all that differentiation without DNA methylation, how can vertebrates use DNA methylation as an important gene regulator during development? This was an often-quoted idea despite the fact that early *Drosophila* embryos with their syncytial development are dramatically different from early vertebrate embryos and have a much smaller genome. Furthermore, it was already clear in the 1970s that despite the many common themes in molecular biology among diverse organisms, considerably different genetic pathways can yield similar biochemical outcomes. For example, very many bacterial strains use *dam* methylation (at the N^6 position of the A in GATC) to direct DNA mismatch repair as well as for regulating gene expression and DNA replication (Langle-Rouault et al. 1987; Palmer and Marinus 1994), but most bacterial strains do not have *dam* methylation (Barbeyron et al. 1984; Ehrlich et al. 1985). These *dam* methylation-negative bacteria can use asymmetrical nicks in the DNA generated during discontinuous DNA replication of one strand to direct mismatch repair (Lacks et al. 1985; Huang et al. 2004). Even the premise that *Drosophila* has no genomic m^5C was disproved. Recently it was clearly shown that *Drosophila* has small amounts of this methylated base in its genome although this methylation is not essential for differentiation (Gowher et al. 2000; Lyko et al. 2000; Kunert et al. 2003).

In contrast to *Drosophila*, vertebrates require DNA methylation for normal development (Li et al. 1992; Beard et al. 1995; Panning and Jaenisch 1996; Okano et al. 1999; Fan et al. 2001; Lee et al. 2001; Yung et al. 2001; Biniszkiewicz et al. 2002). This conclusion comes largely from studies of transgenic mice. Complicating studies of knockout or hypomorphic mice that are mutant in one of the main DNA methyltransferase genes (*DNMT1, DNMT3A,* or *DNMT3B*) are the multiple activities of all studied vertebrate DNA methyltransferases (DNMTs) (Bachman et al. 2001; Liang et al. 2002; Rhee et al. 2002; Datta et al. 2003; Fuks et al. 2001, 2003a; Pradhan and Esteve 2003; Geiman et al. 2004). However, other kinds of gene knockouts in mice also globally affect DNA methylation (Dennis et al. 2001; Yan et al. 2003), and results from those transgenic animals complement the conclusions from the DNA methyltransferase mutants. In addition, naturally occurring mutations in the portion of *DNMT3B* that specifies the catalytic domain. These mutations, which are found in the majority of patients with the ICF (immunodeficiency, centromeric region instability, facial anomalies) syndrome, reinforce the direct connection of decreased DNA methylation and abnormalities in development (see Sect. 7).

Another important methylation analysis tool developed by Peter Jones and colleagues for studies of the functionality of DNA methylation is the use of DNA methylation inhibitors: 5-azacytidine (azaCR) or the more specific 5-azadeoxycytidine (azaCdR) (Constantinides et al. 1977; Taylor and Jones 1979). These inhibitors can activate expression of many genes that had been inactivated by hypermethylation (Jones and Baylin 2002). A caveat for studies with these inhibitors is that their incorporation into DNA leads to DNA-protein cross-links, inhibition of DNA replication, and mutations (Juttermann et al. 1994; Jackson-Grusby et al. 1997), as well as to DNA hypomethylation. Results from studies involving azaCdR or azaCR combined with other experimental approaches to decrease genomic methylation or to increase methylation clearly implicate DNA methylation in the regulation of gene expression and mammalian development (Maier et al. 2003; Moreau et al. 2003; Suzuki et al. 2004; Tsuji-Takayama et al. 2004; and see Sect. 3 above).

5
Global DNA Hypomethylation in Human Cancer

In 1983, altered methylation of DNA in human cancer was first reported as DNA hypomethylation (Feinberg and Vogelstein 1983a, b; Gama-Sosa et al. 1983b). Our study of more than 100 human tumors of various types was done in collaboration with Charles Gehrke and demonstrated that there was global DNA hypomethylation (overall decreases in genomic m^5C) in cancers compared to 15 types of normal somatic tissue by examining DNA digested to deoxynucleosides and subjected to HPLC (Gama-Sosa et al. 1983b). Although there are tissue-specific differences in the genomic m^5C content of

DNA from normal samples (Ehrlich et al. 1982), we could identify, in collaboration with Charles Gehrke and Emerich Fiala, cancer-linked global DNA hypomethylation by lower levels of methylation in many of the cancers compared to a wide variety of control postnatal somatic tissues (Ehrlich et al. 2002, 2005; Widschwendter et al. 2004). This approach provided a lower estimate of this genome-wide hypomethylation because we looked for lower genomic m^5C levels in the tumors than in all the normal somatic controls. For most tumors, DNA hypomethylation is much more closely associated with malignancy with malignancy than with benign tumorigenesis (Ehrlich 2002). Cancer-linked DNA hypomethylation has been confirmed in diverse cancers by other laboratories (Kim et al. 1994; Cheng et al. 1997; Soares et al. 1999; Linn et al. 2001; Wong et al. 2001; Yamamoto et al. 2001; Saito et al. 2002; Tsuda et al. 2002).

Subsequent to the discovery of human cancer-associated DNA hypomethylation, hypermethylation of CpG island-promoters was found (Baylin et al. 1986). Hypermethylation of the promoter or 5′ regions of tumor suppressor genes has been shown to be very important in downregulating expression of these genes in cancer (Costello et al. 2000; Jones and Baylin 2002). A role for both hypomethylation and hypermethylation of DNA in cancer is supported by the findings that decreasing DNA methylation in some model systems enhances tumorigenicity, while increasing it in others does the same (Carr et al. 1984; Denda et al. 1985; Thomas and Williams 1992; Eads et al. 2002; Gaudet et al. 2003).

6 Hypomethylation of Tandem Repeats in Cancer

We found that global cancer-linked DNA hypomethylation is significantly associated with cancer-linked hypomethylation in centromeric satellite α DNA (Satα) and juxtacentromeric (centromere-adjacent) satellite DNA, specifically satellite 2 (Sat2) (Narayan et al. 1998; Qu et al. 1999a, b). Normal postnatal somatic tissues have similar high levels of methylation of these satellite DNAs. Moderate hypomethylation of satellite DNA was present in about 40%–80% of Wilms tumors, breast adenocarcinomas, and ovarian epithelial carcinomas and extensive hypomethylation in about 10%–30% of these cancers when comparing them to a wide variety of the normal somatic standards (Narayan et al. 1998; Qu et al. 1999a, b; Wong et al. 2001; Ehrlich et al. 2003). Also, in human hepatocellular carcinomas, Saito et al. (2001) observed that frequent hypomethylation of Sat2 was highly concordant with that of Sat3, the main DNA component of the long juxtacentromeric heterochromatin of Chr9. Itano et al. (2002) found that hypomethylation of two non-satellite DNA tandem repeats was the best indicator of a poor prognosis in patients with hepatocellular carcinoma. In collaboration with Louis Dubeau, we showed that hypomethylation of satellite DNA is significantly correlated with malignant potential in ovarian epithelial tumors (Qu et al. 1999b). Martin Widschwendter's group and my lab demonstrated that satellite hypomethylation in ovarian epithelial carcinomas is also significantly associated, as an independent marker, with decreased overall survival and disease-free survival (Widschwendter et al. 2004). Therefore, hypomethylation of tandem DNA repeats, as well as hypermethylation of gene regions (Brabender et al. 2001), can be an independent prognostic indicator. Furthermore, in collaboration with Peter Laird, we have recently shown that interspersed DNA repeats are also frequently hypomethylated in cancer, and either interspersed repeats (Alu repeats or LINE1 repeats) or Sat2 can serve as a surrogate marker for global DNA hypomethylation (Weisenberger et al. 2005). Therefore, all of these hypomethylation parameters may be of prognostic significance for certain types of cancer.

7 DNA Hypomethylation and Chromosome Rearrangements

How might satellite DNA hypomethylation contribute to carcinogenesis and tumor progression (Table 1)? We proposed that DNA hypomethylation in the centromeric or juxtacentromeric heterochromatin in cancer helps destabilize the genome by leading to rearrangements in the hypomethylated regions (Qu et al. 1999a; Tuck-Muller et al. 2000). There is an association between hypomethylation of Sat2 at 1qh and 16qh (the juxtacentromeric heterochromatin of chromosomes 1 and 16, respectively) and chromatin decondensation in these regions in lymphoid cells from patients with the above-mentioned ICF syndrome (Jeanpierre 1993; Tuck-Muller et al. 2000). This syndrome always involves immunodeficiency and high frequencies of juxtacentromeric rearrangements in chromosomes 1 and/or 16 (Chr1, Chr16) in lymphoid cells as well as other diverse symptoms that vary with the patient. Sat2 hypomethylation (Jeanpierre et al. 1993), due to ICF-linked DNA methyltransferase 3B *(DNMT3B)* mutations (Hansen et al. 1999; Okano et al. 1999; Xu et al. 1999), is seen in all examined ICF cell populations. Chr1 and Chr16 are the only human chromosomes with long Sat2-containing regions, and both Chr1 Sat2 and Chr16 Sat2 (but not Satα), are selectively hypomethylated in all studied ICF cell populations (Ehrlich et al. 2001). We have shown that there is hypomethylation of only a small portion of the ICF genome, i.e., approximately 7% (Kondo et al. 2000; Tuck-Muller et al. 2000), as a result of the ICF-causing

Table 1 Some possible roles for DNA hypomethylation in carcinogenesis

1.	Increasing DNA recombination in *cis* in certain sequences and tumor types
2.	Activation in *cis* of expression of genes that favor tumorigenesis or tumor progression by effects on promoters, enhancers, silencers, and other transcription regulatory sequences
3.	Alteration of sequestration of DNA binding proteins at repeated DNA sequences to control expression of genes in *trans*
4.	Effects on the subnuclear compartmentalization of chromatin or on euchromatin–heterochromatin interactions

(usually missense) *DNMT3B* mutations (Hansen et al. 1999; Okano et al. 1999; Xu et al. 1999; Gowher and Jeltsch 2002).

Our detailed studies of (1) karyotypes of ICF vs control lymphoblastoid cell lines (LCLs) and, in collaboration with David Gisselsson et al. (2005), (2) chromosome dynamics in these cells suggests that the chromatin decondensation in 1qh and 16qh—which can be associated with hypomethylated Sat2—in turn predisposes to rearrangements and prophase and anaphase anomalies in these regions (Jeanpierre et al. 1993; Tuck-Muller et al. 2000). We demonstrated that a normal pro-B cell line was especially susceptible to ICF-like chromosome abnormalities induced by the DNA inhibitors azaCdR and azaCR, including transient multiradial chromosomes with many Chr1 and/or Chr16 arms joined in the pericentromeric region (Hernandez et al. 1997; Ji et al. 1997). These aberrations occurred at high frequencies in cells in the absence of other chromosomal abnormalities. With respect to cancer, centromeric heterochromatin and the juxtacentromeric heterochromatin regions—whose DNA is frequently hypomethylated in a cancer-linked fashion as described above—often are the sites of unbalanced rearrangements (Mitelman et al. 1997) that could contribute to carcinogenesis by the resulting gene imbalances. Although ICF patients display no increased cancer incidence, only about 50–60 patients (mostly children) have been identified, and their very short average lifespan would preclude detection of a cancer predisposition that was not very high and did not result in tumors sufficiently quickly. There is also evidence for a role of DNA hypomethylation in chromosome instability outside of pericentromeric regions (Chen et al. 1998; Eden et al. 2003; Ehrlich et al. 2003; Gaudet et al. 2003), including in the important cancer-linked loss of heterozygosity (Yamada et al. 2005). Genomic instability and DNA hypomethylation can be observed early during tumorigenesis, and both often increase with tumor progression (Feinberg et al. 1983; Mitelman et al. 1997; Ehrlich 2002).

Despite the above evidence for some associations of DNA hypomethylation with increased chromosome recombination, recent studies suggest that increased recombination is not the main means by which DNA hypomethylation might promote carcinogenesis and tumor progression. While hypomethylation of pericentromeric DNA in heterochromatin probably predisposes certain human cell populations to rearrangements in these regions, e.g., ICF lymphoid cells and liver cancer (Wong et al. 2001), as does hypomethylation of euchromatic DNA sequences elsewhere (Chen et al. 1998, Schulz et al. 2002), exceptions to the relationship of Sat2 DNA hypomethylation with pericentromeric rearrangements have been reported for breast carcinomas (Tsuda et al. 2002). Moreover, our recent study of Wilms tumors involving a detailed karyotype analysis and examination of satellite DNA methylation showed that the frequencies of hypomethylation at *Bst*BI sites in Chr1 Sat2 and at Satα throughout the centromeres (51% and 69% of 35 primary tumors, respectively, compared to various normal postnatal somatic tissues) were much greater than the frequencies of rearrangements in the centromeric or juxtacentromeric regions of any of the chromosomes (20%; Ehrlich et al. 2003). Similarly, the very high frequencies of ovarian and breast cancer-associated hypomethylation at satellite DNA suggest that the probable functional significance of this hypomethylation is not limited to fostering chromosome rearrangements (Jackson et al. 2004). While there seems to be some linkage between DNA hypomethylation and chromosome rearrangements, there is no relationship between centromeric DNA hypomethylation and aneuploidy, as determined from studies of ICF LCLs and Wilms tumors (Tuck-Muller et al. 2000; Ehrlich et al. 2003).

8 DNA Hypomethylation and Possible *trans* Effects on Gene Expression

Other possible roles of DNA hypomethylation in cancer relate to either *cis* or *trans* effects on gene expression (Table 1). Activation of DNA methylation-repressed retrotransposons does not appear to play a major role in cancers (Gaudet et al. 2003). Other than the *MAGE/CAGE/BAGE* superfamily of cancer-testes antigen genes of unknown function and certain imprinted genes (De Smet et al. 1996; Tycko 2000; Ehrlich 2002; Feinberg et al. 2002), there is only a small number of convincing reports (e.g., Sato et al. 2003; Ye et al. 2005; Okada et al. 2005) of hypomethylation of gene regions and associated overexpression as a frequent event in cancer. In contrast, DNA repeats seem to be often hypomethylated in a variety of cancers, and satellite DNA hypomethylation is especially prevalent (Florl et al. 1999; Ehrlich 2002).

We hypothesize that human satellite DNA-rich pericentromeric heterochromatin controls in *trans* the expression of some genes present on various chromosomes in a manner that is sensitive to the level of methylation of the satellite DNA and impacts binding of proteins to chromatin. This protein binding to satellite DNA-rich chromatin might be affected negatively or positively by changes in DNA methylation in *cis* depending on the protein, as we and others have shown (Huang et al. 1984; Ehrlich and Ehrlich 1993; Zhang et al. 1993; Wade 2001). This would provide a novel, possibly tissue-specific pathway by which satellite DNA hypomethylation could affect carcinogenesis and gene expression.

Upon satellite DNA demethylation, there could be a decrease in the binding to centromeric and juxtacentromeric heterochromatin of proteins that are not DNA-sequence specific, such as methylation-specific methyl-CpG-binding protein (MeCP)-type proteins and histone deacetylases (HDACs) that interact with MeCP proteins (Nan et al. 1996; Fuks et al. 2007). This change could release these transcription repressors or corepressors to the nucleoplasm to downregulate expression of tumor-suppressor genes and might act synergistically (Baylin and Herman 2000) with tumor suppressor gene hypermethylation in the promoter regions. In addition, constitutive heterochromatin may act as a reservoir for certain DNA sequence-specific transcription regulators that bind to both heterochromatin regions and promoters or enhancers. Given the highly reiterated state of satellite DNA in heterochromatin and the length of pericentromeric heterochromatin, a large percentage of a given transcription regulator could be sequestered in this heterochromatin if it had a moderate affinity for a site in the approximately 1- to 2-kb higher-order repeat of the satellite DNA sequence. Examples of mammalian transcription regulators that bind selectively to centromeric or juxtacentromeric heterochromatin by DNA–protein and/or protein–protein interactions are accumulating: Ikaros; ATRX; heat shock factor 1; C/EBPα, C/EBPβ, and C/EBPδ; DNMT1, DNMT3A, and DNMT3B (which can repress transcription independently of their methylating activities); the MeCP proteins MBD1, MBD2, and MeCP2; HDAC1 and 2; and HP1-associated proteins like TIF1β (Nan et al. 1996; Tang and Lane 1999; Gibbons et al. 2000; Bachman et al. 2001; Sabbattini et al. 2001; Bozhenok et al. 2002; Cammas et al. 2002; Denegri et al. 2002). Heat shock factor 1 localizes to 9qh and the centromeric regions of chromosomes 12 and 15 upon heat shock (Denegri et al. 2002), and there is induction of transcription of Chr9 Sat3 associated with the colocalization of the transcription factor to the Sat3-rich 9qh region (Jolly et al. 2004).

Alternatively, decreases in satellite DNA methylation might increase sequence-specific binding of certain DNA-binding proteins to DNA (Ehrlich and Ehrlich 1993). In addition, binding of proteins to heterochromatin by protein–

protein interactions could be increased by a looser chromatin structure and by indirect effects on histone modification (Xin et al. 2003). Moreover, constitutive heterochromatin in the vicinity of the centromeres might influence gene expression in *trans* by altering subnuclear compartmentalization (Kosak et al. 2002) or heterochromatin-euchromatin interactions between different chromosomes, as has been proposed for murine centromeric heterochromatin and several early lymphogenesis genes (Gasser 2001; Alcobia et al. 2003).

9
DNA Hypomethylation vs DNA Hypermethylation in Cancer: Distinct Aspects and Overlaps

It has been hypothesized that hypomethylation of DNA in cancer is only a prelude or a reaction to DNA hypermethylation. In that case, there should be a positive association between cancer-linked gene hypermethylation and global DNA hypomethylation or hypomethylation of satellite DNAs. In collaboration with Peter Laird, Louis Dubeau, and Mimi Yu, we showed that there is no such association for Wilms tumors analyzed at 13 CpG-rich 5′ gene regions (Ehrlich et al. 2002) or ovarian carcinomas analyzed at 60 such loci by MethyLight assay (Ehrlich et al. 2006). Therefore, satellite DNA and global DNA hypomethylation probably arise during carcinogenesis independently of gene hypermethylation. Lack of a positive association of DNA hypomethylation and hypermethylation was also seen in studies of the following interrelationships: LINE1 hypomethylation vs *GST1* hypermethylation in prostate carcinomas (Santourlidis et al. 1999); satellite and global DNA hypomethylation vs gene hypermethylation in Wilms tumors (Ehrlich et al. 2002); cancer-testes antigen gene hypomethylation vs gene hypermethylation in gastric cancers (Kaneda et al. 2004); and in vitro DNMT acceptor activity vs gene hypermethylation in colon tumors (Bariol et al. 2003; Table 2). Also consistent with this conclusion, we, as well as others, did not detect gene hypermethylation in ICF cells (M. Ehrlich, unpublished data). However, in three studies of ovarian carcinomas or breast cancers (Widschwendter et al. 2004; Ehrlich et al. 2006), we found one gene, *CDH13*, had a significant inverse association of hypermethylation of its 5′ CpG island with satellite DNA hypomethylation. This suggests an antagonistic relationship between waves of genome demethylation and de novo gene methylation such that promoter hypermethylation in cancer could be subsequently reversed during a tumor progression-linked wave of DNA demethylation that includes satellite DNA.

Opposite types of cancer-linked DNA methylation changes can occur in the same DNA sequence (Nishiyama et al. 2005a, b; Table 2, point 3). This

Table 2 Interrelationships of hypermethylation and hypomethylation in cancer

1. In various cancers, CpG island hypermethylation in promoters and 5′ gene regions is *not* positively associated with hypomethylation of the following
   - Satellite2 and satellite α DNA (in juxtacentromeric and centromeric regions, respectively)
   - LINE1 repeats
   - Cancer-testes antigen genes
   - The overall genome (global DNA hypomethylation)
2. In ovarian carcinomas, less hypomethylation of satellite 2 is significantly associated with more hypermethylation of the following
   - *CDH13*'s 5′ CpG island
   - A tandem DNA repent (*NBL2*) at *Hha*I sites
3. A given DNA sequence (*NBL2*) can undergo opposite cancer-linked epigenetic changes and display the following
   - Hypomethylation at certain sites and hypermethylation at others within the same ~0.2-kb subregion
   - Hypomethylation involving both isolated CpG sites and spreading of demethylation
   - Predominant hypomethylation in some ovarian cancers and predominant hypermethylation in others
   - Much more frequent hypomethylation in certain tumor types than others
4. These findings suggest that there is a common step that makes DNA sequences epigenetically unstable in tumors, which can predispose CpGs to de novo methylation and m^5CpGs to demethylation
   - But that there are subsequent, *independent* steps leading to hypermethylation or hypomethylation
   - And that de novo methylation of some sequences in cancers is reversed by spontaneous demethylation

surprising result came first from Southern blot analysis. We found that a 1.4-kb repeat called *NBL2* (Thoraval et al. 1996; *DMHD-1, CNIC*, Y10752, and U59100) can be hypomethylated at *Hha*I sites (5 CGCG-3 sites) in some cancers (17% of ovarian carcinomas) and hypermethylated at very high frequencies (>70% of ovarian carcinomas and Wilms tumors) in other tumors at the same sites (Nishiyama et al. 2005a). The possibility of artifacts was eliminated by internal controls demonstrating complete digestion. Similar patterns of partial methylation at *Hha*I sites in *NBL2* were seen in various normal samples from nine types of postnatal somatic tissues. *NBL2* hypermethylation at *Hha*I sites in ovarian cancers was significantly associated with less hypomethylation at Sat2. There was no detectable transcription of this sequence in normal tissues and only very low levels of transcripts, probably from run-through transcription, in some of the cancers, independent of *NBL2* hypomethylation (Nishiyama et al. 2005a). *NBL2* is a complex tandem repeat in the pericentromeric regions of the short arms of the acrocentric chromosomes and at 9p11 and 9q21, and it lacks a long open reading frame (Nishiyama et al. 2005a). It was found to be hypomethylated at *Not*I sites, the focus of previous studies, in 83% of neuroblastomas and 75% of liver cancers (Thoraval et al. 1996; Itano et al. 2002). Hypermethylation in cancer was not previously observed with this enzyme because *Not*I cleaves control somatic DNA too infrequently to reveal hypermethylation in cancers. We showed that the ovarian carcinomas that were hypomethylated at *Hha*I sites were also the only ones that were hypomethylated at *Not*I sites.

Southern blot analysis of *NBL2* in digests of DNA from the ovarian carcinomas, Wilms tumors, and somatic control tissues with three other CpG methylation-sensitive restriction endonucleases indicated that the changes in methylation patterns in the cancers were yet more varied than those observed in *Hha*I and *Not*I digests (Nishiyama et al. 2005b). Some cancer DNAs exhibited hypermethylation at one type of restriction site and hypomethylation at another. A few cancers displayed two distinct subfractions of *NBL2* sequences, one hypermethylated and the other hypomethylated, relative to all the somatic controls.

Cancer-linked changes in methylation were studied at the local level by hairpin-PCR bisulfite genomic sequencing (Laird et al. 2004) in a study revealing the methylation status of each cytosine residue on complementary strands of two approximately 0.2-kb subregions of the 1.4-kb repeat *NBL2* unit (Nishiyama et al. 2005b). Ten cancers (ovarian carcinomas or Wilms tumors) exhibiting either hypomethylation or hypermethylation at *Hha*I sites by Southern blot analysis and control tissues were sequenced by the hairpin-bisulfite PCR method in which the covalent cross-linking of the two strands allows the unambiguous determination of symmetrical methylation, symmetrical lack of methylation, or hemimethylation at CpG sites from bisulfite-treated DNA. We observed much site-specificity among various normal tissues in the methylation status of individual CpG sites in these two *NBL2* subregions. This finding allowed us to detect cancer-linked increases or decreases in methylation at most of the examined CpG sites. At the CpG sites with conserved methylation status in normal somatic tissues, there was a significant association between methylation changes from the hairpin-bisulfite PCR sequencing and those deduced from the Southern blot analyses of *Hha*I digests. Nonetheless, molecular clones from the cancer DNAs showed both hypomethylation and hypermethylation at different CpG sites. In some cancers, hypomethylation predominated and in others, hypermethylation. Therefore, the interrelationships between cancer-linked hypermethylation and cancer-linked hypomethylation within a given DNA sequence can be intimate despite our findings that CpG island hypermethylation and satellite or global DNA hypomethylation show no significant positive association (Ehrlich et al. 2002, 2006). Our results indicate that a small region of DNA can be made unstable epigenetically during carcinogenesis so that close CpG sites can undergo opposite changes in DNA methylation (Table 2).

A minor but considerable portion of the CpG sites was hemimethylated in both normal tissues and tumors (Nishiyama et al. 2005b). This indicates that maintenance methylation is less efficient than commonly assumed. Furthermore, the methylation patterns that we observed in individual clones suggest that there is spreading of demethylation in cancer, although most of the cancer-linked methylation changes in *NBL2* involved non-adjacent CpG sites. Previous studies had demonstrated spreading of de novo methylation along DNA sequences (Doerfler 1984, Orend et al. 1991; Singal and van Wert 2001; Kim et al. 2002; Turker 2002).

10
Conclusions

The findings that global DNA, satellite DNA, and gene hypomethylation exhibit no statistically significant positive association with CpG island hypermethylation (Kaneda et al. 1994; Santourlidis et al. 1999; Ehrlich et al. 2002; Bariol et al. 2003) indicate that oncogenesis-linked DNA demethylation is not just a cellular response to or stimulus for de novo methylation. The importance of cancer-linked DNA hypomethylation is evidenced by its widespread occurrence and by the findings that increases in DNA repeat hypomethylation are significantly associated with an increase in aggressiveness of some types of cancers and a decrease in patient survival (Ehrlich 2002; Itano et al. 2002; Widschwendter et al. 2004). Also, some DNA hypomethylation, like DNA hypermethylation, can be detected very early in tumorigenesis or even in abnormal non-neoplastic tissue (Ehrlich 2002; Jackson et al. 2004). Whatever the most important biological targets of tumorigenesis-related DNA hypomethylation, cancer geneticists and clinicians should be aware that decreases in DNA methylation induced as part of a therapeutic regimen might contribute to carcinogenesis or tumor progression. Nonetheless, DNA demethylation therapy clearly may be very useful in cases where better alternatives do not exist. Our limited knowledge of DNA demethylation during carcinogenesis also points to the need to better understand its contribution to tumor formation and tumor progression.

Recent findings that a non-satellite tandem DNA repeat shows predominant hypermethylation in some ovarian carcinomas and hypomethylation in others indicates the surprising plasticity of epigenetic changes in DNA in cancer (Nishiyama et al. 2005). Furthermore, over a 0.3-kb region, there were hypomethylated CpG sites that were normally highly methylated, and hypermethylated CpG sites that were usually unmethylated (Nishiyama et al. 2005b). These results and the previously mentioned ones suggest that an early cancer-linked change in chromatin structure can predispose to both de novo methylation and demethylation during tumorigenesis but that subsequent independent events and the genetic and epigenetic features of the DNA sequence dictate whether there will be losses or gains in methylation in a DNA region (Table 2). Moreover, there are significant inverse associations of *CDH13* hypermethylation and satellite 2 hypomethylation in ovarian and breast cancer (Widschwendter et al. 2004) and of *NBL2* hypermethylation and satellite 2 hypomethylation in ovarian cancer (Nishiyama et al. 2005). All these findings suggest that tumor-associated changes in DNA methylation in a given sequence occur in a stepwise manner, which can consist of multiple rounds of de novo methylation, demethylation, or de novo methylation followed by demethylation. Therefore, CpG-island promoter sequences that become hypermethylated in cancer may similarly undergo subsequent demethylation during waves of DNA demethylation as the tumor develops, and such reversal of de novo methylation may contribute to carcinogenesis (Cheng et al. 2001; Zhu et al. 2003). In conclusion, DNA methylation changes in cancer appear to be dynamic due to an underlying, long-lasting, epigenetic instability.

Acknowledgements This work was supported in part by NIH Grant CA81506 and a Louisiana Research Consortium grant.

References

Alcobia I, Quina AS, Neves H, Clode N, Parreira L (2003) The spatial organization of centromeric heterochromatin during normal human lymphopoiesis: evidence for ontogenetically determined spatial patterns. Exp Cell Res 290:358–369

Anonymous (1985) Biochemistry and biology of DNA methylation. Proceedings of a Fogarty International Center Conference. Bethesda, Maryland, April 17-19, 1985. Prog Clin Biol Res 198:1–324

Attwood JT, Yung RL, Richardson BC (2002) DNA methylation and the regulation of gene transcription. Cell Mol Life Sci 59:241–257

Bachman KE, Rountree MR, Baylin SB (2001) Dnmt3a and Dnmt3b are transcriptional repressors that exhibit unique localization properties to heterochromatin. J Biol Chem 276:32282

Barbeyron T, Kean K, Forterre P (1984) DNA adenine methylation of GTC sequences appeared recently in the Escherichia coli lineage. J Bacteriol 160:586–590

Bariol C, Suter C, Cheong K, Ku SL, Meaghr A, Hawkins N, Ward R (2003) The relationship between hypomethylation and CpG island methylation in colorectal neoplasia. Am J Pathol 162:1361–1371

Baylin SB, Herman JG (2000) Epigenetics and loss of gene function in cancer. In: Ehrlich M (ed) DNA and alterations in cancer: genetic and epigenetic alterations. Eaton Publishing, Natick, pp 293–309

Baylin SB, Hoppener JW, de Bustros A, Steenbergh PH, Lips CJ, Nelkin BD (1986) DNA methylation patterns of the calcitonin gene in human lung cancers and lymphomas. Cancer Res 46:29172922

Beard C, Li E, Jaenisch R (1995) Loss of methylation activates Xist in somatic but not in embryonic cells. Genes Dev 9:2325–2334

Biniszkiewicz D, Gribnau J, Ramsahoye B, Gaudet F, Eggan K, Humpherys D, Mastrangelo MA, Jun Z, Walter J, Jaenisch R (2002) Dnmt1 overexpression causes genomic hypermethylation, loss of imprinting, and embryonic lethality. Mol Cell Biol 22:2124–2135

Bird AP (1984) DNA methylation—how important in gene control? Nature 307:503–504

Bozhenok L, Wade PA, Varga-Weisz P (2002) WSTF-ISWI chromatin remodeling complex targets heterochromatic replication foci. EMBO J 21:2231–2241

Brabender J, Usadel H, Danenberg KD, Metzger R, Schneider PM, Lord RV, Wickramasinghe K, Lum CE, Park J, Salonge D, Singer J, Sidransky D, Holscher AH, Meltzer SJ Danenberg PV (2001) Adenomatous polyposis coli gene promoter hypermethylation in non-small cell lung cancer is associated with survival. Oncogene 20:3528–3532

Bruniquel D, Schwartz RH (2003) Selective, stable demethylation of the interleukin-2 gene enhances transcription by an active process. Nat Immunol 4:235–240

Cairns BR (2001) Emerging roles for chromatin remodeling in cancer biology. Trends Cell Biol 11:S15–21

Cammas F, Oulad-Abdelghani M, Vonesch JL, Huss-Garcia Y, Chambon P, Losson R (2002) Cell differentiation induces TIF1 beta association with centromeric heterochromatin via an HP1 interaction. J Cell Sci 115:3439–3448

Carr BI, Reilly G, Smith SS, Winberg C, Riggs A (1984) The tumorigenicity of 5-azacytidine in the male Fischer rat. Carcinogenesis 5:1583–1590

Chen RZ, Pettersson U, Beard C, Jackson-Grusby L, Jaenisch R (1998) DNA hypomethylation leads to elevated mutation rates. Nature 395:89–93

Cheng CW, Wu PE, Yu JC, Huang CS, Yue CT, Wu CW, Shen CY (2001) Mechanisms of inactivation of E-cadherin in breast carcinoma: modification of the two-hit hypothesis of tumor suppressor gene. Oncogene 20:3814–3823

Cheng P, Schmutte C, Cofer KF, Felix JC, Yu MC, Dubeau L (1997) Alterations in DNA methylation are early, but not initial, events in ovarian tumorigenesis. Br J Cancer 75:396–402

Constantinides PG, Jones PA, Gevers W (1977) Functional striated muscle cells from non-myoblast precursors following 5-azacytidine treatment. Nature 267:364–366

Cooper DN (1983) Eukaryotic DNA methylation. Hum Genet 64:315–333

Costello JF, Fruhwald MC, Smiraglia DJ, Rush LJ, Robertson GP, Gao X, Wright FA, Feramisco JD, Peltomaki P, Lang JC, Schuller DE, Yu L, Bloomfield CD, Caligiuri MA, Yates A, Nishikawa R, Su Huang H, Petrelli NJ, Zhang X, O'Dorisio MS, Held WA, Cavenee WK, Plass C (2000) Aberrant CpG-island methylation has non-random and tumour-type-specific patterns. Nat Genet 24:132–138

Craig JM, Earle E, Canham P, Wong LH, Anderson M, Choo KH (2003) Analysis of mammalian proteins involved in chromatin modification reveals new metaphase centromeric proteins and distinct chromosomal distribution patterns. Hum Mol Genet 12:3109–3121

Cusack SM, Rohn TT, Medeck RJ, Irwin KM, Brown RJ, Mercer LM, Oxford JT (2004) Suppression of MeCP2beta expression inhibits neurite extenson in PC12 cells. Exp Cell Res 299:442–453

Datta J, Ghoshal K, Sharma SM, Tajima S,Jacob ST (2003) Biochemical fractionation reveals association of DNA methyltransferase (Dnmt) 3b with Dnmt 1 and that of Dnmt 3a with a histone H3 methyltransferase and Hdac 1. J Cell Biochem 88:855–864

De Smet C, De Backer O, Faraoni I, Lurquin C, Brasseur F, Boon T (1996) The activation of human gene MAGE-1 in tumor cells is correlated with genome wide demethylation. Proc Natl Acad Sci USA 93:7149–7153

Denda A, Rao PM, Rajalakshmi S, Sarma DS (1985) 5-Azacyidine potentiates initiation induced by carcinogens in rat liver. Carcinogenesis 6:145146

Denegri M, Moralli D, Rocchi M, Biggiogera M, Raimondi E, Cobianchi F, De Carli L, Riva S, Biamonti G (2002) Human chromosomes 9, 12, and 15 contain the nucleation sites of stress-induced nuclear bodies. Mol Biol Cell 13:2069–2079

Dennis K, Fan T, Geiman T, Yan Q, Muegge K (2001) Lsh, a member of the SNF2 family, is required for genome-wide methylation. Genes Dev 15:2940–2944

Doerfler W (1984) DNA-Methylierung: Geninaktivierung durch sequenzspezifische DNA Methylierungen. Angew Chem Weinheim Bergstr Ger 23:919–929

Eads CA, Nickel AE, Laird PW (2002) Complete genetic suppression of polyp formation and reduction of CpG-island hypermethylation in Apc(Min/+) Dnmt 1-hypomorphic mice. Cancer Res 62:1296–1299

Eden A, Gaudet F, Waghmare A, Jaenisch R (2003) Chromosomal instability and tumors promoted by DNA hypomethylation. Science 300:455

Ehrlich M (2000) DNA hypomethylation and cancer. In: Ehrlich M (ed) DNA and alterations in cancer: genetic and epigenetic alterations. Eaton Publishing, Natick, pp 273–291

Ehrlich M (2002) DNA methylation in cancer: too much, but also too little. Oncogene 21:5400–5413

Ehrlich M (2003) Expression of various genes is controlled by DNA methylation during mammalian development. J Cell Biochem 88:899–910

Ehrlich M, Ehrlich KC (1993) Effect of DNA methylation on the binding of vertebrate and plant proteins to DNA. In: Jost JP, Saluz HP (eds) DNA methylation: biological significance. Birkhauser Verlag, Boston, pp 145–168

Ehrlich M, Wang RY (1981) 5-Methylcytosine in eukaryotic DNA. Science 212:1350–1357

Ehrlich M, Gama-Sosa MA, Huang LH, Midgett RM, Kuo KC, McCune RA, Gehrke C (1982) Amount and distribution of 5-methylcytosine in human DNA from different types of tissues of cells. Nucleic Acids Res 10:2709–2721

Ehrlich M, Gama-Sosa MA, Carreira LH, Ljungdahl LG, Kuo KC (1985) DNA methylation in thermophilic bacteria: N4-methylcytosine, 5-methylcytosine, and N6-methyladenine. Nucleic Acids Res 13:1399–1412

Ehrlich M, Buchanan KL, Tsien F, Jiang G, Sun B, Uicker W, Weemaes CM, Smeets D, Sperling K, Belohradsky BH, Tommerup N, Misek DE, Rouillard JM, Kuick R, Hanash SM (2001) DNA methyltransferase 3B mutations linked to the ICF syndrome cause dysregulation of lymphogenesis genes. Hum Mol Genet 10:2917–2931

Ehrlich M, Jiang G, Fiala E, Dome JS, Yu MC, Long TI, Youn B, Sohn OS, Widschwendter M, Tomlinson GE, Chintagumpala M, Champagne M, Parham D, Liang G, Malik K, Laird PW (2002) Hypomethylation and hypermethylation of DNA in Wilms tumors. Oncogene 21:6694–6702

Ehrlich M, Hopkins NE, Jiang G, Dome JS, Yu MC, Woods CB, Tomlinson GE, Chintagumpala NM, Champagne M, Dillerg L, Parham DM, Sawyer J (2003) Satellite DNA hypomethylation in karyotyped Wilms tumors. Cancer Genet Cytogenet 141:97–105

Ehrlich M, Woods C, Yu M, Dubeau L, Yang F, Campan M, Weisenberger D, Long T, Youn B, Fiala E, Laird P (2006) Quantitative analysis of associations between DNA hypermethylation, hypomethylation, and DNMT RNA in ovarian tumors. Oncogene Mar 13; (Epub ahead of print)

Feinberg AP, Vogelstein B (1983a) Hypomethylation distinguishes genes of some human cancers from their normal counterparts. Nature 301:89–92

Feinberg AP, Vogelstein B (1983b) Hypomethylation of ras oncogenes in primary human cancers. Biochem Biophys Res Commun 111:47–54

Feinberg AP, Cui H, Ohlsson R (2002) DNA methylation and genomic imprinting: insights from cancer into epigenetic mechanisms. Semin Cancer Biol 12:389–398

Florl AR, Lower R, Schmitz-Drager BJ, Schulz WA (1999) DNA methylation and expression of LINE-1 and HERV-K provirus sequences in urothelial and renal cell carcinomas. Br J Cancer 80:1312–1321

Fuks F, Burgers WA, Godin N, Kasai M, Kouzarides T (2001) Dnmt3a binds deacetylases and is recruited by a sequence-specific repressor to silence transcription. EMBO J 20:2536

Fuks F, Hurd PJ, Deplus R, Kouzarides T (2003a) The DNA methyltransferases associate with HP1 and the SUV39H1 histone methyltransferase. Nucleic Acids Res 31:2305–2312

Fuks F, Hurd PJ, Wolf D, Nan X, Bird AP, Kouzarides T (2003b) The methyl-CpC-binding protein MeCP2 links DNA methylation to histone methylation. J Biol Chem 278:4035–4040

Gama-Sosa MA, Midgett SM, Slagel VA, Githens S, Kuo KC, Gehrke CW, Ehrlich M (1983a) Tissue-specific differences in DNA methylation in various mammals. Biochim Biophys Acta 740:212–219

Gama-Sosa MA, Slagel VA, Trewyn RW, Oxenhandler R, Kuo KC, Gehrke CW, Ehrlich M (1983b) The 5-methylcytosine content of DNA from human tumors. Nucleic Acids Res 11:6883–6895

Gasser SM (2001) Positions of potential: nuclear organization and gene expression. Cell 104:639–642

Gaudet F, Hodgson JG, Eden A, Jackson-Grusby L, Dausman J, Gray JW, Leonhardt H, Jaenisch R (2003) Induction of tumors in mice by genomic hypomethylation. Science 300:489–492

Geiman TM, Sankpal UT, Robertson AK, Chen Y, Mazumdar M, Heale JT, Schmiesing JA, Kim W, Yokomori K, Zhao Y, Robertson KD (2004) Isolation and characterization of a novel DNA methyltransferase complex linking DNMT3B with components of the mitotic chromosome condensation machinery. Nucleic Acids Res 32:2716–2729

Geyer CB, Kiefer CM, Yang TP, McCarrey JR (2004) Ontogeny of a demethylation domain and its relationship to activation of tissue-specific transcription. Biol Reprod 71:837–844

Gibbons RJ, McDowell TL, Raman S, O'Rourke DM, Garrick D, Ayyub H, Higgs DR (2000) Mutations in ATRX, encoding a SWI/SNF-like protein, cause diverse changes in the pattern of DNA methylation. Nat Genet 24:368–371

Gisselsson D, Shao C, Tuck-Muller C, Sogorovic S, Palsson E, Smeets D, Ehrlich M (2005) Interphase chromosomal abnormalities and mitotic missegregation of hypomethylated sequences in ICF syndrome cells. Chromosoma 114:118–126

Gowher H, Jeltsch A (2002) Molecular enzymology of the catalytic domains of the Dnmt3A and Dnmt3b DNA methyltransferases. J Biol Chem 277:20409–20414

Gowher H, Leismann O, Jeltsch A (2000) DNA of Drosophila melanogaster contains 5-metylcytosine. EMBO J 19:6918–6923

Hansen RS, Wijmenga C, Luo P, Stanek AM, Canfield TK, Weemaes CM, Gartler SM (1999) The DNMT3B DNA methyltransferase gene is mutated in the ICF immunodeficiency syndrome. Proc Natl Acad Sci USA 96:14412–14417

Heard E (2004) Recent advances in X-chromosome inactivation. Curr Opin Cell Biol 16:247–255

Heard E, Clerc P, Avner P (1997) X-chromosome inactivation in mammals. Annu Rev Genet 31:571–610

Hernandez R, Frady A, Zhang XY, Varela M, Ehrlich M (1997) Preferential induction of chromosome 1 multibranched figures and whole-arm deletions in a human proB cell line treated with 5-azacytidine or 5-azadeoxycytudube. Cytogenet Cell Genet 76:196–201

Howell CY, Bestor TH, Ding F, Latham KE, Mertineit C, Trasler JM, Chaillet JR (2001) Genomic imprinting disrupted by a maternal effect mutation in the Dnmt1 gene. Cell 104:829–838

Huang LH, Wang R, Gama-Sosa MA, Shenoy S, Ehrlich M (1984) A protein from human placental nuclei binds preferentially to 5-methylcytosine-rich DNA. Nature 308:293–295

Huang YM, Chen SU, Goodman SD, Wu SH, Kao JT, Lee CN, Cheng WC, Tsai KS,Fang WH (2004) Interaction of nick-directed DNA mismatch repair and loop repair in human cells. J Biol Chem 279:30228–30235

Itano O, Ueda M, Kikuchi K, Hashimoto O, Hayatsu S, Kawaguchi M, Seki H, Aiura K, Kitajima M (2002) Correlation of postoperative recurrence in hepatocellular carcinoma with demethylation of repetitive sequences. Oncogene 21:789–797

Iwano H, Nakamura M, Tajima S (2004) Xenopus MBD3 plays a crucial role in an early stage of development. Dev Biol 268:416–428

Jackson K, Yu MC, Arakawa K, Fiala E, Youn B, Fiegl H, Muller-Holzner E, Widschwendter M, Ehrlich M (2004) DNA hypomethylation is prevalent even in low-grade breast cancers. Cancer Biol Ther 3:1225–1231

Jackson-Grusby L, Laird PW, Magge SN, Moeller BJ, Jaenisch R (1997) Mutagenicity of 5-aza-2'-deoxycytidine is mediated by the mammalian DNA methyltransferase. Proc Natl Acad Sci USA 94:4681–4685

Jeanpierre M, Turleau C, Aurias A, Prieur M, Ledeist F, Fischer A, Viegas-Pequinot E (1993) An embryonic-like methylation pattern of classical satellite DNA is observed in ICF syndrome. Hum Mol Genet 2:731–735

Ji W, Hernandez R, Zhang XY, Qu GZ, Frady A, Varrela M, Ehrlich M (1997) DNA demethylation and pericentromeric rearrangements of chromosome 1. Mutat Res 379:33–41

Jolly C, Metz A, Govin J, Vigneron M, Turner BM, Khochbin S, Vourch C (2004) Stress-induced transcription of satellite III repeats. J Cell Biol 164:25–33

Jones PA, Baylin SB (2002) The fundamental role of epigenetic events in cancer. Nat Rev Genet 3:415–428

Juttermann R, Li E, Jaenisch R (1994) Toxicity of 5-aza-2'deoxycytidine to mammalian cells is mediated primarily by covalent trapping of DNA methyltransferase rather than DNA demethylation. Proc Natl Acad Sci USA 91:11797–11801

Kaneda A, Tsukamoto T, Takamura-Enya T, Watanabe N, Kaminishi M, Sugimura T, Tatematsu M, Ushijima T (2004) Frequent hypomethylation in multiple promoter CpG islands is associated with global hypomethylation, but nor with frequent promoter hypermethylation. Cancer Sci 95:58–64

Kim GD, Ni J, Kelesoglu N, Roberts RJ, Pradhan S (2002) Co-operation and communication between the human maintenance and de novo DNA (cytosine-5) methyltransferases. EMBO J 21:4183

Kim YI, Giuliano A, Hatch KD, Schneider A, Nour MA, Dallal GE, Selhub J, Mason JB (1994) Global DNA hypomethylation increases progressively in cervical dysplasia and carcinoma. Cancer 74:893–899

Klochendler-Yeivin A, Muchardt C, Yaniv M (2002) SWI/SBF chromatin remodeling and cancer. Curr Opin Genet Dev 12:73–79

Kondo T, Bobek MP, Kuick R, Lamb B, Zhu X, Narayan A, Bourchis D, Viegas-Pequignot E, Ehrlich M, Hanash SM (2000) Whole-genome methylation scan in ICF syndrome: hypomethylation of non-satellite DNA repeats D4Z4 and NBL2. Hum Mol Genet 9:597–604

Kosak ST, Skok JA, Medina KL, Riblet R, Le Beau MM, Fisher AG, Singh H (2002) Subnuclear compartmentalization of immunoglobulin loci during lymphocyte development. Science 296:158–162

Kunert N, Marhold J, Stanke J, Stach D, Lyko F (2003) A Dnmt2-like protein mediates DNA methylation in Drosophila. Development 130:5083–5090

Lacks SA, Dunn JJ, Greenberg B (1982) Identification of base mismatches recognized by the heteroduplex-DNA-repair system of Streptococcus pneumoniae. Cell 31:327–336

Laird CD, Pleasant ND, Clrk AD, Sneeden JL, Hassan KM, Manley NC, Vary JC Jr, Morgan T, Hansen RS, Stoger R (2004) Hairpin-bisulfite PCR: assessing epigenetic methylation patterns on complementary strands of individual DNA molecules. Proc Natl Acad Sci USA 101:204–209

Langle-Rouault F, Maenhaut-Michel G, Radman M (1987) GATC sequences, DNA nicks and the MutH function in Escherichia coli mismatch repair. EMBO J 6:1121–1127

Lee PP, Fitzpatrick DR, Beard C, Jessup HK, Lehar S, Makar KW, Perez-Melgosa M, Sweetser MT Schlissel MS, Nguyen S, Cherry SR, Tsai JH, Tucker SM, Weaver WM, Kelso A, Jaenisch R, Wilson CB (2001) A critical role for Dnmt1 and DNA methylation in T cell development, function, and survival. Immunity 15:763–774

Li E, Bestor TH, Jaenisch R (1992) Targeted mutation of the DNA methyltransferase gene results in embryonic lethality. Cell 69:915–926

Liang G, Chan MF, Tomigahara Y, Tsai YC, Gonzales FA, Li E, Laird PW, Jones PA (2002) Cooperativity between DNA methyltransferases in the maintenance methylation of repetitive elements. Mol Cell Biol 22:480–491

Lin CH, Hsieh SY, Sheen IS, Lee WC, Chen TC, Shyu WC, Liaw YF (2001) Genome-wide hypomethylation in hepatocellular carcinogenesis. Cancer Res 61:4238–4243

Lyko F, Ramsahoye BH, Jaenisch R (2000) DNA methylation in Drosophila melanogaster. Nature 408:538–540

Maier H, Colbert J, Fitzsimmons D, Clark DR, Hagman J (2003) Activation of the early B-cell-specific mb-1 (Ig-alpha) gene by Pax-5 is dependent on an unmethylated Ets binding site. Mol Cell Biol 23:1946–1960

Makar KW, Perez-Melgosa M, Shnyreva M, Weaver WM, Fitzpatrick DR, Wilson CB (2003) Active recruitment of DNA methyltransferases regulates interleukin 4 in thymocytes and T cells. Nat Immunol 4:1183–1190

Martinowich K, Hattori D, Wu H, Fouse S, He F, Hu Y, Fan G, Sun YE (2003) DNA methylation-related chromatin remodeling in activity-dependent BDNF gene regulation. Science 302:890–893

Mitelman F, Mertens F, Johansson B (1997) A breakpoint map of recurrent chromosomal rearrangements in human neoplasia. Nat Genet 15 Spec No:417–474

Moreau P, Mouillot G, Rousseau P, Marcou C, Dausset J, Carosella ED (2003) HLA-G gene repression is reversed by demethylation. Proc Natl Acad Sci USA 100:1191–1196

Namihira M, Nakashima K, Taga T (2004) Developmental stage dependent regulation of DNA methylation and chromatin modification in a immature astrocyte specific gene promoter. FEBS Lett 572:184

Nan X, Tate P, Li E, Bird A (1996) DNA methylation specifies chromosomal localization of MeCP2. Mol Cell Biol 16:414–421

Narayan A, Ji W, Zhang XY, Marrogi A, Graff JR, Baylin SB, Ehrlich M (1998) Hypomethylation of pericentromeric DNA in breast adenocarcinomas. Int J Cancer 77:833–838

Nishino K, Hattori N, Tanaka S, Shiota K (2004) DNA methylation-mediated control of Sry gene expression in mouse gonadal development. J Biol Chem 279:22306–22313

Nishiyama R, Qu L, Tsumagari K, Dubeau L, Weissbecker K, Champagne M, Sikka S, Nagai H, Ehrlich MA (2005a) DNA repeat, NBL2, is hypermethylated in some cancers but hypomethylated in others. Cancer Biol Ther 4:440–448

Nishiyama R, Qi L, Lacey M, Ehrlich M (2005b) Both hypomethylation and hypermethylation in a 0.2-kb region of a DNA repeat in cancer. Mol Cancer Res 3:617–626

Okada H, Kimura MT, Tan D, Fujiwara K, Igarashi J, Makuuchi M, Hui AM, Tsurumaru M, Nagase H (2005) Frequent trefoil factor 3 (TFF3) overexpression and promoter hypomethylation in mouse and human hepatocellular carcinomas. Int J Oncol 26:369–377

Okano M, Bell DW, Haber DA, Li E (1999) DNA methyltransferases Dnmt3a and Dnmt3b are essential for de novo methylation and mammalian development. Cell 98:247–257
Orend G, Kuhlmann I, Doerfler W (1991) Spreading of DNA methylation across integrated foreign (adenovirus type 12) genomes in mammalian cells. J Virol 65:4301–4308
Palmer BR, Marinus MG (1994) The dam and dcm strains of Escherichia coli—a review. Gene 143:1–12
Panning B, Jaenisch R (1996) DNA hypomethylation can activate Xist expression and silence X-linked genes. Genes Dev 10:1991–2002
Plass C, Soloway PD (2002) DNA methylation, imprinting and cancer. Eur J Hum Genet 10:6–16
Pradhan S, Esteve PO (2003) Mammalian DNA (cytosine-5) methyltransferases and their expression. Clin Immunol 109:6–16
Qu G, Grundy PE, Narayan A, Ehrlich M (1999a) Frequent hypomethylation in Wilms tumors of pericentromeric DNA in chromosomes 1 and 16. Cancer Genet Cytogenet 109:34–39
Qu G, Dubeau L, Narayan A, Yu M, Ehrlich M (1999b) Satellite DNA hypomethylation vs. overall genomic hypomethylation in ovarian epithelial tumors of different malignant potential. Mutat Res 423:91–101
Rhee I, Bachman KE, Park BH, Jair KW, Yen RW, Schuebel KE, Cui H, Feinberg AP, Lengauer C, Kinzler KW, Baylin SB, Vogelstein B (2002) DNMT1 and DNMT3b cooperate to silence genes in human cancer cells. Nature 416:552–556
Riggs AD (1975) X inactivation, differentiation, and DNA methylation. Cytogenet Cell Genet 14:9–25
Riggs AD, Jones PA (1983) 5-methylcytosine, gene regulation, and cancer. Adv Cancer Res 40:1–30
Romanov GA, Vanyushin BF (1981) Methylation of reiterated sequences in mammalian DNAs. Effects of the tissue type, age, malignancy and hormonal induction. Biochim Biophys Acta 653:204–218
Sabbattini P, Lundgren M, Georgiou A, Chow C, Warnes G, Dillon N (2001) Binding of Ikaros to the lambda5 promoter silences transcription through a mechanism that does not require heterochromatin formation. EMBO J 20:2812–2822
Saito Y, Kanai Y, Sakamoto M, Saito H, Ishii H, Hirohashi S (2001) Expression of mRNA for DNA methyltransferases and methyl-CpG-binding proteins and DNA methylation status on CpG islands and pericentromeric satellite regions during human hepatocarcinogenesis. Hepatology 33:561–568
Saito Y, Kanai Y, Sakamoto M, Saito H, Ishii H, Hirohashi S (2002) Overexpression of a splice variant of DNA methyltransferase 3b, DNMT4b4, associated with DNA hypomethylation on pericentromeric satellite regions during human hepatocarcinogenesis. Proc Natl Acad Sci USA 99:10060–10065
Santourlidis S, Florl A, Ackermann R, Wirtz HC, Schulz WA (1999) High frequency of alterations in DNA methylation in adenocarcinoma of the prostate. Prostate 39:166–174
Sato N, Maitra A, Fukushima N, van Heek NT, Matsubayashi H, Iacobuzio-Donahue CA, Rosty C, Goggins M (2003) Frequent hypomethylation of multiple genes overexpressed in pancreatic ductal adenocarcinoma. Cancer Res 63:4158–4166

Schulz WA, Elo JP, Florl AR, Pennanen S, Santourlidis S, Engers R, Buchardt M, Seifert HH, Visakorpi T (2002) Genomewide DNA hypomethylation is associated with alterations on chromosome 8 in prostate carcinoma. Genes Chromosomes Cancer 35:58–65

Singal R, van Wert JM (2001) De novo methylation of an embryonic globin gene during normal development is strand specific and spreads from the proximal transcribed region. Blood 98:3441–3446

Soares J, Pinto AE, Cunha CV, Andre S, Barao I, Sousa JM, Cravo M (1999) Global DNA hypomethylation in breast carcinoma: correlation with prognostic factors and tumor progression. Cancer 85:112–118

Strathdee G, Sim A, Brown R (2004) Control of gene expression by CpG island methylation in normal cells. Biochem Soc Trans 32:913–915

Suzuki M, Harashima A, Okochi A, Yamamoto M, Nakamura M, Nakamura S, Motoda R, Yamasaki F, Orita K (2004) 5-Azacytidine supports the long-term repopulating activity of cord blood CD34(+) cells. Am J Hematol 77:313–315

Tagoh H, Schebesta A, Lefevre P, Wilson N, Hume D, Busslinger M, Bonifer C (2004) Epigenetic silencing of the c-fms locus during B-lymphopoiesis occurs in discrete steps and is reversible. EMBO J 23:4275–4285

Tang QQ, Lane MD (1999) Activation and centromeric localization of CCAAT/enhancer-binding proteins during the mitotic clonal expansion of adipocyte differentiation. Genes Dev 13:2231–2241

Taylor SM, Jones PA (1979) Multiple new phenotypes induced in 10T1/2 and 3T3 cells treated with 5-azacytidine. Cell 17:771–779

Thomas GA, Williams ED (1992) Production of thyroid tumours in mice by demethylating agents. Carcinogenesis 13:1039–1042

Thoraval D, Asakawa J, Wimmer K, Kuick R, Lamb B, Richardson B, Ambros P, Glover T, Hanash S (1996) Demethylation of repetitive DNA sequences in neuroblastoma. Genes Chromosomes Cancer 17:234–244

Tsuda H, Takarabe T, Kanai Y, Fukutomi T, Hirohashi S (2002) Correlation of DNA hypomethylation at pericentromeric regions of chromosomes 16 and 1 with histological features and chromosomal abnormalities of human breast carcinomas. Am J Pathol 161:859–866

Tsuji-Takayama K, Inoue T, Ijiri Y, Otani T, Motoda R, Nakamura S, Orita K (2004) Demethylating agent, 5-azacytidine, reverses differentiation of embryonic stem cells. Biochem Biophys Res Commun 323:86–90

Tuck-Muller CM, Narayan A, Tsien F, Smets DF, Sayer J, Fiala ES, Sohn OS, Ehrlich M (2000) DNA methylation and unusual chromosome instability in cell lines from ICF syndrome patients. Cytogenet Cell Genet 89:121–128

Turker MS (2002) Gene silencing in mammalian cells and the spread of DNA methylation. Oncogene 21:5388–5393

Tycko B (2000) Genomic imprinting and human neoplasia. In: Ehrlich M (ed) DNA and alterations in cancer: genetic and epigenetic alterations. Eaton Publishing, Natick, pp 333–349

Vanyushin BF, Tkacheva SG, Belozersky AN (1970) Rare bases in animal DNA. Nature 225:948–949

Vanyushin BF, Nemirovsky LE, Klimenko VV, Vasiliev VK, Belozersky AN (1973) The 5-methylcytosine in DNA of rats. Tissue and age specificity and the changes induced by hydrocortisone and other agents. Gerontologia 19:138–152

Wade PA (2001) Methyl CpG binding proteins: coupling chromatin architecture to gene regulation. Oncogene 20:3166–3173

Weaver IC, Cervoni N, Champagne FA, D'Alessio AC, Sharma S, Seckl JR, Dymov S, Szyf M, Meaney MJ (2004) Epigenetic programming by maternal behavior. Nat Neurosci 7:847–854

Weisenberger DJ, Campan M, Long TI, Kim M, Woods C, Fiala E, Ehrlich M (2005) Analysis of repetitive element DNA methylation by MethyLight. Nucleic Acids Res 33:6823–6836

Widschwendter M, Jiang G, Woods C, Muller HM, Fiegl H, Goebel G, Marth C, Muller-Holzner E, Zeimet AG, Laird PW, Ehrlich M (2004) DNA hypomethylation and ovarian cancer biology. Cancer Res 64:4472–4480

Wong N, Lam WC, Lai PB, Pang E, Lau WY Johnson PJ (2001) Hypomethylation of chromosome 1 heterochromatin DNA correlates with q-arm copy gain in human hepatocellular carcinoma. Am J Pathol 159:465–471

Xin H, Yoon HG, Singh PB, Wong J, Qin J (2003) Components of a pathway maintaining histone modification and heterochromatin protein 1 binding at the pericentric heterochromatin in mammalian cells. J Biol Chem 279:9536–9546

Xu GL, Bestor TH, Bourchis D, Hsieh CL, Tommerup N, Bugge M, Hulten M, Qu X, Russo JJ, Viegas-Pequignot E (1999) Chromosome instability and immunodeficiency syndrome caused by mutations in DNA methyltransferase gene. Nature 402:187–191

Yamada Y, Jackson-Grusby L, Linhart H, Meissner A, Eden A, Lin H, Jaenisch R (2005) Opposing effects of DNA hypomethylation on intestinal and liver carcinogenesis. Proc Natl Acad Sci U S A 102:13580–13585

Yamamoto F, Yamamoto M, Soto JL, Kojima E, Wang EN, Perucho M (2001) NotI-MseII methylation-sensitive amplified fragment length polymorphism for DNA methylation analysis of human cancers. Electrophoresis 22:1946–1956

Yan Q, Huang J, Fan T, Zhu H, Muegge K (2003) Lsh, a modulator of CpG methylation, is crucial for normal histone methylation. EMBO J 22:5154

Ye L, Li X, Kong X, Wang W, Bi Y, Hu L, Cui B, Ning G (2005) Hypomethylation in the promoter region of POMC gene correlates with ectopic overexpression in thymic carcinoids. J Endocrinol 185:337–343

Yung R, Ray D, Eisenbraun JK, Dengh C, Attwood J, Eisenbraun MD (2001) Unexpected effects of a heterozygous dnmt 1 null mutation on age-dependent DNA hypomethylation and autoimmunity. J Gerontol A Biol Sci Med Sci 56:B268–B276

Zhang XY, Jabrane-Ferrat N, Aiedu CK, Samac S, Peterlin BM, Ehrlich M (1993) The major histocompatibility complex class II promoter-binding protein RFX (NF-X) is a methylated DNA-binding protein. Mol Cell Biol 13:6810–6818

Zhu X, Leav I, Leung YK, Wu M, Liu Q, Gao Y, McNeal JE, Ho SM (2004) Dynamic regulation of estrogen receptor-beta expression by DNA methylation during prostate cancer development and metastasis. Am J Pathol 164:2003–2012

Subject Index

Current Topics in Microbiology and Immunology

Volumes published since 1989 (and still available)

Vol. 266: **Cooper, Koproski (Eds.):** The Interface Between Innate and Acquired Immunity, 2002. 15 figs. XIV, 116 pp. ISBN 3-540-42894-X

Vol. 267: **Mackenzie, John S.; Barrett, Alan D. T.; Deubel, Vincent (Eds.):** Japanese Encephalitis and West Nile Viruses. 2002. 66 figs. X, 418 pp. ISBN 3-540-42783X

Vol. 268: **Zwickl, Peter; Baumeister, Wolfgang (Eds.):** The Proteasome-Ubiquitin Protein Degradation Pathway. 2002. 17 figs. X, 213 pp. ISBN 3-540-43096-2

Vol. 269: **Koszinowski, Ulrich H.; Hengel, Hartmut (Eds.):** Viral Proteins Counteracting Host Defenses. 2002. 47 figs. XII, 325 pp. ISBN 3-540-43261-2

Vol. 270: **Beutler, Bruce; Wagner, Hermann (Eds.):** Toll-Like Receptor Family Members and Their Ligands. 2002. 31 figs. X, 192 pp. ISBN 3-540-43560-3

Vol. 271: **Koehler, Theresa M. (Ed.):** Anthrax. 2002. 14 figs. X, 169 pp. ISBN 3-540-43497-6

Vol. 272: **Doerfler, Walter; Böhm, Petra (Eds.):** Adenoviruses: Model and Vectors in Virus-Host Interactions. Virion and Structure, Viral Replication, Host Cell Interactions. 2003. 63 figs., approx. 280 pp. ISBN 3-540-00154-9

Vol. 273: **Doerfler, Walter; Böhm, Petra (Eds.):** Adenoviruses: Model and Vectors in VirusHost Interactions. Immune System, Oncogenesis, Gene Therapy. 2004. 35 figs., approx. 280 pp. ISBN 3-540-06851-1

Vol. 274: **Workman, Jerry L. (Ed.):** Protein Complexes that Modify Chromatin. 2003. 38 figs., XII, 296 pp. ISBN 3-540-44208-1

Vol. 275: **Fan, Hung (Ed.):** Jaagsiekte Sheep Retrovirus and Lung Cancer. 2003. 63 figs., XII, 252 pp. ISBN 3-540-44096-3

Vol. 276: **Steinkasserer, Alexander (Ed.):** Dendritic Cells and Virus Infection. 2003. 24 figs., X, 296 pp. ISBN 3-540-44290-1

Vol. 277: **Rethwilm, Axel (Ed.):** Foamy Viruses. 2003. 40 figs., X, 214 pp. ISBN 3-540-44388-6

Vol. 278: **Salomon, Daniel R.; Wilson, Carolyn (Eds.):** Xenotransplantation. 2003. 22 figs., IX, 254 pp. ISBN 3-540-00210-3

Vol. 279: **Thomas, George; Sabatini, David; Hall, Michael N. (Eds.):** TOR. 2004. 49 figs., X, 364 pp. ISBN 3-540-00534X

Vol. 280: **Heber-Katz, Ellen (Ed.):** Regeneration: Stem Cells and Beyond. 2004. 42 figs., XII, 194 pp. ISBN 3-540-02238-4

Vol. 281: **Young, John A. T. (Ed.):** Cellular Factors Involved in Early Steps of Retroviral Replication. 2003. 21 figs., IX, 240 pp. ISBN 3-540-00844-6

Vol. 282: **Stenmark, Harald (Ed.):** Phosphoinositides in Subcellular Targeting and Enzyme Activation. 2003. 20 figs., X, 210 pp. ISBN 3-540-00950-7

Vol. 283: **Kawaoka, Yoshihiro (Ed.):** Biology of Negative Strand RNA Viruses: The Power of Reverse Genetics. 2004. 24 figs., IX, 350 pp. ISBN 3-540-40661-1

Vol. 284: **Harris, David (Ed.):** Mad Cow Disease and Related Spongiform Encephalopathies. 2004. 34 figs., IX, 219 pp. ISBN 3-540-20107-6

Vol. 285: **Marsh, Mark (Ed.):** Membrane Trafficking in Viral Replication. 2004. 19 figs., IX, 259 pp. ISBN 3-540-21430-5

Vol. 286: **Madshus, Inger H. (Ed.):** Signalling from Internalized Growth Factor Receptors. 2004. 19 figs., IX, 187 pp. ISBN 3-540-21038-5

Vol. 287: **Enjuanes, Luis (Ed.):** Coronavirus Replication and Reverse Genetics. 2005. 49 figs., XI, 257 pp. ISBN 3-540-21494-1

Vol. 288: **Mahy, Brain W. J. (Ed.):** Foot-and-Mouth-Disease Virus. 2005. 16 figs., IX, 178 pp. ISBN 3-540-22419X

Vol. 289: **Griffin, Diane E. (Ed.):** Role of Apoptosis in Infection. 2005. 40 figs., IX, 294 pp. ISBN 3-540-23006-8

Vol. 290: **Singh, Harinder; Grosschedl, Rudolf (Eds.):** Molecular Analysis of B Lymphocyte Development and Activation. 2005. 28 figs., XI, 255 pp. ISBN 3-540-23090-4

Vol. 291: **Boquet, Patrice; Lemichez Emmanuel (Eds.)** Bacterial Virulence Factors and Rho GTPases. 2005. 28 figs., IX, 196 pp. ISBN 3-540-23865-4

Vol. 292: **Fu, Zhen F (Ed.):** The World of Rhabdoviruses. 2005. 27 figs., X, 210 pp. ISBN 3-540-24011-X

Vol. 293: **Kyewski, Bruno; Suri-Payer, Elisabeth (Eds.):** CD4+CD25+ Regulatory T Cells: Origin, Function and Therapeutic Potential. 2005. 22 figs., XII, 332 pp. ISBN 3-540-24444-1

Vol. 294: **Caligaris-Cappio, Federico, Dalla Favera, Ricardo (Eds.):** Chronic Lymphocytic Leukemia. 2005. 25 figs., VIII, 187 pp. ISBN 3-540-25279-7

Vol. 295: **Sullivan, David J.; Krishna Sanjeew (Eds.):** Malaria: Drugs, Disease and Post-genomic Biology. 2005. 40 figs., XI, 446 pp. ISBN 3-540-25363-7

Vol. 296: **Oldstone, Michael B. A. (Ed.):** Molecular Mimicry: Infection Induced Autoimmune Disease. 2005. 28 figs., VIII, 167 pp. ISBN 3-540-25597-4

Vol. 297: **Langhorne, Jean (Ed.):** Immunology and Immunopathogenesis of Malaria. 2005. 8 figs., XII, 236 pp. ISBN 3-540-25718-7

Vol. 298: **Vivier, Eric; Colonna, Marco (Eds.):** Immunobiology of Natural Killer Cell Receptors. 2005. 27 figs., VIII, 286 pp. ISBN 3-540-26083-8

Vol. 299: **Domingo, Esteban (Ed.):** Quasispecies: Concept and Implications. 2006. 44 figs., XII, 401 pp. ISBN 3-540-26395-0

Vol. 300: **Wiertz, Emmanuel J.H.J.; Kikkert, Marjolein (Eds.):** Dislocation and Degradation of Proteins from the Endoplasmic Reticulum. 2006. 19 figs., VIII, 168 pp. ISBN 3-540-28006-5

Vol. 301: **Doerfler, Walter; Böhm, Petra (Eds.):** DNA Methylation: Basic Mechanisms. 2006. 24 figs., VIII, 324 pp. ISBN 3-540-29114-8

Vol. 302: **Robert N. Eisenman (Ed.):** The Myc/Max/Mad Transcription Factor Network. 2006. 28 figs. XII, 278 pp. ISBN 3-540-23968-5

Vol. 303: **Thomas E. Lane (Ed.):** Chemokines and Viral Infection. 2006. 14 figs. XII, 154 pp. ISBN 3-540-29207-1

Vol. 304: **Stanley A. Plotkin (Ed.):** Mass Vaccination: Global Aspects -- Progress and Obstacles. 2006. 40 figs. X, 270 pp. ISBN 3-540-29382-5

Vol. 305: **Radbruch, Andreas; Lipsky, Peter E. (Eds.):** Current Concepts in Autoimmunity. 2006. 29 figs. IIX, 276 pp. ISBN 3-540-29713-8

Vol. 306: **William M. Shafer (Ed.):** Antimicrobial Peptides and Human Disease. 2006. 12 figs. XII, 262 pp. ISBN 3-540-29915-7

Vol. 307: **John L. Casey (Ed.):** Hepatitis Delta Virus. 2006. 22 figs. XII, 228 pp. ISBN 3-540-29801-0

Vol. 308: **Honjo, Tasuku; Melchers, Fritz (Eds.):** Gut-Associated Lymphoid Tissues. 2006. 24 figs. XII, 204 pp. ISBN 3-540-30656-0

Vol. 309: **Polly Roy (Ed.):** Reoviruses: Entry, Assembly and Morphogenesis. 2006. 43 figs. XX, 261 pp. ISBN 3-540-30772-9

Printing: Krips bv, Meppel
Binding: Stürtz, Würzburg